Flowers on the Tree of Life

Genetic and molecular studies have recently come to dominate botanical research at the expense of more traditional morphological approaches. This broad introduction to modern flower systematics demonstrates the great potential that floral morphology has to complement molecular data in phylogenetic and evolutionary investigations.

Contributions from experts in floral morphology and evolution take the reader through examples of how flowers have diversified in a large variety of lineages of extant and fossil flowering plants. They explore angiosperm origins and the early evolution of flowers and analyse the significance of morphological characters for phylogenetic reconstructions on the tree of life.

The importance of integrating morphology into modern botanical research is highlighted through case studies exploring specific plant groups where morphological investigations are having a major impact. Examples include the clarification of phylogenetic relationships and an understanding of the significance and evolution of specific floral characters, such as pollination mechanisms and stamen and carpel numbers.

LIVIA WANNTORP currently works as a researcher at the Swedish Museum of Natural History in Stockholm, where she leads several projects involving flower morphology and systematics of many different groups of flowering plants. She is Associate Editor of the journal *Plant Systematics and Evolution* and President of the *Swedish Systematics Association*.

LOUIS P. RONSE DE CRAENE is director of the M.Sc. course on the Biodiversity and Taxonomy of Plants at the Royal Botanic Garden Edinburgh. His research interests include the morphology and evolution of flowers and encompass a broad range of angiosperm families. He is the author of *Floral Diagrams: An Aid to Understanding Flower Morphology and Evolution* (Cambridge University Press, 2010).

The Systematics Association
Special Volume Series

SERIES EDITOR

DAVID J. GOWER

Department of Zoology, The Natural History Museum, London, UK

The Systematics Association promotes all aspects of systematic biology by organizing conferences and workshops on key themes in systematics, running annual lecture series, publishing books and a newsletter, and awarding grants in support of systematics research. Membership of the Association is open globally to professionals and amateurs with an interest in any branch of biology, including palaeobiology. Members are entitled to attend conferences at discounted rates, to apply for grants and to receive the newsletter and mailed information; they also receive a generous discount on the purchase of all volumes produced by the Association.

The first of the Systematics Association's publications *The New Systematics* (1940) was a classic work edited by its then-president Sir Julian Huxley. Since then, more than 70 volumes have been published, often in rapidly expanding areas of science where a modern synthesis is required.

The Association encourages researchers to organize symposia that result in multiauthored volumes. In 1997 the Association organized the first of its international Biennial Conferences. This and subsequent Biennial Conferences, which are designed to provide for systematists of all kinds, included themed symposia that resulted in further publications. The Association also publishes volumes that are not specifically linked to meetings, and encourages new publications (including textbooks) in a broad range of systematics topics.

More information about the Systematics Association and its publications can be found at our website: www.systass.org

Previous Systematics Association publications are listed after the index for this volume.

Systematics Association Special Volumes published by Cambridge University Press:

THE SYSTEMATICS ASSOCIATION SPECIAL

VOLUME 80

Flowers on the Tree of Life

EDITED BY

LIVIA WANNTORP

Department of Phanerogamic Botany, Swedish Museum of Natural History,
Stockholm, Sweden

LOUIS P. RONSE DE CRAENE

Royal Botanic Garden Edinburgh, UK

CAMBRIDGE
UNIVERSITY PRESS

Shaftesbury Road, Cambridge CB2 8EA, United Kingdom

One Liberty Plaza, 20th Floor, New York, NY 10006, USA

477 Williamstown Road, Port Melbourne, VIC 3207, Australia

314–321, 3rd Floor, Plot 3, Splendor Forum, Jasola District Centre, New Delhi – 110025, India

103 Penang Road, #05–06/07, Visioncrest Commercial, Singapore 238467

Cambridge University Press is part of Cambridge University Press & Assessment,
a department of the University of Cambridge.

We share the University's mission to contribute to society through the pursuit of
education, learning and research at the highest international levels of excellence.

www.cambridge.org
Information on this title: www.cambridge.org/9780521765992

First published 2011

A catalogue record for this publication is available from the British Library

Library of Congress Cataloging-in-Publication data
Flowers on the tree of life / [edited by] Livia Wanntorp, Louis P. Ronse De Craene.
    p.   cm. – (Systematics Association special volume series)
Includes bibliographical references and indexes.
ISBN 978-0-521-76599-2
1. Flowers – Morphology.   2. Flowers – Evolution.   3. Angiosperms – Morphology.
4. Angiosperms – Evolution.   I. Wanntorp, Livia.   II. Ronse Decraene, L. P. (Louis Philippe)
QK653.F59 2011
581.3′8–dc23   2011019851

ISBN   978-0-521-76599-2   Hardback

Contents

Colour plate section appears between pages 214 and 215.

Contributors

JULIEN B. BACHELIER Institute of Systematic Botany, University of Zurich,
  Switzerland

WILLIAM J. BAKER Royal Botanic Gardens, Kew, UK

MARIA VON BALTHAZAR Department of Systematic and Evolutionary Botany,
  University of Vienna, Austria

ANDERS BARFOD Department of Biological Sciences, Aarhus University, Denmark

RICHARD M. BATEMAN School of Geography, Earth and Environmental Sciences,
  University of Birmingham, UK

REGINE CLAßEN-BOCKHOFF Institut für Spezielle Botanik und Botanischer Garten,
  University Mainz, Germany

PETER R. CRANE School of Forestry and Environmental Studies, Yale University, New
  Haven, CT, USA

JAMES A. DOYLE Department of Evolution and Ecology, University of California,
  Davis, CA, USA

PETER K. ENDRESS Institute of Systematic Botany, University of Zurich, Switzerland

ELSE MARIE FRIIS Department of Palaeobotany, Swedish Museum of Natural
  History, Stockholm, Sweden

JASON HILTON School of Geography, Earth and Environmental Sciences, University of
  Birmingham, UK

ALEXANDRA C. LEY Department of Evolutionary Biology and Ecology, Université
  Libre de Bruxelles, Belgium

SOPHIE NADOT Laboratoire Ecologie, Systématique et Evolution, Université Paris-
  Sud, Orsay, France

KAJ RAUNSGAARD PEDERSEN Department of Geology, University of Aarhus,
  Denmark

DARIN PENNEYS Department of Botany, California Academy of Sciences, San
  Francisco, CA, USA

GERHARD PRENNER Jodrell Laboratory, Royal Botanic Gardens, Kew, UK

CARMEN PUGLISI Royal Botanic Garden Edinburgh, UK

LOUIS P. RONSE DE CRAENE Royal Botanic Garden Edinburgh, UK

PAULA J. RUDALL Jodrell Laboratory, Royal Botanic Gardens, Kew, UK

JULIE SANNIER Laboratoire Ecologie, Systématique et Evolution, Université Paris-Sud, Orsay, France

LIVIA WANNTORP Department of Phanerogamic Botany, Swedish Museum of Natural History, Stockholm, Sweden

Acknowledgements

We thank the organizers of an excellent meeting in Leiden, especially Professor Erik Smets, Director of the National Herbarium of the Netherlands and Dr Peter Hovenkamp for their welcome and continuous support.

We express our thanks to the Systematics Association, especially Richard Bateman and Alan Warren for supporting this initiative, and Cambridge University Press for publishing this work.

We are grateful to all authors who contributed to this book and colleagues who agreed to take time to review the chapters improving the standard of assembled articles, especially Peter Stevens, Thomas Stuetzel, Favio González, Jürg Schönenberger, Dmitry Sokoloff, Angelica Bello, Rolf Rutishauser, Merran Matthews, Patrick Herendeen, John Dransfield, Mark Newman, Fabian Michelangeli and Erik Smets.

1

Introduction: Establishing the state of the art – the role of morphology in plant systematics

LOUIS P. RONSE DE CRAENE AND LIVIA WANNTORP

1.1 Outlook

Scientific biological research is dominated by genetics and molecular studies nowadays. This research is extremely important and has led to a tremendous advance in the fields of systematic botany and evolutionary developmental genetics. Nevertheless, from the start the molecular approach grew at the expense of more traditional approaches, such as morphology, embryology, palynology and cytology, and today molecular Barcoding and phylogenetic studies often appear to be the dominant, sometimes exclusive, research areas. Despite this, most systematists would agree that morphological and molecular data are complementary and should, when possible, be used together in phylogenetic and evolutionary investigations. A common approach used in systematics, combining the molecular and morphological methods, routinely maps unexplored morphological characters or putative synapomorphies on well-supported phylogenetic trees in order to study the evolution of these characters. There is an important problem with this approach, that morphological characters can be wrongly defined or are often unknown or superficially assessed. However, understanding the characters used for phylogenetic studies is crucial for understanding evolutionary processes in plants.

Flowers on the Tree of Life, ed. Livia Wanntorp and Louis P. Ronse De Craene. Published by Cambridge University Press. © The Systematics Association 2011.

A general appreciation of floral morphology is also becoming difficult to grasp, with the disappearance of generalists, and this is not helped by the fact that there is little or no funding for any PhDs that are non-molecular. With the cutback of traditional botany in university education, lack of interest and funding from decision-making bodies, floral morphology is left increasingly aside. This is tragic, because it represents a loss of knowledge, which needs to be 'rediscovered' (as currently happens with the oblivion of obscure nineteenth century observations in even more obscure journals) and a non-appreciation of the value of morphology in contributing to solving the biodiversity crisis. Alas, morphology and general botany are increasingly scrapped from university curricula in the constricted atmosphere of 'efficient' research funding, with retiring experts not being replaced and with an increased specialization in botany on offer. Very few universities still have a morphology-based, integrative botany course. Recent developments such as genetic Barcoding are undoubtedly useful, but they remove interest and funding from other studies, such as those focusing on floral morphology.

There is an urgent need to stop the continuous erosion of expertise in morphology and systematics that has been going on for several decades. However, researchers in disciplines dealing with living organisms are increasingly confronted with a lack of knowledge of structures, with dire consequences for inaccurate interpretations and lack of perspective to their research. The understanding of the morphology of organisms in a broad range of taxa is essential to make links between observed patterns and helps the researcher to connect with the real world.

Flowers have always been the first pole of attraction to botany. While the beauty of flowers is part of human appreciation, the underlying parts and mechanisms of flowers have become the domain of floral morphology. Floral morphology is the reflection of an ongoing evolutionary process that started some 130 million years ago, and is continuously progressing with time. It is the accomplishment of close co-evolution with pollinating agents or physical mechanisms that lead to a single goal: fertilization and production of offspring. As a science, morphology is highly synthetic, even holistic. It finds unifying features in the great diversity of life (cf. Classen-Bockhoff, 2001; Kaplan, 2001). Floral morphology has a long and rich tradition in botany. Starting with herbals, it evolved into a major science that found its heydays in the nineteenth century, mainly in Germany and France. At the end of the twentieth century, morphology (especially of flowers) had become a well-established component in systematic research, as a major tool to reconstruct phylogeny, with far-reaching speculations about the evolution and significance of organs (e.g. Cronquist, 1981; Takhtajan, 1997). However, misconceptions grew out of a superficial approach to floral morphology, where things may look the same, but are not. Conspicuous examples of these generalizations are the derived nature of inferior ovaries, multistaminate androecia and the nature of petals. The rise of cladistics and molecular systematics was a great milestone in the understanding of relationships of plants, but it did not help in stopping a trend of a progressive demise of morphology. Important shifts were proposed in the classification systems that were originally considered to be sacrosanct. The acceptance of the important principles of parsimony and the understanding that only apomorphies can be considered valid led to increasing suspicions on the legality of floral morphology, as a nest of convergences and homoplasy.

Despite this, the advantages of investigating morphological characters with the help of molecular phylogenetic trees have become more evident during the past years.

The use of molecular methods is undeniably valuable and has helped botany to put order in the system, but its expansion went at the expense of traditional botany, of which floral morphology is a major component. The appreciation that producing molecular phylogenies for the sake of creating trees is hardly practical without 'real' or 'visual' characters, has led to a reconsideration of morphology. It is now generally understood that the support of phylogenetic trees is usually improved by combining molecular data with morphological data. The increased use of fossil evidence in phylogenies makes the use of structural data more important, as fossils possess no retrievable DNA (Endress, this volume; Doyle and Endress, 2010). Morphology is increasingly used as the basis of the molecular phylogenies, to understand how patterns evolve. This is useful to a certain extent but becomes doubtful when the morphological characters are not well understood or even wrongly assessed. When the principle of parsimony is used, some assumptions of morphological change quickly become listed as 'reversals', as characters are reconstructed along the nodes. This can lead in some cases to changes that are difficult to explain on morphological grounds. Especially at the level of their genetic mechanisms, the expression of morphological characters can be switched on and off very easily along the nodes of a tree.

However, with increased stability of the angiosperm tree of life, the value of flower morphology becomes increasingly obvious as a worthy counterpart to molecular characters in phylogenetic studies and as a source of data that help clarifying floral evolution and the underlying mechanisms of flower development. There are still major parts of the tree that need to be explored. For example, there are angiosperm groups whose positions in the system are yet unresolved and genera (even families) whose flower morphology and ontogeny is either poorly or completely unknown. This lack of knowledge is limiting our current understanding of the evolution of flowers and their structures considerably, as well as our general comprehension of the systematics of angiosperms. However, the complexity of floral morphology is increasingly being investigated, thanks to the development of evo-devo research, and our understanding of underlying structural and genetic factors does not fail to impress researchers.

1.2 Contents

Flowers on the Tree of Life aims to be a celebration of floral morphology by representing a compilation of articles from eminent researchers in the field of floral morphology. This volume is the proceedings of a symposium 'Flowers on the tree of life', organized by the editors and supported by the UK Systematics Association. This symposium was part of an international conference, entitled 'Systematics', organized by the Systematics Association and the Federation of European Biological Systematic Societies in Leiden from 10–14 August 2009. The symposium was built around three major themes: use of floral characters in phylogeny reconstructions, theoretical background for morphological characters and case-studies of specific groups of angiosperms whose flower morphology is controversial, unexplored or in need of further research. The symposium has clearly demonstrated that there is a great potential to re-establish floral morphology as centre-stage in botanical research and on the tree of life, putting more emphasis on the definition of sound characters that can be used to clarify relationships of plants and the intrinsic evolution of their flowers. Morphology is hugely important because it is a synthetic science that builds on information from multiple sources.

Chapters 2–4 deal with questions of relationships in angiosperm origins and early evolution of flowers.

A central subject to our understanding of flowers is how and when they diverged in evolution. This question has been pursued from several angles such as molecular phylogenetics and evolutionary developmental genetics, as well as morphology, including extant taxa and fossils (e.g. Bateman, Hilton and Rudall, 2006; Doyle, 2008; Doyle and Endress, 2010; Endress and Doyle, 2009; Friis, Pedersen and Crane, 2010; Frohlich, 2006; Soltis et al., 2009; Soltis and Soltis, 2004; Theissen and Melzer, 2007).

Bateman, Hilton and Rudall (Chapter 2) explore the relationships of bisexual angiosperm flowers with their gymnosperm relatives. A major obstacle to our understanding of flowers comes from the fact that gymnosperms are unisexual. The authors analyse this question from a historical as well as a developmental-genetic perspective and conclude that changes in gender from gymnosperm ancestors are inversely proportional to divergence between male and female developmental programmes.

Maria von Balthazar and co-workers (Chapter 3) explore the diversity of fossil Lauraceae and demonstrate how characters of extant Laurales are represented in the fossil record, but in different combinations.

James Doyle and Peter Endress (Chapter 4) discuss the origin and early evolution of the angiosperm flowers by mapping morphological characters on a phylogeny based on molecular data. They demonstrate that reconstructions of floral evolution depend on the phylogenetic framework, but also on morphological aspects of floral organizations.

Chapters 5 and 6 analyse the significance of morphological characters for phylogenetic reconstructions on the tree of life. Peter Endress (Chapter 5) discusses how the role of flowers has changed away from phylogenetics to a better understanding of evolutionary processes. However, increased use of fossil evidence increases the role of morphological data in phylogenetic reconstructions. The flexibility of changes in morphological characters is illustrated by several examples and highlights the importance of exploring morphology at deeper levels. A genetic predisposal can be treated as a character in its own right, even if it is not always expressed in a clade. The understanding of these underlying genetic systems will undoubtedly clarify our concepts of character evolution. This corresponds to a distinction between cladistic homology and biological homology as discussed by Endress.

Understanding key characters at deep phylogenetic nodes becomes an important part of morphological research. Paula Rudall (Chapter 6) explores characters that are significant at deeper nodes of the angiosperm tree of life. The centrifugal direction of stamen development is shown to be partially significant, but not at the level of Corner's (1946) prediction. This update of our knowledge of stamen development puts a final nail in the coffin of Cronquist's belief of centrifugal stamen development as a basic character for his subclass Dilleniidae. We now have a much better understanding of the relationship of centrifugal stamen development and polyandry (see also Ronse De Craene and Smets, 1992).

The next chapters (7–12) are examples of specific groups of plants where morphological investigations have a major impact, in understanding the phylogenetic relationships of palms with polyandry (Chapter 7) and genera making up the family Nitrariaceae (Chapter 8), and the understanding of the evolution and significance of floral morphological characters (Chapters 9–12).

Sophie Nadot and co-workers (Chapter 7) explore how the optimization of floral characters on a supertree of palms helps in understanding the evolution of the androecium in the diversification of the family.

Julien Bachelier, Peter Endress and Louis Ronse De Craene (Chapter 8) analyse how newly explored floral characters support the recent amalgamation of *Nitraria*, *Peganum* and *Tetradiclis* in a family Nitrariaceae of Sapindales.

Livia Wanntorp and co-workers (Chapter 9) study the variation in merism in *Conostegia* (Melastomataceae) and explore how changes in petal and stamen numbers correlate with increases in carpel numbers.

Alexandra Ley and Regine Classen-Bockhoff (Chapter 10) study the variation in the elaborate pollination mechanism of Marantaceae. By mapping major transformations of characters on a phylogenetic tree they are able to reconstruct the evolution of this unique pollination mechanism throughout the family.

The contribution of Gerhard Prenner (Chapter 11) studies a special case of high stamen and carpel numbers in *Acacia celastrifolia* and how this appears to be a concerted increase, linked with specific pollination strategies.

Finally Louis Ronse De Craene (Chapter 12) demonstrates on floral developmental evidence that petals of *Napoleonaea* (Lecythidaceae) represent a true corolla and not staminodes as previously suggested. This study also shows the value of comparative floral development in interpreting morphological structures.

In conclusion, flower morphological and ontogenetic studies have been and are going to remain an important contribution to systematics. This book clearly demonstrates that there is a great potential to re-establish floral morphology as centre-stage in botanical research and on the tree of life, putting more emphasis on the definition of sound characters that can be used to clarify relationships between plants and the intrinsic evolution of their flowers. New methods for working with flower morphological characters, such as synchrotron X-ray tomographic microscopy and more traditional light and electron microscopy techniques, as well as new ways of storing and making morphological data available to the community, such as the use of databases, are facilitating the use of such characters. This together with the necessity to use morphological characters as a complement to molecular data will ensure, whenever possible, that the morphological tradition will not be forgotten. As the world's flora is under increasing threat, understanding of plant structure becomes increasingly vital and experts with a global view on biodiversity become a necessary breed. Let us hope that there is room for a renaissance in morphology and that the present symposium becomes commonplace in the future.

1.3 References

Bateman, R. M., Hilton, J. and Rudall, P. J. (2006). Morphological and molecular phylogenetic context of the angiosperms: contrasting the 'top-down' and 'bottom-up' approaches used to infer the likely characteristics of the first flowers. *Journal of Experimental Botany*, **57**, 3471–3503.

Classen-Bockhoff, R. (2001). Plant morphology: the historic concepts of Wilhelm Troll, Walter Zimmermann and Agnes Arber. *Annals of Botany*, **88**, 1153–1172.

Corner, E. J. H. (1946). Centrifugal stamens. *Journal of the Arnold Arboretum*, **27**, 423–437.

Cronquist, A. (1981). *An Integrated System of Classification of Flowering Plants*. New York: Columbia University Press.

Doyle, J. A. (2008). Integrating molecular phylogenetic and paleobotanical evidence on origin of the flower. *International Journal of Plant Science*, **169**, 816–843.

Doyle, J. A. and Endress, P. K. (2010). Integrating Early Cretaceous fossils into the phylogeny of living angiosperms: Magnoliidae and eudicots. *Journal of Systematics and Evolution*, **48**, 1–35.

Endress, P. K. and Doyle, J. A. (2009). Reconstructing the ancestral angiosperm flower and its initial specializations. *American Journal of Botany*, **96**, 22–66.

Friis, E. M., Pedersen, K. R. and Crane, P. R. (2010). Diversity in obscurity: fossil flowers and the early history of angiosperms. *Philosophical Transactions of the Royal Society, B*, **365**, 369–382.

Frohlich, M. W. (2006). Recent developments regarding the evolutionary origins of flowers. *Advances in Botanical Research*, **44**, 63–127.

Kaplan, D. R. (2001). The science of plant morphology: definition, history and role in modern biology. *American Journal of Botany*, **88**, 1711–1741.

Ronse De Craene, L. P. and Smets, E. (1992). Complex polyandry in the Magnoliatae: definition, distribution and systematic value. *Nordic Journal of Botany*, **12**, 621–649.

Soltis, P. S., Brockington, S. F., Yoo, M.-J. et al. (2009). Floral variation and floral genetics in basal angiosperms. *American Journal of Botany*, **96**, 110–128.

Soltis, P. S. and Soltis, D. E. (2004). The origin and diversification of angiosperms. *American Journal of Botany*, **91**, 1614–1626.

Takhtajan, A. (1997). *Diversity and Classification of Flowering Plants*. New York: Columbia University Press.

Theissen, G. and Melzer, R. (2007). Molecular mechanisms underlying origin and diversification of the angiosperm flower. *Annals of Botany*, **100**, 603–619.

2

Spatial separation and developmental divergence of male and female reproductive units in gymnosperms, and their relevance to the origin of the angiosperm flower

RICHARD M. BATEMAN, JASON HILTON AND PAULA J. RUDALL

2.1 Introduction: aims and terminology

It is now generally accepted that angiosperms are monophyletic and are derived from a gymnospermous ancestor. It is also widely recognized that, among extant seed-plants, angiosperm reproductive units are typically bisexual (= bisporangiate, hermaphrodite) and are termed flowers, whereas putatively comparable units produced by the four groups of gymnosperms represented in the extant flora are typically unisexual, either functionally dioecious (cycads, *Ginkgo*, gnetaleans) or a more complex admixture of dioecious and monoecious taxa (conifers) (e.g. Tandre et al., 1995). Individual extant gymnosperms are either monoecious, bearing male and female units on separate axes of the same plant, or dioecious, each individual bearing units of only one gender (note: in this chapter, the terms 'male' and 'female' are used consistently as colloquial shorthand for the ovuliferous and (pre)polleniferous conditions, respectively). A positive correlation

Flowers on the Tree of Life, ed. Livia Wanntorp and Louis P. Ronse De Craene. Published by Cambridge University Press. © The Systematics Association 2011.

with dispersal mechanism is evident, monoecious extant gymnosperms typically producing dry, wind-dispersed seeds and dioecious gymnosperms bearing fleshy, often animal-dispersed seeds (cf. Givnish, 1980; Donoghue, 1989).

Further terminological clarifications are needed. Bateman et al. (2006, p. 3472) reviewed relevant definitions before defining a flower as 'a determinate axis bearing megasporangia that are surrounded by microsporangia and are collectively subtended by at least one sterile laminar organ'. Accepting this controversial definition means that the angiosperm flower is not unique; comparable hermaphrodite structures occur in at least one other group of seed-plants, specifically a putatively highly derived clade within bennettites (Crane, 1988). Extending this logic, the term inflorescence could also be applied more widely, to encompass axial systems that bear multiple reproductive units of gymnosperms. However, we have chosen to use throughout this chapter the more phylogenetically neutral term 'truss' to describe any reproductive shoot, unbranched or branched (we recognize that this usage of 'truss' contradicts that employed in the telome theory of Zimmermann, 1952).

Returning to the topic of gymnosperm–angiosperm relationships, the transition from unisexual to bisexual (bisporangiate, dimorphic, hermaphrodite) reproductive units, thereby spatially – though by no means always temporally – unifying male and female expression, has been consistently implicated as a key step in the long-debated origin of the angiosperm flower (reviewed by Bateman et al., 2006; Endress and Doyle, 2009; Specht and Bartlett, 2009). In contrast, the likelihood that the earlier origin of the gymnosperms involved the physical *separation* of male and female genders has received less consideration. In this paper we review the evidence for, and assumptions that underlie, these assertions of contrasting but equally radical shifts in the spatial location of male and female expression. We critically assess several scenarios emerging from recent evolutionary-developmental genetic studies in the light of: (a) the reproductive morphology of fossil gymnosperms and (b) teratological reproductive structures found in extant gymnosperms.

2.2 The enlarged phylogenetic gap between extant angiosperms and extant gymnosperms

Cladistic analyses conducted during the last three decades have consistently revealed vast morphological (e.g. Crane, 1985, 1988; Doyle and Donoghue, 1986; Donoghue and Doyle, 1989; Nixon et al., 1994; Rothwell and Serbet, 1994; Hilton and Bateman, 2006; Endress and Doyle, 2009; Rothwell et al., 2009) and molecular (e.g. Rydin et al., 2002; Burleigh and Mathews, 2004; Rai et al., 2008; Graham and Iles, 2009; Mathews, 2009) disparities between extant angiosperms and extant

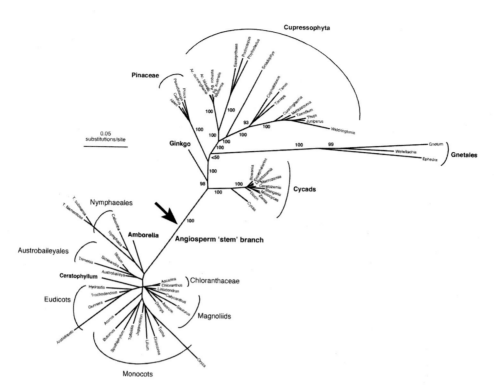

Fig 2.1 Unrooted phylogeny of extant seed-plants (modified after Graham and Iles, 2009, Fig 2). The arrow indicates the most common location of the tree root.

gymnosperms, irrespective of detailed topology (Fig 2.1). This exceptional phylogenetic divergence not only challenges the credibility of root placement in such trees but also represents a potentially insurmountable barrier to the important goal of reconstructing the phenotype and/or genotype of the angiosperm ancestor.

Similar ambiguity surrounds the characteristics of the hypothetical ancestor of the angiosperm crown group (i.e. the next node towards the apex of the seed-plant tree). Although it has not yet been quantified to our satisfaction, there appears to be greater morphological diversity among early-divergent extant angiosperms (termed by many authors the ANITA grade plus the magnoliids) than subsequently diverging and far more species-rich taxa of monocots and eudicots. This interpretation is encouraging in suggesting that the early-divergent extant angiosperms represent a genuine residue of the initial angiosperm radiation. However, the comparatively high level of morphological diversity also brings more negative consequences, as it renders very uncertain any attempted reconstruction of the shared ancestor of the crown group; when addressing this important question we can expect only limited help from 'top-down' angiospermo-centrism (Bateman

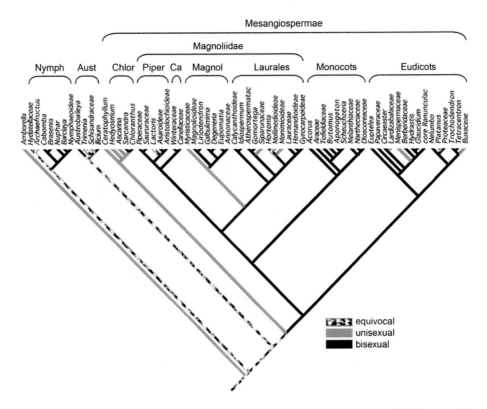

Fig 2.2 Optimization of hermaphrodism across a molecularly constrained phylogeny of the angiosperms, based on the preferred topology and the morphological data matrix of Endress and Doyle (2009, Fig 2b).

et al., 2006). Perhaps the best-informed attempt to achieve this aim using morphological cladistics (Endress and Doyle, 2009, pp. 44–46) concluded that 'Our results imply that the ancestral angiosperm flower had more than two whorls (or series) of tepals, more than two whorls (series) of stamens, probably with adaxial microsporangia, and several ascidiate carpels ..., most likely with one pendent bitegmic ovule, which was probably anatropous. Perianth and androecium phyllotaxis is uncertain. The most striking uncertainty is whether the ancestral flower was bisexual or unisexual.' This last character, critical but highly ambiguous (Fig 2.2), provides the focus of the present paper.

Moreover, the much-discussed Early Cretaceous fossil angiosperm *Archaefructus* (Archaefructaceae) and the recently recognized extant near-basal angiosperm *Trithuria* (Hydatellaceae) are unhelpfully ambivalent in both senses (Rudall et al., 2008; Rudall and Bateman, 2010). The reproductive truss of *Archaefructus* has been interpreted as being either a single flower lacking axial condensation (Sun et al., 2002) or an inflorescence of several unisexual flowers, the males concentrated basally and the females apically (Friis et al., 2003; Doyle, 2008; Endress and Doyle, 2009). As with all credible fossil angiosperms, the most likely phylogenetic placements of *Archaefructus* are within the angiosperm crown group (e.g. Friis et al., 2003; Bateman et al., 2006). Within the genus *Trithuria*, despite the fact that only a dozen extant species are currently known, four strongly contrasting reproductive strategies have been documented: dioecious annual plant; monoecious annual plant with bisexual reproductive units; monoecious annual plant with large unisexual reproductive units (both male and female on the same individual) and perennial apomicts bearing only reproductive units female (Remizova et al., 2008; Sokoloff et al., 2008). Thus, in this genus also the ancestral condition is ambiguous with regard to the gender of fertile trusses (Iles et al., 2008), an ambiguity reinforced by a combination of aquatic adaptations and putatively paedomorphic features (Bateman, 1996; Endress and Doyle, 2009). During optimization, the ambiguity caused by taxa such as *Archaefructus* and *Trithuria* is transferred downwards to the node that subtends the entire angiosperm crown group, with nihilistic consequences.

2.3 Evolutionary-developmental genetic models of the origin of the (hermaphrodite) flower

The classic ABC(D)E model of floral organ identity controlled through key developmental genes (dominantly MADS-box genes) has become hugely influential during the last two decades, though there is increasing recognition that the power of the model wanes in near-basal lineages of extant angiosperms. The evolutionarily derived and taxonomically dominant eudicots are adequately (if not perfectly) represented by the combinatorial ABC model (Bowman et al., 1989; Coen and Meyerowitz, 1991), wherein A-function genes specify sepals, A+B specify petals, B+C specify pollen-bearing stamens, and C alone specify ovule-bearing carpels. It was later recognized that D-function genes are required for successful ovule development (Colombo et al., 1995) and that A, B and C factors operate correctly only within a metabolic environment created by E factors (Pelaz et al., 2000). Also, Theissen (2001) and Theissen and Saedler (2001) advanced the quartets model, which postulated that A, B, C and E transcription factors reliably interact in aggregates of four protein molecules organized as dimers to determine contrasting floral organ identities. Noting that the distinction between sepals and petals is less evident in, or even absent from, pre-eudicot angiosperms, the Shifting Border (Bowman, 1997), Sliding Boundary (Kramer et al., 2003) and Fading Borders (Buzgo et al., 2004; Kim et al., 2005; Soltis et al., 2007, 2009) models were proposed to explain the apparently less rigid control of organ development in early-divergent angiosperms and the weaker applicability of the eudicot-centred ABC model; evidence for A function is especially poor.

Several evolutionary-developmental genetic ('evo-devo') theories of the origin of the angiosperm flower were published during the early 2000s, all relying primarily on the same observations of two sets of key developmental genes: B- and C-class MADS-box genes (e.g. Theissen and Becker, 2004) and *LEAFY* (e.g. Frohlich and Parker, 2000). Ultimately, the two classes of genes are functionally linked, as *LEAFY* binds with *UFO* to activate B-class genes in angiosperms, though gymnosperms lack *UFO* orthologues (e.g. Meyroud et al., 2010).

2.3.1 B- and C-class MADS-box genes

C-class genes, necessary for expression of both maleness and femaleness in angiosperms, diverged from B-class genes, necessary for expression of maleness in angiosperms, before angiosperms separated from the lineage leading to extant gymnosperms. In contrast, A-class genes controlling petal–sepal divergence are present only within angiosperms. In gymnosperms, B-class genes are reputedly expressed only in male cones (as shown by evidence from several conifer families and the gnetalean *Gnetum*) alongside C-class genes, whereas female cones show expression of C-class genes only (reviewed by Becker et al., 2000; Theissen et al., 2002; Theissen and Becker, 2004; Melzer et al., 2010). At least two lineages of B-class genes occur in conifers and Gnetales, though neither is orthologous with the two main groups of B-class genes found in angiosperms.

2.3.2 The *LEAFY* gene family

In angiosperms, *LEAFY* is arguably the most 'networked' of all the key developmental genes, controlling the switch from indeterminate to determinate meristem development and directly influencing the expression of A-, B- and C-class MADS-box organ identity genes through co-activation with the *SEPALLATA* class of MADS-box genes (e.g. Theissen et al., 2002). Two *LEAFY*-like gene lineages, *LEAFY s.s.* and *NEEDLY*, occur in (almost?) all gymnosperms, whereas only *LEAFY s.s.* occurs in angiosperms; *NEEDLY* is absent (Frohlich and Parker, 2000; Frohlich, 2002, 2006). Both *LEAFY* and *NEEDLY* are up-regulated in axillary meristems irrespective of the organs subsequently developed, and they apparently act in combination with *SEPALLATA*-related genes to control the expression of B- and C-class genes, mirroring the system documented more precisely in model angiosperms (e.g. Zahn et al., 2005). More importantly, expression evidence from *Pinus radiata* (Pinaceae: Mellerowicz et al., 1998; Mouradov et al., 1998) was prematurely extrapolated by most authors into a general assumption that, throughout gymnosperms, *LEAFY s.s.* is expressed only in male organs and early expression of *NEEDLY* occurs primarily in female organs.

The *LEAFY* and *NEEDLY* lineages are said, on molecular phylogenetic evidence, to represent a duplication event that occurred early in seed-plant evolution, prior to the separation of the angiosperm and extant gymnosperm lineages, followed by a presumed loss of *NEEDLY* from the angiosperm lineage prior to crown-group divergence (Frohlich and Parker, 2000).

These two pairs of groups of key developmental genes – B- versus C-class and *LEAFY* versus *NEEDLY* – provided much of the basis of the following four scenarios designed to explain the origin of the angiosperm flower.

2.3.3 'Mostly Male' theory

In this scenario, developed by Frohlich and colleagues (Frohlich and Parker, 2000; Frohlich, 2002, 2006), the 'feminizing' gene *NEEDLY* was lost from the angiosperm stem lineage after it separated from the lineage hypothesized to lead to all extant gymnosperm groups. The consequent absence of the NEEDLY protein shortens the reproductive axes and permits ectopic expression of ovules towards the apex of formerly male (i.e. simple rather than compound) cones, with possible downstream effects on ovule-forming D-function genes. This model was partly inspired by the observation that occasional *Ginkgo biloba* trees produce teratological leaves; these bear along their margins one or more ectopic ovules that ultimately attain a range of developmental stages, some eventually approximating maturity (Frohlich, 2002). The occurrence of sterile ovules within functionally male cones of Gnetales, notably *Welwitschia* (e.g. Mundry and Stützel, 2004a), was used to argue that there is a selective advantage in having female expression in close proximity to male (e.g. Endress, 1996; but see Section 2.4). The Mostly Male theory requires that the resulting hermaphrodite flower more closely resembles the ancestral male structure than the ancestral female structure. When seeking potential ancestors among the more derived pteridosperms that apparently possessed such characteristics, Frohlich (2002) selected as the most likely candidates the Triassic corystosperm *Pteruchus* (Yao et al., 1995; Taylor and Taylor, 2009) and Jurassic corystosperm *Pteroma* (Harris, 1964; Frohlich, 2002). Interestingly, it has recently become evident that the corystosperms survived into the Early Cretaceous and thus grew alongside angiosperms during their early radiation (Stockey and Rothwell, 2009). However, the adaxial expression of ovules assumed by Frohlich (2002) to characterize the corystosperms was challenged by Klavins et al. (2002).

2.3.4 'Pleiotropy Constraint' model

The originators of this model (Albert et al., 2002) did not name it; the term 'pleiotropy constraint' is here suggested. Following duplication (most likely of the entire ancestral genome), either *LEAFY s.s.* or *NEEDLY* should, under most circumstances, have been rapidly lost from the gymnosperm lineage – a phenomenon that occurs immediately following recent duplications of *LEAFY* within angiosperms. However, both copies could in theory be preserved due to pleiotropy (a single gene fulfilling multiple morphogenetic roles) if: (a) the roles of the two copies only partly overlap and (b) *LEAFY* is more resistant to inactivation than *NEEDLY* but is also less effective at controlling the reproduction-inducing C-class genes. Such a 'balance of power' would mirror interactions observed between class

B and C genes in angiosperms that are gynodioecious (individual plants produce either exclusively female or exclusively bisexual flowers: e.g. Barrett, 2002). In this context, gymnosperm-sourced copies of *LEAFY* are more effectively expressed when transgenically inserted into model angiosperms such as *Arabidopsis* than are copies of the gymnosperm-only *NEEDLY* clade.

We note that this theory would be applicable even if *LEAFY s.s.* and *NEEDLY* did not control maleness and femaleness; there simply needs to be some kind of divergence of functional role between them via neofunctionalization (cf. Lynch, 2002). Increase in the effectiveness of C-function genes controlled by *LEAFY s.s.* could easily lead to the loss of the relatively vulnerable *NEEDLY*, thereby allowing the immediate (and potentially saltational: Bateman and DiMichele, 2002) formation of bisexual cones as flower precursors. The Pleiotropy Constraint model helpfully allows co-expression of both the male and female programmes in a single determinate meristem; in contrast with the Mostly Male theory, there is no requirement for ectopic expression of either the female or the male developmental programmes.

2.3.5 'Out of Male' and 'Out of Female' co-hypotheses

Developed by Theissen and colleagues (Theissen et al., 2002; Theissen and Becker, 2004), this scenario relies heavily on the premise that B-class genes are expressed only in the male cones of gymnosperms – a premise based on evidence from several conifer groups plus the gnetalean *Gnetum*. Of three B genes found in the pine-relative *Picea*, two show partly divergent expression patterns (subfunctionalization): *DAL11* in the cone axis and *DAL13* in the sporophylls. The model hypothesizes that modification of a B-gene promoter enhances the apical–basal gradient in one or more phytohormones or transcription factors, thereby inducing a bisexual cone (basally male, distally female). This could occur through either (a) reduction of B-gene expression in the apical region of the male cone (termed the 'Out of Male' hypothesis) or (b) ectopic expression of B genes in the basal region of the female cone (termed the 'Out of Female' hypothesis).

2.3.6 'Further Out of Male' hypothesis

Again, we have chosen the name for this model in the absence of a previous descriptor. It constitutes the first phase of a more comprehensive, four-stage evolutionary-developmental genetic scenario advanced by Baum and Hileman (2006) to explain not only the origin, but also the early evolution of the angiosperm flower:

(1) Evolution of a bisexual axis via a gynomonoecious intermediate; homeotic conversion of apical microsporophylls (stamens) into megasporophylls (carpels) within the gymnosperm pollen cone, via differences in maximal expression levels of B- and C-class organ identity genes and competition among their gene products (a mechanism echoing the earlier 'Out of Male' hypothesis).

(2) Evolution of floral axis compression and determinacy, caused by C-class genes becoming negative regulators of the meristem maintenance gene *WUSCHEL* (in our opinion, this is a crucial step that is surprisingly overlooked in most flower-origin scenarios).

(3) Evolution of a petaloid perianth by sterilization of the outer zone of stamens, caused by co-option of *WUSCHEL* as co-regulator of C-class genes (note that evidence also exists in some lineages for the origin of petals as bracteopetals, derived from bracts: e.g. Zanis et al., 2003).

(4) Evolution of the dimorphic perianth (i.e. petal–sepal differentiation) that characterizes the core eudicots, caused by B-class function becoming dependent on *UFO* co-regulation.

2.3.7 Selecting among the competing hypotheses

Theissen and Becker (2004) argued that the best test of the competing hypotheses is to make genetically (and taxonomically?) broad comparisons between extant angiosperms and extant gymnosperms. More specifically, they stated that the best test of the hypotheses is to carefully compare gene networks in bisexual angiosperm flowers with male and female gymnosperm cones.

Thus, if the gene networks in the flower prove to be most similar to those of the male gymnosperm cone, then the Mostly Male Theory, Out of Male (and Further Out of Male) hypotheses are supported. If the gene networks in the flower most closely resemble those of the female gymnosperm cone, then the Out of Female hypothesis is supported. And lastly, if the gene networks in the flower are approximately equally similar to both the male and female gymnosperm cones, then the 'Pleiotropy Constraint' hypothesis is supported. Although we agree with the broad sweep of this logic, available data inevitably prompt several codicils, a few of which we will now explore.

2.4 Spanners in the works: erroneous assumptions and a predominance of clinal expression patterns

Additional lines of research have yielded valuable results since the above scenarios were developed. Here, we focus on just two contrasting research areas: the location of key gene expression with gymnospermous reproductive structures, and the ability for radical shifts in gender expression that are evident in teratological gymnosperm cones.

2.4.1 Location of expression

Current evidence suggests that the origin and/or early evolution of the seed-plant clade entailed greater localization of expression of *LEAFY*-group genes inherited from its pteridophytic progymnospermous ancestor and the establishment of co-activity with *SEPALLATA*-group to control newly emerged B- and C-class MADS-box genes, together forming a reproductive gene network. Increased complexity and specificity of *LEAFY*-group expression is indicated, most notably up-regulation in axillary meristems (e.g. Moyroud et al., 2010).

However, a spanner (arguably a monkey wrench) was thrown in the works of the *LEAFY*-based evolutionary hypotheses when evo-devo studies expanded outwards from Pinaceae into other groups of extant gymnosperms (Fig 2.1). Specifically, Vazquez-Lobo et al. (2007) demonstrated that, contrary to expectations, both *LEAFY* and *NEEDLY* are actually meaningfully expressed in both male *and* female cones. *LEAFY* is in fact highly expressed, consistently and through much of their ontogeny, in female cones of not only *Pinus* (Dornelas and Rodriguez, 2005) and the closely related *Picea* (Carlsbecker et al., 2004; Vazquez-Lobo et al., 2007) but also the more distantly related, fleshy-fruited conifers *Podocarpus* and *Taxus* (Vazquez-Lobo et al., 2007) and even the gnetalean *Gnetum* (Shindo et al., 2001) (Table 2.1). Similar results were reported by Guo et al. (2005) for *Ginkgo*, wherein both *LEAFY* and *NEEDLY* are expressed in trees of both genders but expression of *NEEDLY* appears more localized within the tree architectures than that of *LEAFY*. Vazquez-Lobo et al. (2007) suggested that the apparent absence of *LEAFY* expression from male (but not female) cones of *Pinus* erroneously reported earlier by Mellerowicz et al. (1998) could be due either to a technical flaw resulting in undetectable expression levels or, less likely, an upstream regulatory peculiarity restricted to *Pinus*. These expanded observations require varying degrees of modification of the previous theories, prompting greater emphasis of the differential expression of *LEAFY s.s.* and *NEEDLY within* rather than *between* cones.

The altered interpretation of *LEAFY* expression provides at least two valuable lessons. Firstly, if it is technically difficult to conclusively demonstrate the expression of key developmental genes in conifers, it is even more challenging to conclusively demonstrate the *non*-expression of genes. Secondly, typological observations are suspect; it is important to expand expression studies into multiple related species. A further crucial factor is the scientific 'sociology' of these studies. The papers that presented the general ABC model of organ expression are very heavily cited, those that established the competing hypotheses of angiosperm flower origin are fairly heavily cited, but the benchmark contribution that convincingly contradicted critical elements of those hypotheses (Vazquez-Lobo et al., 2007) has thus far received remarkably little attention.

Although the more recent observations require modifications of earlier scenarios, they do not wholly undermine them. *LEAFY* and *NEEDLY* evidently play subtly different roles in the reproductive organs across a wide taxonomic range of extant conifers, especially in the female cone (Table 2.1). According to our reading of Vazquez-Lobo et al. (2007), in the female 'dry cones' of *Picea*, *LEAFY* is expressed in the ovule and parts of the ovuliferous scale (largely coinciding with the location

Table 2.1 Comparison of *LEAFY* and *NEEDLY* expression in the female and male cones of three morphologically disparate genera of extant conifers (data from Vazquez-Lobo et al., 2007; note that they interpreted the epimatium and aril as components of the ovuliferous scale).

Genus/GENDER	*LEAFY*	*NEEDLY*
Female		
Picea	Initially in the peripheral zone of the apical meristem; later in the ovule and ovuliferous scale primordia	Initially in the bract primordia; later in the ovuliferous scale but excluding the ovule
Podocarpus	Initially throughout the meristem; later in the central portion of the bract, the epimatium and the nucellus	Initially in the bract primordia; later in the ovuliferous scale but excluding the ovule
Taxus	Ovule nucellus	Aril primordium (interpreted as part of the ovuliferous scale) and arguably the ovule nucellus
Male		
All three genera	Microsporangia	Microsporophylls (synangia in *Taxus*)

of expression of the C-class gene *DAL2*), whereas *NEEDLY* is expressed in the extra-ovular vasculature. Moving on to consider fleshy-seeded taxa, in *Podocarpus*, *LEAFY* is expressed mainly in the ovule and *NEEDLY* in the extra-ovular vasculature, while in *Taxus*, *LEAFY* is expressed in the ovule and *NEEDLY* mainly in the surrounding aril. It is perhaps significant that, although there is also strong divergence of expression patterns between *LEAFY* and *NEEDLY* in the male cones of these three genera, there is little divergence in male expression among the genera; *LEAFY* is expressed in the microsporangia and *NEEDLY* in the microsporophylls (synangia in the case of *Taxus*). Thus, irrespective of phylogenetic position within conifers, location of expression of these genes appears substantially more consistent among conifer families in the male cone than in the female cone.

Further insights are gained when the ontogeny of conifer cones is considered in greater detail. *LEAFY* expression is located in shoot apices and organ primordia and decreases towards maturity. In contrast, *NEEDLY* expression is not confined to primordia; it persists in organs still actively growing, but is excluded from the immediate vicinity of the ovules and pollen grains. Placing these observations in an evolutionary-developmental context, heterochronic shifts in timing and location of *LEAFY s.s.* expression in female cones may help to explain radical morphological divergences among extant conifers. *NEEDLY* may confer late-stage modifications on size and shape of the ovule-bearing structures, permitting both heterochronic (shape change) and allometric (size change only) shifts between putative ancestor and descendant species (Bateman, 1996).

Overall, these evo-devo studies have given us a more dynamic perspective in both space and (ontogenetic) time, emphasizing the need to consider in more detail the subtleties of gene expression and interaction during ontogeny. If we look beneath this pattern-based evolutionary terminology for potential causal processes, spatial gradients in phyto-hormones (perhaps gibberellins) and/or transcripts provide attractive prospects, perhaps operating in tandem with variable gene expression reflecting differential levels of methylation (Adams and Wendel, 2005; Madlung et al., 2005).

Lastly, we note that although the comparative approach employed by Vazquez-Lobo et al. (2007) and others in place of the more conventional and commonplace reliance on a single model organism has given us a better understanding of the role of the *LEAFY* class of developmental genes in gymnosperms, thus far it has done so only in conifers. Even then, Meyroud et al. (2010, p. 399) concluded that 'it is still unknown whether modifications in protein stability, affinity for DNA, sequence-specific recognition, or interaction with coregulators account for differences between the [*LEAFY* and *NEEDLY*] paralogs'.

2.4.2 Teratological phenotypes

Much of our knowledge of modern genetics has been derived through comparison of conspecific wildtype and mutant morphs. Such comparisons are increasingly controlled in the laboratory, with the primary objective of identifying the precise (epi)genetic and biochemical processes that dictate the relevant phenotypic divergence. Comparative surveys of naturally occurring teratologies are far fewer and confined to angiosperms (e.g. those on orchid flowers by Bateman and Rudall, 2006, and on Brassicaceae flowers by Endress, 1992), but fulfil a useful function by indicating genuine evolutionary potential (Bateman and DiMichele, 2002; Ziermann et al., 2009). Despite the complaints by Meyroud et al. (2010, p. 349) of a 'lack of gymnosperm [*LEAFY*] mutants' and by Groth (2010, p. 24) that 'there are no mutants available for any of the gymnosperm MADS-box genes', occasional papers have addressed particular terata observed within particular conifer species, such as the mutant cones of *Picea abies* discussed by Theissen and Becker (2004). Here, we conduct a brief survey of reproductive teratologies occurring naturally in gymnosperms, both living and fossil (Figs 2.3–2.5); past and present reports are summarized in Table 2.2 (see also Rudall et al., 2011).

The most frequently observed category of teratological conifer cones encompasses those that revert from determinate reproductive growth to indeterminate

Fig 2.3 Teratological female cones of extant conifers showing reversion from reproductive to vegetative growth (a–e), in one case (c) followed by a switch to male expression. (a, b) *Cunninghamia lanceolata*, (c) *Cryptomeria japonica*, (d) *Sequoia sempervirens*, (e) *Abies koreana*. Images by Richard Bateman (a–c, e), Julien Bachelier (d). For colour illustration see plate section.

vegetative growth (often termed 'proliferated' cones), thereby suggesting that the 'determinizing' effect of *LEAFY* and associated genes is reversible. Examples of cone axes that revert to production of substantial numbers of wildtype needle-leaves are frequently observed in extant Pinaceae (*Picea, Abies*: Fig 2.3e), Sciadopityaceae (*Sciadopitys*), Cephalotaxaceae (*Cephalotaxus*) and Cupressaceae *s.l.* (*Cunninghamia, Cryptomeria*: Fig 2.3a–c; *Sequoia*: J. Bachelier, pers. comm., 2009: Fig 2.3d) (Table 2.2). Indeed, Masters (1890, p. 314) noted that it is 'rare in some seasons to meet with a tree of *Cryptomeria japonica* in which some of the

Fig 2.4 Bisporangiate cone-aggregates and cones of extant conifers. (a, b) Single terminal female cone of *Cunninghamia lanceolata* closely subtended by a pseudowhorl of lateral male cones. (c) Artificially coloured scanning electron micrographs of two teratological conifer cones that show a transition from basal male to apical female expression in *Tsuga dumosa* (right) and from basal female to indeterminate to male to apical female in *Tsuga caroliniana* (left). Colours: green = vegetative, blue = male, pink = female, purple = equivocal gender. (d) Cone of *Araucaria bidwillii* that is closer in size (10 cm long) and gross morphology to a typical male cone of the species but bears a longitudinal 'feminized' zone along the convex margin. Images by Richard Bateman (a, b), Paula Rudall (c), Raymond van der Ham (d). For colour illustration see plate section.

cones are not so altered', a comment reinforced by our own observations on both *Cryptomeria* and *Cunninghamia*. In the majority of cases, the re-instituted vegetative growth is weak and ceases within months (Fig 2.3a), but in other cases the fresh growth is more vigorous and longer lasting (Fig 2.3b).

In addition, branches of the Pennsylvanian species *Barthelia furcata*, a taxon that diverged earlier than all living and most other fossil conifers, may have routinely

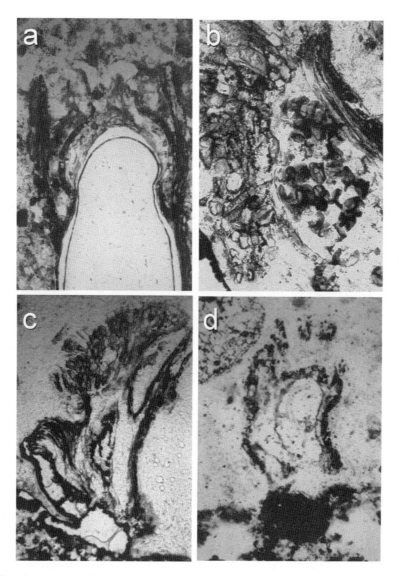

Fig 2.5 Light micrographs highlighting a putative teratos of the hydrasperman pteridosperm cupule *Pullaritheca longii* from the Mississippian of southeast Scotland that together suggest clinal control of gender. (d) Portion of one of 38 recorded dehisced specimens showing one of several abortive ovules attached to the placenta below. (b, c) Two teratological specimens first reported by Long (1977a) at the margin of a single atypical cupule; (c) largely resembles the abortive ovules, but has undergone atypical proliferation of the apical nucellar tissue responsible for capturing pre-pollen grains, (b) shows similar nucellar proliferation, but contains many microspores rather than the expected single megaspore. (a) Shows the wildtype pollen-receiving apparatus of the ovule, compressed by expansion of the ovum following successful pollination (modified after Bateman and DiMichele, 2002, Fig 7.4). Images by Richard Bateman. For colour illustration see plate section.

Table 2.2 Reports of teratological conifer cones that are either bisexual or proliferated (reversion from fertile to vegetative development). Note that some citations to Masters (1890) and Worsdell (1900) represent earlier references cited by these authors.

Proliferated seed-cones	Bisexual cones
Araucariaceae	
	Araucaria bidwillii (present study, pp. R. van der Ham)
Cephalotaxaceae	
Cephalotaxus fortunei, proliferated (leafy) primary and secondary axes (Worsdell, 1900); present study	
Cupressaceae	
Cryptomeria japonica (Masters, 1890; present study)	*Cupressus lawsoniana* (Masters, 1890)
Sequoia sempervirens (Masters, 1890; present study, pp. J. Bachelier)	*Sequoia sempervirens* (Shaw, 1896)
Cunninghamia lanceolata (present study)	
	Juniperus communis (Renner, 1904)
Pinaceae	
Abies koreana (present study)	*Abies balsamea* (Schooley, 1967)
Larix laricina (Tosh and Powell, 1986)	Several *Larix* species, including *L. europaea, L. laricina, L. microcarpa, L. occidentalis* (Masters, 1890; Bartlett, 1913; Kirkwood, 1916; Tosh and Powell, 1986)
Several *Picea* species, including *P. alba, P. glauca, P. mariana* (Masters, 1890; Worsdell, 1900; Santamour, 1959; Owens and Molder, 1977; Caron and Powell, 1991; present study)	Several *Picea* species, including *P. abies, P. excelsa* (as *Abies excelsa*), *P. mariana, P. nigra* (Dickson, 1860; Masters, 1890; Caron and Powell, 1991; M. Fladung *in* Theissen and Becker, 2004)
Pinus (Masters, 1890)	Several *Pinus* species and hybrids, including *P. rigida, P. thunbergii, P. laricio, P. montana, P. johannis* (Masters, 1890; Fisher, 1905; Steil, 1918; Righter, 1932; Mergen, 1963; Flores-Renteria, 2010)
Pseudotsuga (Masters, 1890)	*Pseudotsuga taxifolia* (Littlefield, 1931)

Table 2.2 (*cont.*)

Proliferated seed-cones	Bisexual cones
Tsuga brunoniana (Masters, 1890)	*Tsuga canadensis* (Holmes, 1932)
Sciadopityaceae	
Sciadopitys (Masters, 1890)	

reverted to vegetative growth after cone formation (Rothwell and Mapes, 2001); similar observations have been made on other early-divergent fossil conifers such as *Thucydia mahoningensis* (Hernandez-Castillo et al., 2001) and *Parasciadopitys aequata* (Yao et al., 1997). The apparent fixation of 'partial indeterminacy' in these fossils indicates that the similar teratologies observed in extant conifers are atavistic. The transition from vegetative to reproductive growth (and back again) was evidently subtle in even the earliest conifers.

Moreover, reversion to vegetative growth has, to our knowledge, been reported convincingly in the male cones of only one conifer, specifically in *c.* 0.5% of male cones in a plantation-grown sample of 15-year-old plants of the unusually developmentally labile *Picea mariana* (Caron and Powell, 1991, Figs 9–11). In our opinion, the near-complete restriction of vegetative reversion to female conifer cones provides some circumstantial support for the argument that each female cone scale plus ovuliferous scale in Pinaceae is homologous with an entire laterally borne female cone in cordaites (cf. Worsdell, 1900; Chamberlain, 1935; Florin, 1939, 1951; Wilde, 1944; Thomson, 1949), since this would imply that the determinacy evident in most extant conifer cones has been imposed on a previously indeterminate organ. This phenomenon is likely to be confined to cone-bearing, dry-seeded species, as most fleshy seeds are borne in isolation rather than in more complex trusses.

The basally divergent, dioecious cycad *Cycas* is also relevant in this context, as only the male sporophylls are aggregated into a classic cone. The female sporophylls are produced by the apical meristem as pseudowhorls; these are interspersed with pseudowhorls of leaves, providing an even more striking example of reversion from ovuliferous to vegetative growth in individuals (e.g. Zhang et al., 2004).

The other main category of teratological conifer cones encompasses those that are hermaphrodite (sometimes rather misleadingly termed 'dimorphic') rather than unisexual. Earlier reviews, most brief and morphologically focused (e.g. Littlefield, 1931; Holmes, 1932), identified cases in several genera of Pinaceae (*Pinus, Picea, Abies, Larix, Pseudotsuga, Tsuga*) and a few Cupressaceae s.l. (*Sequoia, Juniperus*) (Table 2.2). Examples explored more recently in the context of evolutionary-developmental genetics were selected exclusively from among Pinaceae, including *Picea abies* (the so-called 'var. *acrocona*': Theissen and Becker, 2004), *Tsuga dumosa* and *T. caroliniana* (F. Vergara-Silva, unpubl. obs., 2007).

Most (but by no means all: Caron and Powell, 1991) bisexual cones produce pollen at the base and ovules toward the apex, thereby mirroring the spatial arrangement of the two genders on the determinate axes of almost all angiosperm flowers (cf. Rudall, 2008; Rudall et al., 2011). The resulting hermaphrodite cones resemble more closely male than female cones and thus are less consistent with the 'Out of Female' hypothesis of Theissen et al. (2002). However, although in many cases the boundary between male and female expression domains is transverse to the cone axis and sharp (Fig 2.4c, right), in others there occurs a zone of hybrid sporophylls that express neither gender clearly (Fig 2.4c, left). In such cases, the 'fundamental gender' of the cone is much less obvious, as is any analogy with the apparently more canalized development of a typical angiosperm flower.

Lastly, the developmentally 'feminized' 'hybrid zone' is most often transverse to the cone axis (Fig 2.4c), and would thus be consistent with a cline established along the axis of the cone in one or more gender-determining phytohormones, perhaps dictated by reduced expression of 'masculinizing' B-class genes towards the apex of the fundamentally male cone. However, in at least one case involving *Araucaria bidwillii* (Araucariaceae), the presumed 'feminized' zone is oriented longitudinally along the axis (Fig 2.4d: R. van der Ham, pers. comm., 2009).

Also relevant is a remarkable hermaphrodite teratos recognized in the Mississippian hydrasperman pteridosperm *Pullaritheca longii* by Long (1977a, b: Fig 2.5) and subsequently re-interpreted in an evolutionary-developmental genetic context by Bateman and DiMichele (2002). This anatomically preserved fossil is interpreted as a fundamentally female ovulate cupule that suffered a developmental perturbation along a small peripheral area of the ovule-generating placenta; one that disturbed the normal ontogenetic trajectory from unfertilized (Fig 2.5d) to fertilized (Fig 2.5a) ovule. Here, imperfectly formed microsporangia (Fig 2.5b) appear to have heterotopically replaced the 'intended' ovules, while the border between the male and female zones is occupied by imperfectly formed ovules (Fig 2.5c). This intriguing cupule specimen suggests that gender control was at least as unstable in the earliest gymnosperms as in extant conifers.

Returning to extant conifers, but focusing on cone aggregates rather than cones per se, a few conifers that possess female cones that are highly compressed and obtusely conical in shape, reflecting ovulate scales that decrease rapidly in size and eventually become sterile, show exceptional developmental lability. In addition to routinely extending the growth of vegetative axes beyond female cones (Fig 2.3a, b), plants of *Cunninghamia lanceolata* (Cupressaceae) frequently cluster several male cones on determinate lateral branches immediately below a single female cone (Fig 2.4a, b), echoing the arrangement of multiple stamens surrounding one or more carpels in angiosperms. Moreover, this arrangement of the conifer 'inflorescence' constitutes circumstantial evidence in support of Florin's decision to homologize an entire male cone with each female bract–scale complex. However, female cones of *Cryptomeria japonica* (Cupressaceae) occasionally exhibit the converse arrangement; they revert to vegetative growth, but then revert yet again, this time to male expression, producing on that distal shoot several poorly developed and laterally expressed male cones (Fig 2.3C).

Finally, entire plants in several extant gymnosperm groups change gender as they become more mature. For example, individuals of *Cupressus sempervirens* (Cupressaceae) typically begin their reproductive lives as females, but later become masculinized (Lev-Yadun and Liphschitz, 1987). In such cases, the entire individual is bisexual if assessed across its entire life-span, but unisexual at certain points in its life history, a condition that might usefully be termed 'transformational dioecy'. It has also proven possible to induce precocious reproduction in juvenile conifers of many extant families through injection with the phytohormone GA3 (e.g. Longman, 1985).

Taken together, these observations suggest that gender control in gymnosperms is both subtle and metastable. Clinal expression that is interrupted by thresholds that constitute tipping-points in gender, suggesting mediation via phytohormones, appears consistent with the behaviour inferred for both the *LEAFY* and B-class gene families among gymnosperms. In this context, we return to the Fading Borders model advanced for extant early-divergent angiosperms. As described by Soltis et al. (2009, p. 119), 'the fading borders model suggests that gradual transitions in organ morphology result from a gradient in levels of expression of floral regulators across the meristem. Overlapping expression of floral regulators would impose some features of adjacent organs onto each other and thus produce morphologically intergrading rather than distinct floral organs.'

2.5 Gender separation is implicated in the origin and early diversification of the (gymnospermous) seed-plants

2.5.1 Did male–female segregation permit the origin of the seed?

All morphological cladistic analyses have identified (or, more correctly, assumed) the sister-group of the seed-plants to be at least one progymnosperm. Where multiple progymnosperms have been included and they have not been constrained to monophyly, they have generally been depicted as paraphyletic with one or more heterosporous species placed as sister to the seed-plants (reviewed by Hilton and Bateman, 2006). The primary candidates are *Archaeopteris halliana* (Upper Devonian: Phillips et al., 1972), *Protopitys scotica* (Mississippian: Walton, 1957) and *Cecropsis luculenta* (Pennsylvanian: Stubblefield and Rothwell, 1989).

Interestingly, although these partially or wholly reconstructed species are sufficiently morphologically disparate that each has been used in many classifications as the basis of a distinct taxonomic order of progymnosperms (e.g. Stewart and Rothwell, 1993), all three bear microsporangia and megasporangia in close proximity. To complicate matters further, some authors argue that the thus far ubiquitously homosporous aneurophyte progymnosperms provide a closer vegetative comparison with the earliest known seed-plants than do the heterosporous species. Given that a direct transition from homospory to the seed habit is unlikely, current evidence suggests that none of the progymnosperms that has so far been reconstructed constitutes a credible ancestor of the gymnosperms.

Yet the earliest fossil seed-plants from the Upper Devonian, such as *Elkinsia polymorpha* (Fig 2.6b: Rothwell and Serbet, 1992) and *Moresnetia zalesskyi* (Fairon-Demaret and Scheckler, 1987), not only have unisexual reproductive units but may also be dioecious. If so, the onset of strong segregation of male and female reproductive organs coincides with, and hence may be implicated in, the origin of the seed (Bateman and DiMichele, 1994).

2.5.2 Architectural ambiguities obscure sexual systems in fossils

Unfortunately, palaeobotanists cannot be certain whether an apparently dioecious fossil plant is actually monoecious, due to inevitable fragmentation of the plant body and the consequent difficulty encountered, when attempting to conceptually reconstruct the plant (Bateman and Hilton, 2009), of detecting sequential monoecy. Such successive maturation of male and female reproductive units within the same individual makes it appear unisexual at any one moment in ontogenetic time (e.g. Longman, 1985). Both male pollen organs and female cupules were illustrated as being borne on different fronds (male above female) of the same individual in the Mississippian pteridosperm *Calathospermum scoticum* (Mississippian), but the 'reconstruction' by Andrews (1948) shown in Fig 2.6a is actually a restoration; the relative positions allocated to male and female organs on the overall bauplan are only educated guesswork, probably based on architectural templates provided primarily by extant conifers (Bateman and Hilton, 2009). The most complete early pteridosperm specimen known is another Mississippian species, *Diplopteridium holdenii* (Fig 2.6c); Rowe (1988) described a potentially near-complete branched specimen that bears five fully or partially expanded ovulate fronds, thereby providing a directly observable (if incomplete) bauplan.

2.5.3 Divergence of developmental programming in male and female organs

Although reproductive units undoubtedly develop at different locations in different pteridosperm groups (reviewed by Retallack and Dilcher, 1988), present evidence

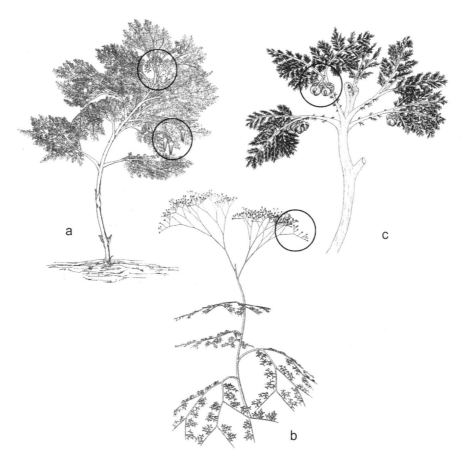

Fig 2.6 Degrees of confidence in architectures inferred for selected Late Devonian (b) and Mississippian (a, c) pteridosperms. Comparison of (a) a highly hypothetical restoration of *Calathospermum scoticum* (Andrews, 1948), (b) a more rigorous reconstruction of *Elkinsia polymorpha* (Rothwell and Serbet, 1992), and (c) an actual compression specimen of *Diplopteridium holdenii* (Rowe, 1988). The rings highlight reproductive trusses.

suggests that male and female units consistently develop in similar locations on the plant's architecture. Examples include the bearing of either multiple synangia or multiple cupules on the fertile median rachis of the trifurcating fronds of some Mississippian lyginopteridalean pteridosperms (Fig 2.7A, B: Long, 1979; see also *Nystroemia pectiniformis sensu:* Hilton and Li, 2003; Wang and Pfefferkorn, 2009), the supposed development of large uniovulate cupules and fused 'megasynangia' directly on the large fronds of some Pennsylvanian medullosan pteridosperms (Fig 2.7E, F: Halle, 1933; Stewart and Delevoryas, 1956; Retallack and Dilcher, 1988), and the distichous rows of either cupules or typically quadripartite (occasionally tripartite or pentapartite) synangia borne laterally on apparently naked

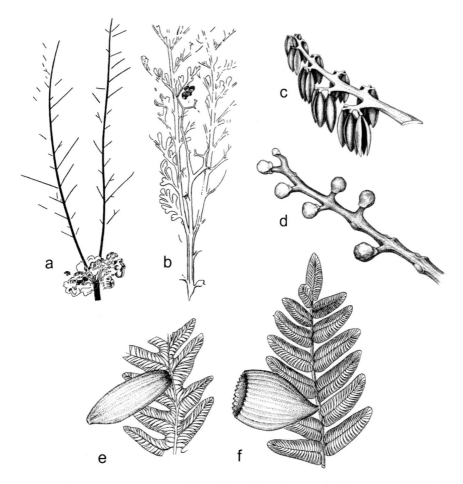

Fig 2.7 Line drawings comparing the male and female reproductive trusses of selected pteridosperms, showing their positional equivalence within the vegetative architecture of the respective families: Mississippian lyginopteridalean female (a) and male (b); Pennsylvanian medullosan putative female (e) and male (f); Jurassic caytonialean male (c) and female (d). (a) *Sphenopteris bifida* (Long, 1979), (b) *Diplopteridium teilianum* (Walton, 1931), (c) *Caytonanthus arberi* (Crane, 1985, after Harris, 1941), (d) *Caytonia nathorstii* (Crane, 1985, after Thomas, 1925), (e) putative *Pachytesta* ovule attached to *Alethopteris* foliage (Stewart and Rothwell, 1993, after Halle, 1933), (f) *Whittleseya* pollen organ attached to *Neuropteris* foliage (Stewart and Rothwell, 1993, after Stewart and Delevoryas, 1956).

axes by some Jurassic caytonialean pteridosperms (Fig 2.7c, d; Jurassic–Lower Cretaceous: Thomas, 1925; Harris, 1964) – an interpretation recently challenged by Wang (2010).

The development of male and female reproductive units directly on the large fronds traditionally attributed to the conceptual whole-plant *Medullosa* is confidently illustrated in most palaeobotany textbooks, but appeared developmentally improbable to the present authors. Such scepticism encouraged Seyfullah and Hilton (2009) to re-examine one of the key compression specimens from the Lower Permian of China and hence to issue two cautionary notes. Ovules supposedly attached to the leaf-genus *Pecopteris* were actually superimposed on the leaves of marattialean ferns, and an 'ovule' supposedly attached to another leaf-genus, *Alethopteris* (Fig 2.7e: cf. Halle, 1933), did indeed prove to be attached but was re-interpreted as a pollen-organ. Despite this misidentification of gender, this specimen confirmed the ability of medullosan pteridosperms to bear fertile organs directly on the frond. We continue to believe that developmental credibility should have an important part to play in assessing the likely accuracy of conceptually reconstructed fossil plants.

The obvious exceptions to the rule of structurally and positionally similar male and female organs are found among the extant gymnosperms: specifically, the conifers, characterized by their supposed compound female cone and simple male cone (Florin, 1931; Wilde, 1944; Meyen, 1987), and arguably *Cycas*, which bears pseudowhorls ('zones') of isolated megasporophylls (Stevenson, 1990; Zhang et al., 2004) and is placed as sister to the remaining cycads in both morphological (Crane, 1988; Stevenson, 1992) and molecular (Hill et al., 2003; Rai et al., 2003) phylogenetic analyses. Most phylogenies show the characteristically dioecious cycads and *Ginkgo* as being separated phylogenetically from angiosperms by fewer nodes than are conifers, or in some cases than Gnetales (e.g. the maximum likelihood tree shown in Fig 2.1, generated from 17 coding and 6 non-coding plastid regions by Graham and Iles, 2009). Indeed, some molecular phylogenetic studies still place particular extant gymnosperm lineages as sister to angiosperms. For example, the 'slow nucleotides only' analysis of the *PHYC/PHYN* coding nuclear regions by Mathews (2009) placed cycads as sister to angiosperms.

Superficially, such topologies appear improbable, but if the lineages leading to extant conifers and extant angiosperms really did diverge as early as the Late Carboniferous (*c.* 290 Ma), placement of various pteridosperm groups on the stems of both lineages would be an expectation rather than a surprise. In this context, extant cycads such as *Cycas* are best regarded as being pteridosperms whose morphology in general, and reproductive morphology in particular, were established as early as the end of the Carboniferous but which happened to persist to the present day (e.g. Hilton and Bateman, 2006).

2.6 Are hermaphrodite bennettites a phylogenetic red herring?

We offer this section as a reminder that the angiosperms are not the only seed-plant lineage to have evolved hermaphrodite reproductive units (i.e. flowers); plant morphology textbooks routinely feature the remarkable hermaphrodite flowers of some (gymnospermous) bennettites. Although bennettites of family Williamsoniaceae bore monosporangiate reproductive units, at least some and possibly all species within family Cycadeoidaceae bore bisexual units, including the textbook classics *Cycadeoidea* (e.g. Crepet, 1974; Rothwell and Stockey, 2002) and *Williamsoniella coronata* (e.g. Harris, 1964). The presence of flowers has helped to embed the bennettites deeply in the infamous, highly derived 'anthophyte' clade that contains extant angiosperms and Gnetales together with extinct bennettites, *Pentoxylon* and arguably also the putatively highly derived pteridosperm *Caytonia* (Crane, 1985; Doyle and Donoghue, 1986, *et seq.*). Bisexual bennettite flowers also helped to raise the profile of heterotopy as a potential mechanism for generating hermaphrodite organs, via the gamoheterotopic model of Meyen (1982, 1988).

Rothwell et al. (2009) recently used an experimental morphological cladistic analysis of seed-plants to attempt to scotch several sacred cows in seed-plant phylogenetics. Character comparison was used to reject previous individualistic sister-group relationships advocated for bennettites + cycads (Nixon et al., 1994) and bennettites + Gnetales (Friis et al., 2007), as well as the more commonly recovered sister-group relationship between *Caytonia* and angiosperms (e.g. Doyle, 1996; Hilton and Bateman, 2006). The preferred matrix of Rothwell et al. (analysis 3) yielded a clade of 'higher seed-ferns' *sensu* Hilton and Bateman (2006), consisting of corystosperms plus *Glossopteris* plus the 'conservatively' re-coded *Caytonia*, that was placed even lower in the tree than the divergences of cordaites, various conifer lineages and *Ginkgo*. In contrast, the remainder of the anthophyte taxa – bennettites, *Pentoxylon* and Gnetales – remained in their 'classical' positions immediately below basal extant angiosperms. Although Rothwell et al. (2009) rejected several previously suggested homologies between the reproductive structures of bennettites and Gnetales, these two groups nonetheless continued to occupy adjacent nodes on the resulting tree.

However, Rothwell et al. (2009) followed most if not all previous broad-brush morphological cladistic studies in using a generalized composite reconstruction as a placeholder for bennettites (cf. Rothwell and Serbet, 1991; Nixon et al., 1994; Doyle, 1996; Hilton and Bateman, 2006; Rothwell et al., 2009). The only morphological phylogenetic study to explore relationships among bennettites (Crane, 1988) recognized only a minority of bennettites as bearing hermaphrodite flowers and resolved those hermaphrodite taxa as a single highly derived clade of Cycadeoidaceae *s.s.* If the character codings and resulting topology are correct, the flowers of *Williamsoniella* and *Cycadeoidea* are pertinent to the origin of the angiosperms not as a potentially ancestral condition but rather as a parallel origin of developmentally fixed bisexuality. In this context, it is unfortunate that the fossil status of bennettites renders them genetically untestable.

Accepting the traditional view that there existed monoecious or dioecious unisexual bennettites and that they were comparatively primitive (Crane, 1988) strongly suggests that, despite their 'centrifugal' zonation of multiple ovules, multiple pollen sacs and tepal-like structures, the hermaphrodite flowers of derived bennettites such as *Williamsoniella* and *Cycadeoidea* are convergent on, rather than synapomorphic with, those of early-divergent angiosperms (Bateman et al., 2006; Rudall and Bateman, 2010). Recent studies show that suitable pollinators were available to early bennettites from the late Middle Jurassic onward, in the form of brachyceran flies that possessed long prosces (Ren et al., 2009). This observation implies that similar co-evolutionary selective pressures could have affected the gnetaleans, bennettitaleans and angiosperms that formed well-known Lower Cretaceous floras such as Jehol, China (e.g. Duan, 1988; Zhou et al., 2002; Rydin et al., 2006) and Crato, Brazil (e.g. Kunzmann et al., 2009).

We conclude that the phylogenetic position of bennettites has not been adequately tested due to their routine treatment as a single aggregate phylogenetic unit (a conclusion reached as early as the mid-1980s by Doyle and Donoghue, 1986). The resulting placement of the bennettites within two nodes of the base of the angiosperms in most morphological cladograms may be incorrect, and the scoring of derived character states within bennettites is likely to cause misleading optimizations (Bateman et al., 2006; Rudall and Bateman, 2010). We are inclined to agree with Rothwell et al. (2009) and earlier authors (e.g. Doyle, 1994) that, on present evidence, bennettite flowers originated independently of those of angiosperms and thus constitute phylogenetic 'red herrings'. In theory, bennettite reproductive units could still prove invaluable by functioning as an independent test of how the architecture of the hermaphrodite flower could originate. For example, strong axial compression is needed to generate the compact bennettite flower, and such compression can be traced back at least as far as Mississippian hydrasperman pteridosperms, where the formation of compact 'megacupules' greatly improved protection of multiple ovules (Long, 1975, 1977b).

2.7 Knowledge of gender control in extant gymnosperms and angiosperms remains inadequate

Despite the ecological success enjoyed by many autogamous angiosperms in the present-day flora, the supposedly great selective advantage of allogamy in most ecological contexts remains one of the fundamental tenets of modern genetic and evolutionary thinking (e.g. Levin, 2000; Coyne and Orr, 2004). Avoidance of genetic stagnation through cross-fertilization provides the obvious explanation for the determined rejection of hermaphrodism by gymnosperms. Available (albeit surprisingly limited) data suggest that intrinsic (biochemical) sterility barriers are absent from extant gymnosperms (e.g. Husband and Schemske, 1996; Owens et al., 1998, but see the account of overdominant lethal factors in *Pinus* embryos by Williams et al., 2003), so they must rely on spatial or temporal separation to at least restrict self-pollination. This 'adaptive goal' is made even more challenging by the fact that the majority of extant gymnosperms produce vast quantities of pollen grains that are transported to the ovules not by animal but rather by comparatively indiscriminate physical vectors, most commonly wind.

It is possible that the increased use of animal pollen vectors inferred for some fossil gymnosperms from the Late Jurassic–Early Cretaceous onwards (e.g. Labandeira, 1997; Ren et al., 2009) and for the great majority of angiosperms, both extinct and extant, greatly reduced the frequency of self-fertilization among hermaphrodite reproductive structures. This inference applies most strongly to taxa such as *Archaefructus*, where the lack of axial compression maintained considerable distance between carpels and stamens, and the former were located well above the latter, thus precluding self-pollination through purely gravitational effects (admittedly, water transport of pollen is feasible in this putative aquatic).

Doyle (pers. comm., 2010) suggested that the earliest angiosperms inherited either protandry or protogyny from their (most likely monoecious) gymnosperm ancestor as a necessary pre-requisite for compression of lateral organs in the determinate reproductive axes of angiosperm flowers (i.e. for phase 2 of the hypothesis of the origin of the angiosperm advanced by evo-devo specialists Baum and Hileman, 2006). Although temporally separated maturation of the two genders might have been sufficient to mitigate the worst effects of autogamy, we speculate that it was the advent of intrinsic (biochemical) sterility barriers that encouraged the predominance of hermaphrodism in angiosperms. This interpretation places emphasis on the respective timings attributed to the origins of sporophytic and gametophytic self-incompatibility in the angiosperm lineage (e.g. Zavada and Taylor, 1986).

The genetics of gender control in angiosperms varies greatly, from a short gene region (most common) to multiple sex chromosomes (Vyskot and Hobza, 2004; Meagher, 2007). Sex chromosomes apparently cause differential expression of B-class 'maleness' genes. Unfortunately, information on the genetics and epigenetics of gender control is available for surprisingly few angiosperms and especially gymnosperms. Although extant *Gingko* has long been reputed to have heterogametic 'XY' males (e.g. Lee, 1954), this conclusion failed to survive a more recent critical review of the available evidence (Hizumu, 1997). This general ignorance precludes serious exploration of key elements in the competing evo-devo hypotheses, not least the widely predicted transition from dioecy to gynodioecy. Moreover, if gymnosperms and basally divergent angiosperms do indeed share a gender-control mechanism that is simple and metastable, and most likely relies on threshold effects across clines of phytohormones with at least some of the mediation of expression being epigenetic, the potential structural and developmental homologies require much greater study.

2.8 Conclusions

(1) Fossil evidence of gender expression and underlying sexual systems in seed-plants will continue to be ambiguous, but nonetheless important, because of the crucial phylogenetic positions occupied by some wholly extinct groups. In contrast, it is self-evident that processes controlling gender can only be studied in extant species.

(2) At present, most knowledge of the developmental-genetic control of seed-plants relates to angiosperms in general and a small number of 'model' angiosperms (mostly derived eudicots) in particular. Within gymnosperms, conifers (notably Pinaceae) have been the preferred study group – a trend prompted more by the economic importance of the family than by its phylogenetic position or its credibility as a source of model organisms (cf. Lev-Yadun and Sederoff, 2000).

(3) We suspect that the likelihood of translocation of female (or male) function to a cone of the opposite gender is inversely proportional to the degree of divergence in the developmental programming of the two genders of cone. Conifers are not only phylogenetically highly divergent (Fig 2.1) but also have the most structurally contrasting male and female reproductive units of any seed-plant group; this developmental disparity is predicted by the textbook hypothesis that the female cone is compound (polyaxial) whereas the male cones have been evolutionarily 'thinned' to leave only a single simple uni-axial unit (cf. Rudall and Bateman, 2010).

(4) Ergo, we are likely to learn more about reproductive conditions pertaining in the gymnospermous ancestor of angiosperms by studying extant gymnosperms exhibiting stronger similarities in their male and female developmental programming, namely *Ginkgo* and especially cycads (Mundry and Stützel, 2004b). These lineages are probably phylogenetically less distant from, and certainly more developmentally similar to, angiosperms than are conifers, and have less disparate male and female developmental programmes; they could legitimately be viewed as living pteridosperms.

(5) Control of gender in gymnosperms is at best metastable when viewed developmentally. Its subtle and gradational nature suggests that it is likely to depend on clines of phytohormones and/or transcription factors that could in turn reflect very small genetic (or epigenetic) changes. One possible control mechanism would be clines of gibberellins such as GA3 established via differential methylation of the relevant genes.

(6) The transition from dioecy to gynodioecy predicted by the majority of the evo-devo hypotheses addressing the origin of the angiosperms has not even

been explored on theoretical grounds, let alone tested empirically. Moreover, epigenetic effects (e.g. methylation) liaising between genome and environment are likely to be crucial elements of any such transition, yet these too have been seriously under-researched. We conclude that it has become essential to reformulate for the twenty-first century the formerly vibrant discipline of phytohormone-focused physiology (e.g. Heslop-Harrison, 1959; Ross and Pharis, 1987), which was wrongly judged passé by 'progressive' physiologists operating towards the end of the twentieth century.

(7) Homology determinations surrounding the origin of the angiosperms remain problematic, irrespective of whether the comparison involves data on static morphology, dynamic morphogenesis or gene expression. The requirement for discrete unitary organs implicit in both the palaeobotanically inspired telome theory (e.g. Kenrick, 2002) and the evo-devo-inspired ABC model of floral organ zonation (e.g. Coen and Meyerowitz, 1991) may have negatively constrained thinking on the evolutionary implications of developmental change. Similarly, the supposedly fundamental distinction in gene phylogenies between orthology (separation of gene lineages during speciation) and paralogy (simple gene duplication) appears less clear-cut on closer inspection, especially as the proportion of speciation events considered to have been *driven by* genome duplication, is increasing rapidly.

(8) Phylogenetic evidence suggests that the origin of seed-plants coincided with maximal spatial and/or temporal separation of male and female reproductive structures. The converse pattern is evident among basal extant angiosperms, though the occurrence of several monoecious and dioecious groups among these lineages means that it is unclear whether hermaphrodism is implicated in the origin of the angiosperms or merely in their early diversification. Protandry and/or protogyny, most likely followed by the acquisition of intrinsic sterility barriers, probably reinforced any earlier selective advantages of hermaphrodism in angiosperms.

(9) Evolutionary-developmental genetics has contributed substantially to development of models of the origins of major groups such as the angiosperms, but the discipline has proven less successful when attempting to test those models (cf. Frohlich, 2006; Kellogg, 2006). Changes in our understanding of *LEAFY* and *AGAMOUS* expression patterns in conifers, notably the radical revisions offered by Vazquez-Lobo et al. (2007) and Groth et al. (2010), provide a useful reminder of the dangers of becoming too wedded to one particular evolutionary model; any model is easily ruined by irritating little facts. Models – either process-based or model organisms – make good servants, but poor masters.

(10) Only fossils can bridge the vast morphological gap that separates extant gymnosperms from extant angiosperms. However, identifying the first angiosperm

to appear in the fossil record essentially becomes a question of definitions – of deciding which of several synapomorphies suggested as delimiting the angiosperms, each presently surrounded by some form of ambiguity, should be chosen as pre-eminent (Bateman and DiMichele, 2003; Bateman et al., 2006). It remains questionable whether, if we were fortunate enough to find in the fossil record the first *bona fide* angiosperm, we would recognize it as such.

Acknowledgements

We thank Paco Vergara-Silva, Raymond van der Ham and Julien Bachelier for providing valuable additional observations on teratological conifer cones, and Jim Doyle, Peter Endress, Margarita Remizova and Dmitry Sokoloff for commenting on the manuscript. This contribution was part of a broader project funded by NERC (NE/E004369/1).

2.9 References

Adams, K. L. and Wendel, J. F. (2005). Polyploidy and genome evolution in plants. *Current Opinion in Plant Biology*, **8**, 135–141.

Albert, V. A., Oppenheimer, D. G. and Lindqvist, C. (2002). Pleiotropy, redundancy and the evolution of flowers. *Trends in Plant Sciences*, **7**, 297–301.

Andrews, H. N. (1948). Some evolutionary trends in the pteridosperms. *Botanical Gazette*, **110**, 13–31.

Barrett, S. C. H. (2002). The evolution of plant sexual diversity. *Nature Reviews Genetics*, **3**, 274–284.

Bartlett, A. W. (1913). Note on the occurrence of an abnormal bisporangiate strobilus of *Larix europaea*, DC. *Annals of Botany*, **27**, 575–576.

Bateman, R. M. (1996). Nonfloral homoplasy and evolutionary scenarios in living and fossil land plants. pp. 91–130 in Sanderson, M. J. and Hufford, L. (eds.), *Homoplasy: The Recurrence of Similarity in Evolution*. San Diego: Academic Press.

Bateman, R. M. and DiMichele, W. A. (1994). Heterospory: the most iterative key innovation in the evolutionary history of the plant kingdom. *Biological Reviews*, **69**, 345–417.

Bateman, R. M. and DiMichele, W. A. (2002). Generating and filtering major phenotypic novelties: neoGoldschmidtian saltation revisited. pp. 109–159 in Cronk, Q. C. B., Bateman, R. M., and Hawkins, J. A. (eds.), *Developmental Genetics and Plant Evolution*. London: Taylor and Francis.

Bateman, R. M. and DiMichele, W. A. (2003). Genesis of phenotypic and genotypic diversity in land plants: the present as the key to the past. *Systematics and Biodiversity*, **1**, 13–28.

Bateman, R. M. and Hilton, J. (2009). Palaeobotanical systematics for the phylogenetic age: applying organ-species, form-species and phylogenetic species concepts in a framework of reconstructed fossil and extant whole-plants. *Taxon*, **58**, 1254–1280.

Bateman, R. M., Hilton, J. and Rudall, P. J. (2006). Morphological and molecular phylogenetic context of the angiosperms: contrasting the 'top-down' and 'bottom-up' approaches used to infer the likely characteristics of the first flowers. *Journal of Experimental Botany*, **57**, 3471–3503.

Bateman, R. M. and Rudall, P. J. (2006). The Good, the Bad and the Ugly: using naturally occurring terata to distinguish the possible from the impossible in orchid floral evolution. *Aliso* (Monocot Special Volume), **22**, 481–496.

Baum, D. A. and Hileman, L. C. (2006). A developmental genetic model for the origin of the flower. pp. 3–27 in Ainsworth, C. (ed.), *Flowering and its Manipulation*. Sheffield: Blackwell.

Becker, A., Winter, K.-U., Meyer, B., Saedler, H. and Theissen, G. (2000). MADS-box gene diversity in seed plants 300 million years ago. *Molecular Biology and Evolution*, **17**, 1425–1434.

Bowman, J. L. (1997). Evolutionary conservation of angiosperm flower development at the molecular and genetic levels. *Journal of Biosciences*, **22**, 515–527.

Bowman, J. L., Smyth, D. R. and Meyerowitz, E. M. (1989). Genes directing flower development in *Arabidopsis*. *Plant Cell*, **1**, 37–52.

Burleigh, J. G. and Mathews, S. (2004). Phylogenetic signal in nucleotide data from seed plants: implications for resolving the seed plant tree of life. *American Journal of Botany*, **91**, 1599–1613.

Buzgo, M., Soltis, P. S. and Soltis, D. E. (2004). Floral developmental morphology of *Amborella trichopoda* (Amborellaceae). *International Journal of Plant Sciences*, **165**, 925–947.

Carlsbecker, A., Tandre, K., Johanson, U., Englund, M. and Engstrom, P. (2004). The MADS-box gene *DAL1* is a potential mediator of the juvenile to adult transition in Norway Spruce (*Picea abies*). *Plant Journal*, **40**, 546–557.

Caron, G. E. and Powell, G. R. (1991). Proliferated seed cones and pollen cones in young black spruce. *Trees*, **5**, 65–74.

Chamberlain, C. J. (1935). *Gymnosperms: Structure and Function*. Chicago: University of Chicago Press.

Coen, E. and Meyerowitz, E. M. (1991). The war of the whorls: genetic interactions controlling flower development. *Nature*, **353**, 31–37.

Colombo, L., Franken, J., Koetje, E. et al. (1995). The petunia MADS box gene *FBP11* determines ovule identity. *Plant Cell*, **7**, 1859–1868.

Coyne, J. A. and Orr, J. A. (2004). *Speciation*. Sunderland, MA: Sinauer.

Crane, P. R. (1985). Phylogenetic analysis of seed plants and the origin of angiosperms. *Annals of the Missouri Botanical Garden*, **72**, 716–793.

Crane, P. R. (1988). Major clades and relationships in the higher gymnosperms. pp. 218–272 in Beck, C. B. (ed.), *Origin and Evolution of Gymnosperms*. New York, NY: Columbia University Press.

Crepet, W. L. (1974). Investigations of North American cycadeoids: the reproductive biology of *Cycadeoidea*. *Palaeontographica B*, **148**, 144–169.

Dickson, A. (1860). Observations on bisexual cones in spruce fir (*Abies excelsa*). *Transactions of the Edinburgh Botanical Society*, **6**, 418–422.

Donoghue, M. and Doyle, J. A. (1989). Phylogenetic analysis of angiosperms and the relationships of Hamamelidae. pp. 17–45 in Crane, P. R. and Blackmore, S. (eds.), *Evolution, Systematics, and Fossil History of the Hamamelidae*, Vol. 1. Oxford: Clarendon Press.

Donoghue, M. J. (1989). Phylogenies and the analysis of evolutionary sequences, with examples from seed plants. *Evolution*, **43**, 1137–1156.

Donoghue, M. J. and Doyle, J. A. (2000). Seed-plant phylogeny: demise of the anthophyte hypothesis? *Current Biology*, **10**, R106–R109.

Dornelas, M. C. and Rodriguez, A. P. M. (2005). A *FLORICAULA/LEAFY* gene homolog is preferentially expressed in developing female cones of the tropical pine *Pinus caribaea* var. *caribaea*. *Genetics and Molecular Biology*, **28**, 299–307.

Doyle, J. A. (1994). Origin of the angiosperm flower: a phylogenetic perspective. *Plant Systematics and Evolution* (supplement), **8**, 7–29.

Doyle, J. A. (1996). Seed plant phylogeny and the relationships of Gnetales. *International Journal of Plant Sciences*, **157** (supplement), S3–S39.

Doyle, J. A. (2008). Integrating molecular phylogenetic and palaeobotanical evidence on origin of the flower. *International Journal of Plant Sciences*, **169**, 816–843.

Doyle, J. A. and Donoghue, M. J. (1986). Seed plant phylogeny and the origin of angiosperms: an experimental cladistic approach. *Botanical Review*, **52**, 321–431.

Doyle, J. A. and Donoghue, M. J. (1992). Fossils and seed plant phylogeny reanalyzed. *Brittonia*, **44**, 89–106.

Doyle, J. A. and Endress, P. K. (2000). Morphological phylogenetic analysis of basal angiosperms: comparison and combination with molecular data. *International Journal of Plant Sciences*, **161** (supplement), S121–S153.

Duan, S.Y. (1988). The oldest angiosperm – a tricarpous female reproductive fossil from western Liaoning province, NE China. *Science in China D*, **41**, 14–20.

Endress, P. K. (1992). Evolution and floral diversity: the phylogenetic surroundings of *Arabidopsis* and *Antirrhinum*. *International Journal of Plant Sciences*, **153** (supplement), S106–S122.

Endress, P. K. (1996). Structure and function of female and bisexual organ complexes in Gnetales. *International Journal of Plant Sciences*, **157** (supplement), S113–S125.

Endress, P. K. and Doyle, J. A. (2009). Reconstructing the ancestral angiosperm flower and its initial specializations. *American Journal of Botany*, **96**, 22–66.

Fairon-Demaret, M. and Scheckler, S. E. (1987). Typification and redescription of *Moresnetia zalesskyi* Stockmans, 1948, an early seed plant from the Upper Famennian of Belgium. *Bulletin de l'Institut Royal des Sciences Naturelles de Belgique*, **57**, 183–199.

Flores-Renteria, L., Whipple, A., Vazquez-Lobo, A. and Pinero, D. (2010) Are bisexual structures an innovation of angiosperms? Evidence of common mechanisms to produce bisexual structures in seed plants. p. 180 in *Abstracts of the Fourth Euro Evo Devo Conference* (Paris).

Florin, R. (1931). Untersuchungen zur Stammesgeschichte der Coniferales und Cordaiten. I. *Kungl Svenska Vertenskansapsakademians Handlingar*, **10**, 1–588.

Florin, R. (1939). The morphology of the female fructifications in cordaites and conifers of Palaeozoic age. *Botaniska Notiser*, **36**, 547–565.

Florin, R. (1951). Evolution in *Cordaites* and conifers. *Acta Horti Bergiani*, **15**, 285–388.

Friis, E. M., Crane, P. R., Pedersen, K. R. et al. (2007). Phase contrast X-ray microtomography links Cretaceous seeds with Gnetales and Bennettitales. *Nature*, **450**, 549–552.

Friis, E. M., Doyle, J. A., Endress, P. K. and Leng, Q. (2003). *Archaefructus*: angiosperm precursor or specialized early angiosperm? *Trends in Plant Science*, **8**, 369–373.

Frohlich, M. W. (2002). The Mostly Male theory of flower origins: summary and update regarding the Jurassic pteridosperm *Pteroma*. pp. 85–108 in Cronk, Q. C. B., Bateman, R. M. and Hawkins, J. A. (eds.), *Developmental Genetics and Plant Evolution*. London: Taylor and Francis.

Frohlich, M. W. (2006). Recent developments regarding the evolutionary origin of flowers. pp. 63–127 in Soltis, D. E., Leebens-Mack, J. and Soltis, P. S. (eds.), *Developmental Genetics of the Flower*. San Diego: Academic Press.

Frohlich, M. W. and Parker, D. S. (2000). The mostly male theory of flower evolutionary origins. *Systematic Botany*, **25**, 155–170.

Givnish, T. J. (1980). Ecological constraints of the evolution of breeding systems in seed plants: dioecy and dispersal in gymnosperms. *Evolution*, **34**, 959–972.

Graham, S. W. and Iles, W. J. D. (2009). Different gymnosperm outgroups have (mostly) congruent signal regarding the root of flowering plant phylogeny. *American Journal of Botany*, **96**, 216–227.

Groth, E., Tandre, K., Engstrom, P. and Vergara-Silva, F. (2010). The expression of patterns of *AGAMOUS* subfamily MADS-box genes in developing seed cones of different conifer families challenge old ideas about organ identity and organ homology. p. 185 in *Abstracts of the Fourth Euro Evo Devo Conference* (Paris).

Guo, C. L., Chen, L. G., He, X. H., Dai, Z. and Yuan, N. Y. (2005). Expression of *LEAFY* homologous genes in different organs and stages of *Ginkgo biloba*. *Hereditas*, **27**, 241–244. [In Chinese]

Halle, T. G. (1933). The structure of certain fossil spore-bearing organs believed to belong to pteridosperms. *Kungl Svenska Vertenskansapsakademians Handlingar*, **12**, 1–103.

Harris, T. M. (1941). *Caytonanthus*, the microsporophyll of *Caytonia*. *Annals of Botany*, **5**, 47–58.

Harris, T. M. (1964). *The Yorkshire Jurassic Flora. II. Caytoniales, Cycadales and Pteridosperms*. London: British Museum (Natural History).

Hernandez-Castillo, G. R., Rothwell, G. W. and Mapes, G. (2001). Thucydiaceae fam. nov., with a review and reevaluation of Paleozoic walchian conifers. *International Journal of Plant Science*, **162**, 1155–1158.

Heslop-Harrison, J. (1959). Growth substances and flower morphogenesis. *Journal of the Linnean Society, Botany*, **56**, 269–281.

Hill, K. D., Chase, M. W., Stevenson, D. W., Hills, H. G. and Schutzman, B. (2003). The families and genera of cycads: a molecular phylogenetic analysis of Cycadophyta based on nuclear and plastid DNA sequences. *International Journal of Plant Sciences*, **146**, 933–948.

Hilton, J. and Bateman, R. M. (2006). Pteridosperms are the backbone of seed-plant evolution. *Journal of the Torrey Botanical Society*, **133**, 119–168.

Hilton, J. and Li, C.-S. (2003). Reinvestigation of *Nystroemia pectiniformis* Halle, an enigmatic seed plant from the Late Permian of China. *Palaeontology*, **46**, 29–51.

Hizumu, M. (1997). Chromosomes of *Ginkgo biloba*. pp. 109–118 in Hari, T., Ridge, R. W., Twecke, W., Del Tredici, P., Trémouillaux-Guiller, J. and Tobe, H. (eds.), *Ginkgo biloba: A Global Treasure*. Tokyo: Springer.

Holmes, S. (1932). A bisporangiate cone of *Tsuga canadensis*. *Botanical Gazette*, **93**, 100–102.

Husband, B. C. and Schemske, D. W. (1996). Evolution of the magnitude and timing of inbreeding depression in plants. *Evolution*, **50**, 54–70.

Iles, W., Rudall, P. J., Sokoloff, D. D. et al. (2008). Phylogenetics of Hydatellaceae. *Botany 2008 (Botanical Society of America) Abstracts* (Vancouver, BC): Abstract #473.

Kellogg, E. A. (2006). Progress and challenges in studies of the evolution of development. *Journal of Experimental Botany*, **57**, 3505–3516.

Kenrick, P. (2002). The telome theory. pp. 365–387 in Cronk, Q. C. B., Bateman, R. M. and Hawkins, J. A. (eds.), *Developmental Genetics and Plant Evolution*. London: Taylor and Francis.

Kim, S., Koh, J., Yoo, M.-J. et al. (2005). Expression of floral MADS-box genes in basal angiosperms: implications for the evolution of floral regulators. *Plant Journal*, **43**, 724–744.

Kirkwood, J. E. (1916). Bisporangiate cones of *Larix [occidentalis]*. *Botanical Gazette*, **61**, 256–257.

Klavins, S. D., Taylor, T. N. and Taylor, E. L. (2002). Anatomy of *Umkomasia* (Corystospermales) from the Triassic

of Antarctica. *American Journal of Botany*, **89**, 664–676.

Kramer, E. M., Di Stilio, V. S. and Schluter, P. M. (2003). Complex patterns of gene duplication in the *APETALA3* and *PISTILLATA* lineages of the Ranunculaceae. *International Journal of Plant Sciences*, **164**, 1–11.

Kunzmann, L., Mohr, B. A. R. and Bernardes-de-Oliveira, M. E. C. (2009). *Cearania heterophylla* gen. et sp. nov., a fossil gymnosperm with affinities to the Gnetales from the Early Cretaceous of Northern Gondwana. *Review of Palaeobotany and Palynology*, **158**, 193–212.

Labandeira, C. C. (1997). Insect mouthparts: ascertaining the paleobiology of insect feeding strategies. *Annual Review of Ecology and Systematics*, **28**, 153–193.

Lee, C. L. (1954). Sex chromosomes in *Ginkgo biloba* L. *American Journal of Botany*, **41**, 545–549.

Levin, D. A. (2000). *The Origin, Expansion, and Demise of Plant Species*. Oxford: Oxford University Press.

Lev-Yadun, S. and Liphschitz, N. (1987). The ontogeny of gender of *Cupressus sempervirens* L. *Botanical Gazette*, **146**, 407–412.

Lev-Yadun, S. and Sederoff, R. (2000). Pines as model gymnosperms to study evolution, wood formation and perennial growth. *Journal of Plant Growth Regulation*, **19**, 290–305.

Littlefield, E. W. (1931). Bisporangiate inflorescences in *Pseudotsuga*. *Ohio Journal of Science*, **31**, 416–417.

Long, A. G. (1975). Further observations on some Lower Carboniferous seeds and cupules. *Transactions of the Royal Society of Edinburgh B*, **69**, 267–293.

Long, A. G. (1977a). Some Lower Carboniferous pteridosperm cupules bearing ovules and microsporangia. *Transactions of the Royal Society of Edinburgh B*, **70**, 1–11.

Long, A. G. (1977b). Lower Carboniferous pteridosperm cupules and the origin of the angiosperms. *Transactions of the Royal Society of Edinburgh B*, **70**, 13–35.

Long, A. G. (1979). The resemblance between the Lower Carboniferous cupules *Hydrasperma* cf. *tenuis* Long and *Sphenopteris bifida* Lindley & Hutton. *Transactions of the Royal Society of Edinburgh B*, **70**, 111–127.

Longman, K. A. (1985). Effects of growth substances on male and female cone initiation in conifers. *Biologia Plantarum*, **27**, 402–407.

Lynch, M. (2002). Gene duplication and evolution. *Science*, **297**, 945–947.

Madlung, A., Tyagi, A. P., Watson, B. et al. (2005). Genome changes in synthetic *Arabidopsis* polyploids. *Plant Journal*, **41**, 221–230.

Masters, M. T. (1890). Review of some points in the comparative morphology, anatomy, and life-history of the *Coniferae*. *Journal of the Linnean Society of London, Botany*, **27**, 226–328.

Mathews, S. (2009). Phylogenetic relationships among seed plants: persistent questions and the limits of molecular data. *American Journal of Botany*, **96**, 228–236.

Meagher, T. R. (2007). Linking the evolution of gender variation to floral development. *Annals of Botany*, **100**, 1–12.

Mellerowicz, E. J., Horgan, K., Walden, A., Coker, A. and Walter, C. (1998). *PRFLLF*, a *Pinus radiata* homologue of *FLORICAULA* and *LEAFY* is expressed in buds containing vegetative shoot and undifferentiated male cone primordia. *Planta*, **206**, 619–629.

Melzer, R., Wang, Y.-Q. and Thiessen, G. (2010). The naked and the dead: the ABCs of gymnosperms reproduction and the origin of the angiosperm flower. *Seminars in Cell and Developmental Biology*, **21**, 118–128.

Mergen, F. (1963). Sex transformation in pine hybrids. *Forest Science*, **9**, 258–262.

Meyen, S. V. (1982). Gymnosperm fructifications and their evolution as evidenced by palaeobotany. *Zhurnal Obscher Biologi*, **50**, 303–323. [In Russian]

Meyen, S. V. (1987). *Fundamentals of Palaeobotany*. London: Chapman and Hall.

Meyen, S. V. (1988). Origin of the angiosperm gynoecium by gamoheterotopy. *Botanical Journal of the Linnean Society*, **97**, 171–178.

Mouradov, A., Glassick, T., Hamdorf, B. et al. (1998). *NEEDLY*, a *Pinus radiata* ortholog of *FLORICAULA/LEAFY* genes, expressed in both reproductive and vegetative meristems. *Proceedings of the National Academy of Sciences USA*, **95**, 6537–6542.

Moyroud, E., Kusters, E., Monnieux, M., Koes, R. and Parcy, F. (2010). *LEAFY* blossoms. *Trends in Plant Science*, **15**, 346–352.

Mundry, M. and Stützel, T. (2004a). Morphogenesis of the reproductive shoots of *Welwitschia mirabilis* and *Ephedra distachya* (Gnetales), and its evolutionary implications. *Organisms, Diversity and Evolution*, **4**, 91–108.

Mundry, M. and Stützel, T. (2004b). Morphogenesis of leaves and cones of male short-shoots of *Ginkgo biloba* L. *Flora*, **199**, 437–452.

Nixon, K. C., Crepet, W. L., Stevenson, D. and Friis, E. M. (1994). A reevaluation of seed plant phylogeny. *Annals of the Missouri Botanical Garden*, **81**, 484–533.

Owens, J. N. and Molder, M. (1977). Bud development in *Picea glauca*. II. Cone differentiation and early development. *Canadian Journal of Botany*, **55**, 2746–2760.

Owens, J. N., Takaso, T. and Runions, C. J. (1998). Pollination in conifers. *Trends in Plant Science*, **3**, 479–485.

Pelaz, S., Ditta, G. S., Baumann, E., Wisman, E. and Yanofsky, M. F. (2000). B and C floral organ identity functions require *SEPALLATA* MADS-box genes. *Nature*, **405**, 200–203.

Phillips, T. L., Andrews, H. N. and Gensel, P. G. (1972). Two heterosporous species of *Archaeopteris* from the Upper Devonian of West Virginia. *Palaeontographica B*, **139**, 47–71.

Rai, H. S., O'Brien, H. E., Reeves, P. A., Olmstead, R. G. and Graham, S. G. (2003). Inference of higher-order relationships in the cycads from a large chloroplast data set. *Molecular Phylogenetics and Evolution*, **29**, 350–359.

Rai, H. S., Reeves, P. A., Peakall, R., Olmstead, R. G. and Graham, S. W. (2008). Inference of higher order conifer relationships from a multilocus plastid data set. *Botany*, **86**, 658–669.

Remizova, M. V., Sokoloff, D. D., Macfarlane, T. D. et al. (2008). Comparative pollen morphology in the early-divergent angiosperm family Hydatellaceae reveals variation at the infraspecific level. *Grana*, **47**, 81–100.

Ren, D., Labandeira, C. C., Santiago-Blay, J. A. et al. (2009). A probable pollination mode before angiosperms: Eurasian, long-proboscid scorpionflies. *Science*, **326**, 840–847.

Renner, O. (1904). Über Zwitterblüthen bei *Juniperus communis*. *Flora*, **93**, 297–300.

Retallack, G. J. and Dilcher, D. L. (1988). Reconstructions of selected seed ferns. *Annals of the Missouri Botanical Garden*, **75**, 1010–1057.

Righter, F. I. (1932). Bisexual flowers among the pines. *Journal of Forestry*, **30**, 873.

Ross, S. D. and Pharis, R. P. (1987). Control of sex expression in conifers. *Plant Growth Regulation*, **6**, 37–60.

Rothwell, G. W., Crepet, W. L. and Stockey, R. A. (2009). Is the anthophyte hypothesis alive and well? New evidence from the reproductive structures of Bennettitales. *American Journal of Botany*, **96**, 296–322.

Rothwell, G. W. and Mapes, G. (2001). *Barthelia furcata* gen. et sp. nov., with a review of Paleozoic coniferophytes and a discussion of coniferophyte systematics. *International Journal of Plant Sciences*, **162**, 637–667.

Rothwell, G. W. and Serbet, R. (1992). Pollination biology of *Elkinsia polymorpha*: implications for the origin of the gymnosperms. *Courier Forschungs-Institut Senckenburg*, **147**, 225–231.

Rothwell, G. W. and Serbet, R. (1994). Lignophyte phylogeny and the evolution of spermatophytes: a numerical cladistic analysis. *Systematic Botany* **19**, 443–482.

Rothwell, G. W. and Stockey, R. A. (2002). Anatomically preserved *Cycadeoidea* (Cycadeoidaceae), with a reevaluation of systematic characters for the seed cones of Bennettitales. *American Journal of Botany*, **89**, 1447–1458.

Rowe, N. P. (1988). New observations on the Lower Carboniferous pteridosperm *Diplopteridium* Walton and an associated synangiate organ. *Botanical Journal of the Linnean Society*, **97**, 125–158.

Rudall, P. J. (2008). Fascicles, filamentous structures and inside-out flowers: comparative ontogeny supports reinterpretation of morphological novelties in the mycoheterotrophic family Truridaceae. *International Journal of Plant Sciences*, **169**, 1023–1037.

Rudall, P. J. and Bateman, R. M. (2003). Evolutionary change in flowers and inflorescences: evidence from naturally occurring terata. *Trends in Plant Science*, **8**, 76–82.

Rudall, P. J. and Bateman, R. M. (2010). Defining the limits of flowers: the challenge of distinguishing between the evolutionary products of simple versus compound strobili. *Philosophical Transactions of the Royal Society B*, **365**, 397–409.

Rudall, P. J., Hilton, J., Vergara-Silva, F. and Bateman, R. M. (2011). Recurrent abnormalities in conifer cones and the evolutionary origins of flower-like structures. *Trends in Plant Sciences*, **16**, 151–159.

Rudall, P. J., Remizova, M. V., Beer, A. et al. (2008). Comparative ovule and megagametophyte development in Hydatellaceae and water lilies reveal a mosaic of features among the earliest angiosperms. *American Journal of Botany*, **101**, 941–956.

Rudall, P. J., Remizowa, M. V., Prenner, G. et al. (2009). Non-flowers near the base of extant angiosperms? Spatiotemporal arrangement of organs in reproductive units of Hydatellaceae, and its bearing on the origin of the flower. *American Journal of Botany*, **96**, 67–82.

Rydin, C., Källersjö, M. and Friis, E. M. (2002). Seed plant relationships and the systematic position of Gnetales based on nuclear and chloroplast DNA: conflicting data, rooting problems, and the monophyly of conifers. *International Journal of Plant Sciences*, **163**, 197–214.

Rydin, C., Wu, S.-Q. and Friis, E.-M. (2006). *Liaoxia* (Gnetales): ephedroids from the Early Cretaceous Yixian Formation in Liaoning, northeastern China. *Plant Systematics and Evolution*, **262**, 293–265.

Santamour, F. S. (1959). Bisexual conelets in spruce. *Morris Arboretum Bulletin*, **10**, 10–11.

Schooley, H. O. (1967). Aberrant ovulate cones in balsam fir. *Forest Science*, **13**, 102–104.

Serbet, R. and Rothwell, G. W. (1992). Characterizing the most primitive seed ferns. 1. A reconstruction of *Elkinsia polymorpha*. *International Journal of Plant Science*, **153**, 602–621.

Seyfullah, L. J. and Hilton, J. (2009). Re-evaluation of Halle's fertile pteridosperms from the Permian floras of Shanxi Province, China. *Plant Systematics and Evolution*, **279**, 191–218.

Shaw, W. R. (1896). Contributions to the life-history of *Sequoia sempervirens*. *Botanical Gazette*, **21**, 297–300.

Shindo, S., Sakakibara, K., Sano, R., Uedo, K. and Hasebe, M. (2001). Characterization of a *FLORICAULA/LEAFY* homologue of *Gnetum parviflorum* and its implications for the evolution of reproductive organs in seed plants. *International Journal of Plant Sciences*, **162**, 1199–1209.

Shiokawa, T., Yamada, S., Futamura, N. et al. (2008). Isolation and functional analysis of the *CjNdly* gene, a homolog in *Cryptomeria japonica* of *FLORICAULA/LEAFY* genes. *Tree Physiology*, **28**, 21–28.

Sokoloff, D. D., Remizowa, M. V., Macfarlane, T. D. and Rudall, P. J. (2008). Classification of the early-divergent angiosperm family Hydatellaceae: one genus instead of two, four new species, and sexual dimorphism in dioecious taxa. *Taxon*, **57**, 179–200.

Soltis, D. E., Chanderbali, A. S., Kim, S., Buzgo, M. and Soltis, P. S. (2007). The ABC model and its applicability to basal angiosperms. *Annals of Botany*, **100**, 155–163.

Soltis, P. S., Brockington, S. F., Yoo, M.-J. et al. (2009). Floral variation and floral genetics in basal angiosperms. *American Journal of Botany*, **96**, 110–128.

Specht, C. D. and Bartlett, M.E. (2009). Flower evolution: the origin and subsequent diversification of the angiosperm flower. *Annual Review of Ecology and Systematics*, **40**, 217–243.

Steil, W. N. (1918). Bisporangiate cones of *Pinus montana*. *Botanical Gazette*, **66**, 68.

Stevenson, D. W. (1990). Morphology and systematics of Cycadales. *Memoirs of the New York Botanical Garden*, **57**, 8–55.

Stevenson, D. W. (1992). A formal classification of the extant cycads. *Brittonia*, **44**, 220–223.

Stewart, W. N. and Delevoryas, T. (1956). The medullosan pteridosperms. *Botanical Review*, **22**, 45–80.

Stewart, W. N. and Rothwell, G. W. (1993). *Paleobotany and the Evolution of Plants.* Cambridge: Cambridge University Press.

Stockey, R. A. and Rothwell, G. W. (2009). Distinguishing angiophytes from the earliest angiosperms: a Lower Cretaceous (Valanginian-Hauterivian) fruitlike reproductive structure. *American Journal of Botany*, **96**, 323–335.

Stubblefield, S. P. and Rothwell, G. W. (1989). *Cecropsis luculentum* gen. et sp. nov., evidence for heterosporous progymnosperms in the Upper Pennsylvanian of North America.

American Journal of Botany, **76**, 1415–1428.

Sun, G., Ji, Q., Dilcher, D. L., Zheng, S., Nixon, K. C. and Wang, X. (2002). Archaefructaceae, a new basal angiosperm family. *Science*, **296**, 899–904.

Tandre, K., Albert, V. A., Sundas, A. and Engstrom, P. (1995). Conifer homologues to genes that control floral development in angiosperms. *Plant Molecular Biology*, **27**, 69–78.

Taylor, E. L. and Taylor, T. N. (2009). Seed ferns from the late Paleozoic and Mesozoic: any angiosperm ancestors lurking there? *American Journal of Botany*, **96**, 237–251.

Theissen, G. (2001). Development of floral identity: stories from the MADS house. *Current Opinion in Plant Biology*, **4**, 75–85.

Theissen, G. and Becker, A. (2004). Gymnosperm orthologues of class B floral homeotic genes and their impact on understanding flower origin. *Critical Reviews in Plant Sciences*, **23**, 129–148.

Theissen, G., Becker, A., Winter, K. U. et al. (2002). How the land plants learned their floral ABCs: the role of MADS-box genes in the evolutionary origin of flowers. pp. 173–205 in Cronk, Q. C. B., Bateman, R. M. and Hawkins, J. A. (eds.). *Developmental Genetics and Plant Evolution.* London: Taylor and Francis.

Theissen, G. and Saedler, H. (2001). Floral quartets. *Nature*, **409**, 469–471.

Thomas, H. H. (1925). The Caytoniales, a new group of angiospermous plants from the Jurassic rocks of Yorkshire. *Philosophical Transactions of the Royal Society of London B*, **213**, 299–363.

Thomson, R. B. (1949). The structure of the cone in the Coniferae. *Botanical Review*, **6**, 49–84.

Tosh, K. J. and Powell, G. R. (1986). Proliferated, bisporangiate, and other atypical cones occurring on young, plantation-grown *Larix laricina*. *Canadian Journal of Botany*, **64**, 469–475.

Vazquez-Lobo, A., Carlsbecker, A., Vergara-Silva, F. et al. (2007). Characterization of the expression patterns of *LEAFY/FLORICAULA* and *NEEDLY* orthologs in female and male cones of the conifer genera *Picea*, *Podocarpus*, and *Taxus*: implications for current evo-devo hypotheses for gymnosperms. *Evolution and Development*, **9**, 446–459.

Vyskot, B. and Hobza, R. (2004). Gender in plants: sex chromosomes are emerging from the fog. *Trends in Genetics*, **20**, 432–438.

Walton, J. (1931). Contributions to knowledge of Lower Carboniferous plants. III. Teilia Quarry. *Philosophical Transactions of the Royal Society B*, **219**, 347–379.

Walton, J. (1957). On *Protopitys* (Göppert): with a description of a fertile specimen *Protopitys scotica* sp. nov. from the Calciferous Sandstone Series of Dumbartonshire. *Transactions of the Royal Society of Edinburgh*, **63**, 333–340.

Wang, J. and Pfefferkorn, H. W. (2009). Nystroemiaceae, a new family of Permian gymnosperms from China with an unusual combination of features. *Proceedings of the Royal Society B*, **277**, 301–309.

Wang, X. (2010). Axial nature of the cupule-bearing organ in Caytoniales. *Journal of Systematics and Evolution*, **48**, 207–214.

Wilde, M. H. (1944). A new interpretation of coniferous cones: 1. Podocarpaceae (*Podocarpus*). *Annals of Botany*, **8**, 1–41.

Williams, C. G, Auckland, L. D., Reynolds, M. M and Leach, K. A. (2003). Overdominant lethals as part of the conifer embryo lethal system. *Heredity*, **91**, 584–592.

Worsdell, W. C. (1900). The structure of the female 'flower' in Coniferae: an historical study. *Annals of Botany*, **14**, 39–82.

Worsdell, W. C. (1901). The morphology of the 'flowers' of *Cephalotaxus*. *Annals of Botany*, **15**, 637–652.

Yao, X., Taylor, T. N. and Taylor, E. L. (1995). The corystosperm pollen organ *Pteruchus* from the Triassic of Antarctica. *American Journal of Botany*, **82**, 535–546.

Yao, X., Taylor, T. N. and Taylor, E. L. (1997). A taxodiaceous seed cone from the Triassic of Antarctica. *American Journal of Botany*, **84**, 343–354.

Yu, H., Ito, T., Wellmer, F. and Meyerowitz, E. M. (2004). Repression of *AGAMOUS-LIKE 24* is a crucial step in promoting flower development. *Nature Genetics*, **36**, 157–161.

Zahn, L. M., Kong, H., Leebens-Mack, J. H. et al. (2005). The evolution of the *SEPALLATA* subfamily of MADS-box genes: a preangiosperm origin with multiple duplications throughout angiosperm history. *Genetics*, **169**, 2209–2223.

Zanis, M. J., Soltis, P. S., Qiu, Y.-L., Zimmer, E. and Soltis, D. E. (2003). Phylogenetic analysis and perianth evolution in basal angiosperms. *Annals of the Missouri Botanical Gardens*, **90**, 129–150.

Zavada, M. S. and Taylor, T. N. (1986). The role of self-incompatibility and sexual selection in the gymnosperm-angiosperm transition: a hypothesis. *American Naturalist*, **128**, 538–550.

Zhang, P., Tan, H. T. W., Pwee, K. H. and Kumar, P. P. (2004). Conservation of class C function of floral organ development during 300 million years of evolution from gymnosperms to angiosperms. *Plant Journal*, **37**, 566–577.

Zhou, Z., Barrett, P. M. and Hilton, J. (2002). An exceptionally preserved Lower Cretaceous ecosystem. *Nature*, **421**, 807–814.

Ziermann, J., Ritz, M. S., Hameister, S. et al. (2009). Floral visitation and reproductive traits of *Stamenoid petals*, a naturally occurring homeotic mutant of *Capsella bursa-pastoris* (Brassicaceae). *Planta*, **230**, 1239–1249.

Zimmermann, W. (1952). Main results of the 'Telome Theory.' *Palaeobotanist*, **1**, 456–470.

2.10 Note added in proof

The topicality of comparative gender control in gymnosperms and angiosperms has been demonstrated by several important papers published since this chapter went to press; collectively, they provide both new insights and useful bibliographies.

Groth et al. (2011; see also Groth, 2010) extended past expression studies of the C-class *AGAMOUS* subfamily of MADS-box genes (Section 2.4.1) from 'model' Pinaceae to several genera of Cupressaceae s.l. and, albeit with less confidence, to Podocarpaceae and Taxaceae. All genera studied apparently contained just a single *AGAMOUS* orthologue that is active in both male and female reproductive units, despite supposedly determining 'femaleness'. Expression patterns among extant conifer families are sufficiently different to challenge structural homologies (notably those relating to the ovuliferous scale) previously assumed within the ovulate cones of Pinaceae relative to those of other coniferous families. Such family-specific expression of C-class genes parallels that reported by Vazquez-Lobo et al. (2007) for *LEAFY* and *NEEDLY*, complicating any attempt to erect a gymnosperm-wide hypothesis of gender control.

Flores-Renteria et al. (2011) added to our own roster of conifer cone terata (Section 2.4.2) by examining in detail many naturally occurring bisexual (teratological) cones of *Pinus johannis*. Both light microscopy and pollen viability assessments demonstrated fertility levels of both the ovules and pollen of bisexual cones equivalent to those of typical unisexual cones. Within bisexual cones, basal sporophylls reliably generated pollen and distal sporophylls produced ovules, though the proportions of each gender differed substantially between cones. Late developmental stages of bisexual cones mirrored those of the dominant gender; male-dominated bisexual cones rapidly shrivelled and were shed before the modest numbers of ovules in the cone could reach maturity, whereas female-dominated bisexual cones resembled wholly female cones in being sufficiently persistent to allow maturation of both pollen and ovules. Developmentally, at least, there remains a major distinction between male and female cones, though

Flores-Renteria et al. provided further evidence that the underlying controls are most likely clinal.

Benefiting from recent technological advances, Tavares et al. (2010) generated several thousand expressed sequence tags from three extant gymnosperms (*Ginkgo* representing the ginkgoaleans, *Zamia* representing the cycads, *Welwitschia* representing the gnetaleans) for comparison with *Arabidopsis* and rice. Although interpretation of the resulting EST data is problematic, it is clear that for these gymnosperms, the male and female cones share expression of statistically equal numbers of orthologous genes with the hermaphrodite angiosperm flowers. Tavares et al. (2010) argued that this outcome falsified Frohlich's (2002) Mostly Male theory but was consistent with (though by no means demonstrated) Theissen and Becker's (2004) Out of Male/Out of Female hypotheses (Section 2.3). However, Tavares et al. rightly noted that the key difference may lie in relative degrees of and/or timing of expression, rather than in the mere presence or apparent absence of expression. Constraints on quantification meant that Tavares et al. did not explore the implications of their data for the widely assumed nonhomology of the male versus female cones of gymnosperms. Presumably, this theory would have predicted at least a modest surplus of genes expressed in relatively complex, supposedly compound female cones relative to simple male cones, at least in conifers (cf. Rudall and Bateman, 2010).

Considered together, these recent neontological studies reinforce several of the trends identified, and predictions made, in the present chapter (Section 2.8). In particular, they usefully highlight the dangers of generalizing limited gene expression data across all extant conifers, and increase the phenotypic disparity perceived to separate Pinaceae from other extant families of conifers, thereby further challenging the wisdom of using Pinaceae as the sole 'model' conifer. All of the evo-devo hypotheses briefly critiqued in Section 2.3 remain credible; although the balance of power has arguably shifted recently, it is proving to be easier to falsify specific aspects of these hypotheses than to reject any hypothesis *in toto*. Also, genetic control of gender among gymnosperms appears increasingly consistent at the most fundamental level but increasingly diverse between and within the major groups of extant gymnosperms. Lastly, the supposed certainties provided by early interpretations of organ-identity genes are giving way to a far more complex and equivocal picture that offers greater roles to hormonal clines and other epigenetic phenomena.

Additional References

Flores-Renteria, L., Vazquez-Lobo, A., Whipple, A. V., et al. (2011). Functional bisporangiate cones in *Pinus johannis* (Pinaceae): implications for the evolution of bisexuality in seed plants. *American Journal of Botany*, **98**, 130–139.

Groth, E. (2010). *Functional Diversification Among MADS-Box Genes and the Evolution of Conifer Seed Cone Development*. Doctoral thesis, University of Uppsala.

Groth, E., Tandre, K., Engström, P. and Vergara-Silva, F. (2011). *Agamous* subfamily MADS-box genes and the evolution of seed cone morphology in Cupressaceae and Taxodiaceae.

Evolution & Development, **13**, 159–170.

Tavares, R., Cagnon, M., Negriutiu, I. and Mouchiroud, D. (2010). Testing the recent theories for the origin of the hermaphrodite flower by comparison of the transcriptomes of gymnosperms and angiosperms. *BMC Evolutionary Biology*, **10**, 240.

3

New flowers of Laurales from the Early Cretaceous (Early to Middle Albian) of eastern North America

Maria von Balthazar, Peter R. Crane,
Kaj Raunsgaard Pedersen and Else Marie Friis

3.1 Introduction

The increasing number of fossil angiosperm reproductive structures described from Cretaceous strata (e.g. Friis et al., 2006) has provided a wealth of new data for understanding aspects of early flowering-plant evolution. In particular, flowers retrieved from many newly discovered mesofossil floras are often three-dimensionally preserved, which permits detailed morphological and systematic analyses. They have thereby provided information on the phylogenetic diversity and reproductive biology of Cretaceous angiosperms (e.g. Friis et al., 2006, 2010). However, an important feature of the angiosperm fossil record from the Cretaceous is that many fossils, particularly from the Early Cretaceous, cannot readily be accommodated in living taxa at the family or genus level, either because they are too poorly preserved to show the diagnostic features needed for secure systematic placement, or because they show a mosaic of features found in several living groups, indicating that they represent extinct lineages on internal branches of the angiosperm phylogenetic tree. The focus of this paper is on two early fossils of the second kind. While

Flowers on the Tree of Life, ed. Livia Wanntorp and Louis P. Ronse De Craene. Published by Cambridge University Press. © The Systematics Association 2011.

their relationships to extant Laurales are secure, they show features indicating that they fall outside the circumscription of extant families in the order.

Studies of relationships among living angiosperms based on analyses of DNA sequences support the recognition of the Laurales as a monophyletic group of seven extant families (Calycanthaceae, Siparunaceae, Gomortegaceae, Atherospermataceae, Hernandiaceae, Monimiaceae, Lauraceae; Renner, 1999, 2005; Renner and Chanderbali, 2000). The Laurales are the sister group to Magnoliales and include between 2840 and 3340 species in about 92 genera (Renner, 2005). The Calycanthaceae are the well-supported sister group to the remainder of the order, the core Laurales (Fig 3.1), within which Atherospermataceae, Gomortegaceae and Siparunaceae also form a well-supported clade (e.g. Renner, 1999, 2005). Relationships among Lauraceae, Monimiaceae and Hernandiaceae are currently not settled securely (Renner and Chanderbali, 2000). Morphological data strongly support a sister relationship of Hernandiaceae and Lauraceae (e.g. Doyle and Endress, 2000; Endress and Doyle, 2009), as do some molecular analyses (e.g. Qiu et al., 1999, 2006). However, in other molecular analyses the pattern of relationships among these three families is not well resolved (e.g. Renner, 1999, 2005; Chanderbali et al., 2001; Soltis et al., 2007).

Fossil flowers related to Laurales are particularly significant in Cretaceous mesofossil floras and first appear around the Albian. Such flowers were already widely distributed by the mid-Cretaceous, and fossils of Laurales remain common through the Late Cretaceous. The Cretaceous fossil record of Laurales includes *Virginianthus calycanthoides* from the Early-Middle Albian Puddledock locality, Virginia, USA (Friis et al., 1994), and *Jerseyanthus*

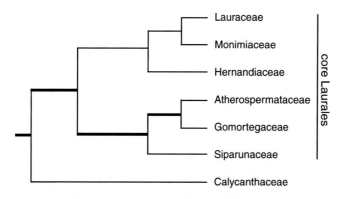

Fig 3.1 Phylogenetic relationships among extant Laurales based on molecular data, re-drawn from Renner (1999, 2005). Branches in bold are well supported, the others have moderate support. Morphological data support a slightly different pattern of relationships, in which Lauraceae and Hernandiaceae are sister taxa, with Monimiaceae sister to both (e.g. Doyle and Endress, 2000).

calycanthoides from the Turonian of New Jersey, USA (Crepet et al., 2005), which both closely resemble extant Calycanthaceae. *Araripia florifera* from the Late Aptian-Early Albian Crato Formation, Brazil (Mohr and Eklund, 2003) is also similar to extant Calycanthaceae in aspects of gross morphology, but it is much less well preserved and its assignment is less secure. *Lovellea wintonensis* from the mid-Cretaceous of Australia shows closer structural similarities to extant Gomortegaceae and Monimiaceae (Dettmann et al., 2009). The fossil flowers *Detrusandra amystagoaga* and *Cronquistiflora sayrevillensis* (Crepet and Nixon, 1998) show various combinations of characters seen in both Magnoliales and Laurales.

Most lauralean fossils from the Cretaceous are considered to be more closely related to extant Lauraceae than to other families within the order. They include *Potomacanthus lobatus* from Early-Middle Albian Puddledock locality, Virginia, USA (von Balthazar et al., 2007), as well as several different taxa described from many localities in North America, Europe and Eastern Asia that are of mid- and Late Cretaceous age (e.g. Retallack and Dilcher, 1981; Drinnan et al., 1990; Crane et al., 1994; Herendeen et al., 1994; Eklund and Kvaček, 1998; Eklund, 2000; Takahashi et al., 2001; Kvaček and Eklund, 2003; Frumin et al., 2004; von Balthazar et al., 2007; Viehofen et al., 2008). Especially distinctive and wide-spread is the extinct genus *Mauldinia*, for which four species have been described from North America, Europe and Central Asia (Drinnan et al., 1990; Eklund and Kvaček, 1998; Frumin et al., 2004; Viehofen et al., 2008). *Mauldinia* shares many characters with Lauraceae, although a sister-group relationship to Lauraceae and Hernandiaceae, or to the Lauraceae/Monimiaceae/Hernandiaceae clade is also possible, depending on certain assumptions regarding the characters of the fossils and the pattern of relationships among extant taxa (Doyle and Endress, 2010).

In this chapter, we present a new flower from the Early Cretaceous (Early-Middle Albian) Puddledock locality, Virginia, USA. We also formally describe a second lauralean floral structure from the same locality that was briefly considered in an earlier publication (Crane et al., 1994). Both taxa have features characteristic of extant Laurales, but provide new evidence of the floral diversity among lauralean lineages from the Early Cretaceous and thereby contribute to a better understanding of the early evolution of the order.

3.2 Materials and methods

3.2.1 Collection and preparation of the fossil material

The material was collected from the Tarmac Lone Star Industries sand and gravel pit (Puddledock locality), located south of Richmond, and east of the Appomattox

River, in Prince George County, Virginia, USA (see Friis et al., 1994, 1995). Palynological analyses (Christopher in Dischinger, 1987) indicate that the sediments at this locality can be referred to the basal part of subzone IIB in the palynological zonation established for the Potomac Group by Brenner (1963) and others (Doyle, 1969; Doyle and Hickey, 1976; Doyle and Robbins, 1977; Hickey and Doyle, 1977). Subzone IIB is of Middle Albian age, but may extend down into the Early Albian (Doyle, 1992).

One of the two taxa described here is known from a single bisexual flower bud (PP53716). The second taxon is known from three fragments of flowers (PP43735, PP43751, PP53825). All specimens are deposited in the palaeobotanical collections of the Field Museum of Natural History, Chicago (PP). Specimen PP53716 was isolated from sediment sample 083, while the specimens PP43735, PP43751, PP53825 are from sediment sample 082. Both samples were collected by P. R. Crane, A. N. Drinnan, E. M. Friis and K. R. Pedersen in 1988 and are two of several samples from organic-rich clay horizons exposed in a part of the Puddledock pit, which have now been lost through further quarrying and flooding of the pit. Sample 082 was collected immediately above sample 083 in the northern wall of the clay pit from the uppermost part of the fossil-bearing horizon, which at this location was about 10 metres above the bottom of the pit.

Plant fossils were extracted from bulk sediment samples by sieving in water followed by treatment with 40% HF, 10% HCl, and thorough rinsing in water. The fossils are typically small, charcoalified or lignitized, and comprise fragments of wood and twigs, as well as detached reproductive organs. The Puddledock flora contains a very diverse and abundant assemblage of angiosperm reproductive structures. Most are fruits and seeds, but there are also many well-preserved flowers, often with their three-dimensional form intact, as well as isolated stamens (Crane et al., 1994; Friis et al., 1994, 1995, 1997; von Balthazar et al., 2007, 2008).

For scanning electron microscopy (SEM) specimens were mounted on aluminium stubs, sputter coated with Au for 60 s, and studied using a Hitachi S-4300 field emission scanning electron microscope at 1–5 KV. Specimen PP53716 was also investigated using synchrotron-radiation X-ray tomographic microscopy (SRXTM) at the TOMCAT beamline of the Swiss Light Source at the Paul Scherrer Institute, Switzerland. For these X-ray studies, the specimen was re-mounted without further treatment on a brass stub, 3 mm in diameter, and examined using the technique outlined by Donoghue et al. (2006). Slice data derived from the scans were then analysed and manipulated using Avizo™ <www.tgs.com> software for computed tomography.

3.2.2 Phylogenetic analyses and interpretations

To evaluate the phylogenetic position of the fossil taxon *Cohongarootonia* (PP53716), the specimen was scored according to the characters and character states in the morphological data set for recent basal angiosperms developed by Endress and

Doyle (2009) and Doyle and Endress (2010). This data set also included placeholders for eudicots and monocots. Thirty-nine reproductive characters were scored as follows for the fossil (numbering according to Endress and Doyle, 2010): 47 sex of flower (bisexual, unisexual), bisexual 0; 48 hypanthium (absent and superior ovary, present and superior ovary, inferior ovary), present and superior ovary 1; 49 receptacle (female) (short, elongate), short 0; 52 floral apex (used up, protruding, one-carpellate taxa scored as unknown), ?; 53 perianth (present, absent), present 0; 54 perianth phyllotaxis (spiral, whorled), whorled 1; 55 perianth merism (trimerous, dimerous, polymerous), trimerous 0; 56 'perianth whorls (series)' (one, two, more than two), two 1; 57 tepal differentiation (all sepaloid, outer sepaloid and inner petaloid, all petaloid), all sepaloid 0; 58 petals (absent, present), absent 0; 59 inner perianth nectaries (absent, present), absent 0; 60 outer perianth fusion (free, at least basally fused), at least basally fused 1; 61 bracteate calyptra (absent, present), absent 0; 62 stamen number (more than one, one), more than one 0; 63 androecium phyllotaxis (spiral, whorled), whorled 1; 64 androecium merism (trimerous, dimerous, polymerous), trimerous 0; 65 'androecium whorls (series)' (one, two, more than two), more than two 2; 66 stamen positions (single, double), single 0; 67 stamen fusion (free, connate), free 0; 68 inner staminodes (absent, present), present 1; 69 food bodies (absent, on stamens or staminodes), absent 0; 70 stamen base (short, 'long, wide', 'long, narrow'), short 0; 71 basal stamen glands (absent, present), present 1; 72 connective apex (extended, truncated or smoothly rounded, peltate), truncated or smoothly rounded 1; 73 pollen sacs (protruding, embedded), embedded 1; 74 microsporangia (four, two), two 1; 75 orientation of dehiscence (distinctly introrse, latrorse-slightly introrse, extrorse), latrorse to slightly introrse 1; 76 anther dehiscence (longitudinal slit, 'H-valvate', 'upward-opening flaps'), upward-opening flaps 2; 96 carpel number (one carpel, '2–5, one whorl (series)', 'more than 5, one whorl', more than one whorl), one 0; 101 style (absent [sessile stigma], present), present 1; 102 stigma (extended, restricted), restricted 1; 103 stigma protuberances (absent, present), absent 0; 106 carpel fusion (apocarpous, parasyncarpous, eusyncarpous, one-carpellate taxa scored as unknown), ?; 108 long hairs on carpels (absent, present), present 1; 109 curved hairs on carpels (absent, present), absent 0; 110 abaxial carpel nectaries (absent, present), absent 0; 111 septal nectaries (absent, present), absent 0; 112 ovule number (one, 'two [or one and two]', more than two), one 0; 113 placentation (ventral, laminar or dorsal), ventral 0; 114 ovule direction (pendant, horizontal, ascendent), pendant 0.

Character states were optimized on a topology (the backbone tree) based on the combined analysis of morphology, 18S nrDNA, *rbcL* and *atpB* by Doyle and Endress (2000), but also incorporating changes and updates based on more recent analyses (e.g. Doyle et al., 2008; Endress and Doyle, 2009; Doyle and Endress, 2010). Ancestral states were reconstructed using the Ancestral State Reconstruction Package implemented in Mesquite (Maddison and Maddison, 2007) and the standard parsimony model (all characters treated as unordered). The fossil was linked iteratively to all possible branches of the backbone tree and the number of required parsimony changes (steps) was calculated for each resulting tree.

3.3 Description of fossil flowers

3.3.1 Bisexual flower bud (PP53716)

Systematics

Order – Laurales
Family Relationships – Lauraceae, Hernandiaceae
Genus – *Cohongarootonia* gen. nov.
Derivation of generic name – From the native American Algonquian name Cohongarooton for the Potomac River, after which the Potomac Group is named.

Generic diagnosis – Flower small, bisexual and actinomorphic, subtended by a bract. Perianth of six free tepals arranged in two alternating trimerous whorls. Organs of the outer perianth whorl broader than those of inner whorl and enclosing all inner floral organs. Long trichomes are present on the adaxial and abaxial surfaces of all tepals and oil cells are present in all tepals. Androecium of six stamens, followed by six staminodes, arranged in four alternating trimerous whorls. Stamens with distinct filaments; anthers basifixed without distinct joint between anther and filament. Anthers apparently disporangiate, dehiscing by two apically hinged flaps. Trichomes present on filaments; oil cells are present in filaments and anthers. Staminodes with short paired lateral appendages, adnate to the base of the filaments; appendages spade-shaped. Gynoecium of one superior, unilocular carpel containing a single pendant ovule. Trichomes present on style and ovary.

Type species – *Cohongarootonia hispida* sp. nov.

Derivation of species name – Latin *hispidus* for hairy and rough, referring to the trichomes present on all organs.

Specific diagnosis – As for the genus.

Dimensions – Flower *c.* 1.2 mm long, *c.* 0.6 mm wide; outer perianth organs *c.* 0.8 mm long, 0.5–0.6 mm wide; filament *c.* 0.2 mm long; anther *c.* 0.5 mm long; staminodes *c.* 0.15–0.2 mm long; carpel *c.* 0.6 mm long; style *c.* 0.4 mm long, ovary *c.* 0.2 mm long.

Holotype – PP53716 (illustrated in Figs 3.2, 3.3, 3.4 and 3.6B).

Type locality – Puddledock locality, Tarmac Lone Star Industries sand and gravel pit, located south of Richmond and east of Appomattox River in Prince George County, Virginia, USA.

Type stratum – Patapsco Formation, Potomac Group.

Age – Early Cretaceous (Basal part of Subzone IIB; Early-Middle Albian).

Description

Flower morphology – The flower is subtended by a bract. The flower is bisexual, *c.* 1.2 mm long and 0.6 mm broad (Fig 3.2A–B). The floral organs are organized in alternating trimerous whorls (Figs 3.4, 3.6B). The flower appears to be fossilized in a late bud stage, which is indicated by the well-developed reproductive organs that are still completely enclosed by the tepals (Fig 3.2).

Perianth – The perianth consists of six tepals in two alternating whorls (Figs 3.2A–E, 3.4). Tepals of both whorls are ovate. Tepals of the outer whorl are slightly longer and broader than the tepals of the inner whorl. The bases of the tepals of the outer whorl abut one another and thus enclose the remainder of the flower (Figs 3.2C, 3.4). Long trichomes are abundant on both the adaxial and abaxial sides of all perianth organs (Fig 3.2D–E). Oil cells are present in both outer and inner tepals.

Androecium – The androecium consists of six fertile stamens and six staminodes arranged in four alternating whorls. The stamens of the outer whorl alternate with the inner tepals (Figs 3.2E–G, 3.3A–D, 3.4). All stamens and staminodes are free from each other and free from the perianth. Stamens are *c.* 0.7 mm long and differentiated into anther and filament, but without a distinct joint at the base of the anther (Figs 3.2F–H, 3.3A–B). Filaments are broad and *c.* 0.2 mm long. Anthers are basifixed and *c.* 0.5 mm long. Anthers have a distinct rounded apical connective protrusion. The number of sporangia per theca appears to be one and anthers are thus disporangiate. Dehiscence of the thecae is latrorse to slightly introrse and the shape of the stomium indicates flap-like dehiscence through an apically hinged valve (Fig 3.3B). Anthers have extensive tissue between the pollen sacs and in the dorsal region of the connective (Fig 3.4). Staminodes are *c.* 0.15–0.2 mm long and characterized by paired, most likely nectariferous, appendages that are attached at the filament base (Fig 3.3B–D). The appendages are spade-like in shape and protrude beyond the short central staminodial tip. Trichomes are present on the filaments of both stamens and staminodes. Oil cells are present in the filaments and in the connective tissue of the anthers.

Pollen – No pollen grains were observed on the specimen.

Gynoecium – The gynoecium consists of a single carpel, *c.* 0.6 mm long (Figs 3.2I, 3.3E–I). The style is slender, curved and *c.* 0.4 mm long. The stigmatic area appears small. The ovary is *c.* 0.2 mm long and superior. Details of carpel form and carpel sealing are not preserved. A stylar canal is present for most of the length of the style (Fig 3.3H). Whether this canal was open or filled with secretion at anthesis is unknown. A single ovule is present in the ovary (Figs 3.2I, 3.3G–I). Insertion of the pendant ovule is median and ventral-apical; ovule curvature is unclear. The ovule does not fill the locule. Trichomes are present over the entire external surface of the carpel.

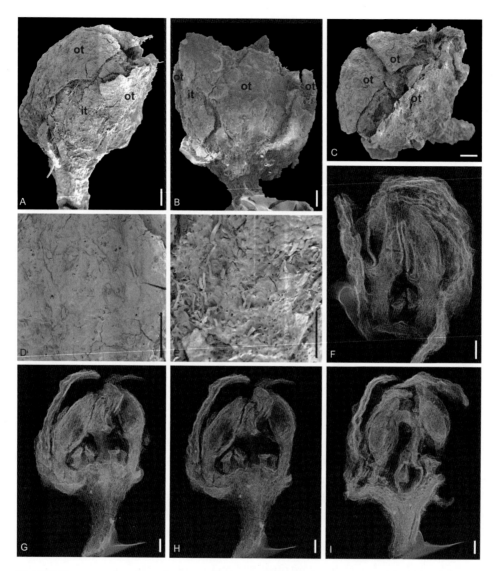

Fig 3.2 Flower and androecium morphology of *Cohongarootonia hispida* (PP53716). SEM micrographs and SRXTM micrographic reconstructions. (A–B) Lateral views of flower showing a whorl of three outer tepals. Only two of the inner tepals are visible. (C) Apical view of flower, showing the whorled arrangement of the three outer tepals. (D) Close-up of outer tepal surface scattered with trichomes and trichome bases. (E) Close-up of inner tepal surface showing dense trichomes and trichome bases. (F) Semi-transparent SRXTM reconstruction of flower showing three tepals, two stamens and two staminodes. (G) Semi-transparent SRXTM reconstruction showing tepals, stamens and staminodes. (H) Semi-transparent SRXTM reconstruction showing size and position of carpel and ovule. Abbreviations: it = inner tepal, ot = outer tepal. Scale bars = 100 μm.

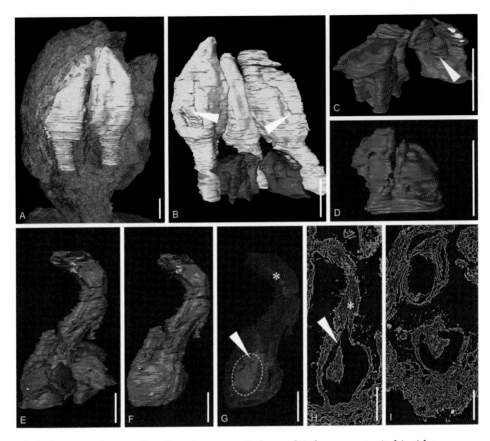

Fig 3.3 Gynoecium and androecium morphology of *Cohongarootonia hispida* (PP53716). SRXTM micrographic reconstructions and SRXTM sections. (A) Three-dimensional SRXTM reconstructions showing size and position of androecium in flower. (B) Three-dimensional SRXTM reconstructions showing three stamens (yellow) and three staminodes (pink and purple), lateral view. Arrowheads indicate apically hinged flaps. (C) Three-dimensional SRXTM reconstructions showing three staminodes with paired spade-like appendages, ventral view. Arrow head indicates short central staminodial tip. (D) Three-dimensional SRXTM reconstructions showing staminode with paired spade-like appendages, dorsal view. (E) Three-dimensional SRXTM reconstruction showing lateral view of carpel (blue) and three staminodes. (F) Three-dimensional SRXTM reconstructions showing lateral view of carpel. (G) Three-dimensional semi-transparent SRXTM reconstructions showing size and position of single ovule and stylar canal (asterisk). Arrowhead indicates point of ovule insertion. (H) Tangential SRXTM section of carpel, showing insertion of ovule (arrow head) and stylar canal (asterisk). (I) Transverse SRXTM section of carpel showing median and ventral-apical insertion of ovule. Scale bars: (A–H) = 100 μm. For colour illustration see plate section.

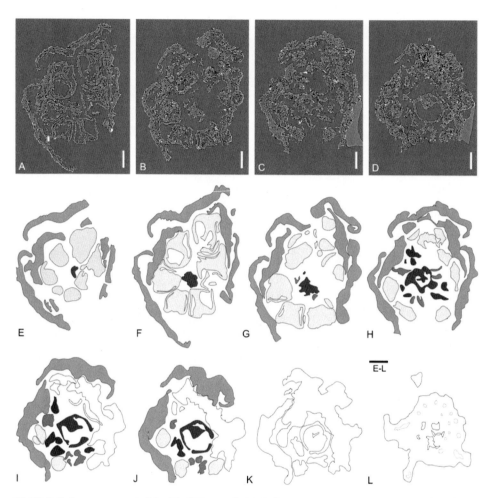

Fig 3.4 *Cohongarootonia hispida* (PP53716). Digital transverse SRXTM sections (A–D) and line drawing series of digital transverse sections (E–L), beginning at the apex and progressing to the base of the flower, indicating arrangement of organs (green = tepals, yellow = stamens, pink and purple = staminodes, blue = carpel). Scale bars = 100 μm. For colour illustration see plate section.

3.3.2 Flower fragments (PP43735, PP43751, PP53734)

Systematics

Order – Laurales
Family relationships – Lauraceae, Hernandiaceae, Monimiaceae
Genus – *Powhatania* gen. nov.
Derivation of generic name – From the native American name 'Powhatan' for the James River in Virginia, USA, to which watershed the Appomattox River with the Puddledock locality belongs.

Generic diagnosis – Flowers small (presumed actinomorphic). Perianth of three broad outer tepals and at least three narrow inner tepals (details of their arrangement unclear). Androecium of at least three stamens or staminodes (only filaments/filament bases preserved) and six stalked nectariferous structures. Arrangement of androecial organs in groups of three with the central filament base opposite each outer perianth organ, flanked by two associated stalked nectariferous structures. Filaments and nectariferous structures fused to the perianth base. Stamens with long and thin filaments (anthers not preserved). Nectariferous structures with short stalks and a peltate or semi-peltate head.

Type species – *Powhatania connata* sp. nov.

Derivation of species name – Latin *connatus* for united/fused, referring to the union of perianth organs and stamens.

Specific diagnosis – As for the genus.

Dimensions – Outer tepal *c*. 1. 4 mm wide; inner tepal *c*. 0.5 mm wide; associated nectariferous structure *c*. 0.5 mm long; semi-peltate/peltate head of nectariferous structure *c*. 0.3 mm wide; filament *c*. 0.5 mm long.

Holotype – PP43735 (illustrated in Figs 3.5, 3.6C).

Other specimens – PP43751, PP53734.

Type locality – Puddledock locality, Tarmac Lone Star Industries sand and gravel pit, located south of Richmond and east of the Appomattox River in Prince George County, Virginia, USA.

Type stratum – Patapsco Formation, Potomac Group.

Age – Early Cretaceous (Basal part of Subzone IIB, Early-Middle Albian).

Description

Flower morphology – The flowers are only partially preserved with the floral centre missing in all fragments. It is therefore unclear whether the flowers had additional male or female organs and thus whether the flowers were unisexual or bisexual (Fig 3.5). Flowers appear actinomorphic (PP43735) with the floral organs arranged in three trimerous whorls: two alternating perianth whorls and at least one alternating stamen/staminode whorl, all three whorls are fused basally to form a floral cup (Fig 3.5A–E).

Perianth – The perianth appears to consist of two alternating, trimerous whorls that are fused into a short floral tube at the base. The outer whorl consists of broad, possibly five-veined, tepals. The inner whorl consists of narrower, three-veined tepals (Fig 3.5G). Scattered gland-like structures, presumably oil glands, are present on the inside of the floral cup (not shown).

Androecium – The androecium consists of at least six structures that are interpreted as nectariferous, and three stamens or staminodes of which only the filaments/filament bases are preserved. The nectariferous structures are arranged in pairs and each pair is associated with a more centrally placed filament/filament

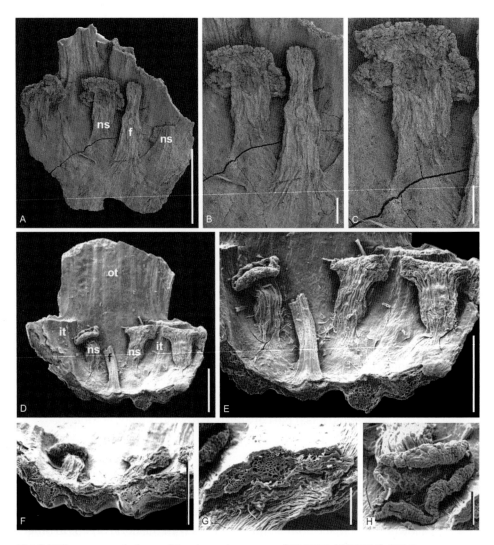

Fig 3.5 Flower morphology of *Powhatania connata* (PP43751, PP53734). SEM micrographs. (A–C) Flower fragment PP53734. (A) Ventral view of fragmentary flower showing part of the perianth and androecium. (B) Detail of broken stamen filament and one of the associated stalked nectariferous structures. (C) Detail of stalked nectariferous structure. (D–H) Flower fragment PP43751. (D) Ventral view of flower fragment showing parts of one outer tepal and two inner tepals with a filament and two nectariferous structures inserted on the floral cup. (E) Detail of part of the androecium, showing one broken filament flanked by two associated stalked structures and one additional stalked nectariferous structure. (F) Transverse section through the lower portion of the flower cup showing ribs. (G) Transverse section through the upper portion of inner perianth part showing three vascular bundles. (H) Apical view of peltate head of stalked nectariferous structure. Abbreviations: f = filament; it = inner tepal; ns = nectariferous structure; ot = outer tepal. Scale bars: (A,D–F) = 500 µm; (B,C,G,H) = 100 µm.

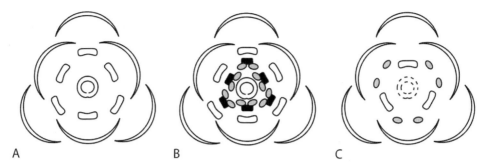

Fig 3.6 Androecium diversity among flowers of Laurales from the Early Cretaceous (Early to Middle Albian) Puddledock locality. Floral diagrams. (A) *Potomacanthus lobatus* floral diagram (PP44882). (B) *Cohongarootonia hispida* (PP53716; staminodes indicated as black rectangles with two spade-like appendages (shaded). (C) Reconstructed floral diagram based on three flower fragments of *Powhatania connata* (PP43735, PP43751, PP53734; shading indicates nectariferous structures). Partly redrawn from Figs 3a and 4 in Crane et al. (1994) and von Balthazar et al. (2007).

base that is positioned in front of each outer perianth organ (Figs 3.5A–E, 3.6). The nectariferous structures have short and broad stalks and peltate or semi-peltate heads; the edges of the heads are curved upwards; the edges appear further incised on one side and may thus be semi-peltate (Fig 3.5E–H).

Pollen – No pollen was observed on the specimens.

Gynoecium – No carpels were preserved in the specimens.

3.4 Discussion

3.4.1 General systematic relationships of the fossils

Flowers of extant Laurales, and in particular those of taxa included in core Laurales, are characterized by a series of floral features that are unique or otherwise rare among other extant angiosperms. These distinctive floral features include the conspicuous paired nectariferous structures, attached or associated with stamens/staminodes, and anther dehiscence by apically hinged flaps (e.g. Endress and Hufford, 1989; Endress, 1994; Rohwer, 1994, 2009; Renner, 1999). Paired nectariferous stamen/staminode structures apparently originated in the order after the divergence of Calycanthaceae. They arose either once, with a reversal in Siparunaceae and Mollinedioideae (Monimiaceae), or twice, in the Atherospermataceae/Gomortegaceae and Monimiaceae/Lauraceae/Hernandiaceae lineages (Endress and Doyle, 2009) (see Fig 3.11B). Similarly, apically hinged anther flaps have either arisen once, after the divergence of Calycanthaceae with a reversal in Monimiaceae, or have originated separately in the Atherospermataceae/Gomortegaceae/

Siparunaceae clade and in the Lauraceae/Hernandiaceae clade (Endress and Doyle, 2009; Doyle and Endress, 2010) (see Fig 3.10B).

The two fossil taxa presented here share many characters with flowers of extant core Laurales (Figs 3.8–3.13), especially the paired, presumably nectariferous, structures associated with the stamens/staminodes and the anther dehiscence by apically hinged flaps (Figs 3.10B, 3.11B). More formal assessment of the possible phylogenetic relationship of *Cohongarootonia* (Fig 3.7) shows that placing it as the sister group to the Lauraceae and Hernandiaceae clade requires the fewest additional character state changes compared to the number of character state changes with extant taxa alone (total tree length: 1018 compared to 1015 steps). Placing *Cohongarootonia* within core group Laurales also generally results in short trees, independent of its exact position (Fig 3.7, 1018–1024 steps). Positions outside core Laurales are less parsimonious and require at least 1025 steps (Fig 3.7). These results suggest that *Cohongarootonia* is more closely related to core Laurales than to any other extant group of angiosperms (Fig 3.7) and that the most parsimonious position of the fossil, in relation to the backbone tree based on molecular and morphological data, is as sister group to both Lauraceae and Hernandiaceae (Fig 3.7).

The flower of *Powhatania connata* is much less completely understood than the *Cohongarootonia* flower, but nevertheless, its perianth and androecial characters suggest an association with the extant families Monimiaceae/Lauraceae/Hernandiaceae among core Laurales.

3.4.2 Interpretation of the fossil *Cohongarootonia hispida* and its systematic affinity

Based on the regular trimerous flower organization, two whorls of similarly differentiated perianth organs, two outer whorls of stamens with disporangiate anthers that dehisce by apically hinged flaps, two inner whorls of staminodes with paired nectariferous appendages, as well as a single carpel with a single apically inserted ovule (see floral diagram Fig 3.6), *Cohongarootonia* can be assigned unequivocally to the order Laurales (Figs 3.7–3.13). A simple whorled trimerous organization of perianth and androecium is characteristic of extant Lauraceae (e.g. Rohwer, 1993), but also occurs in the closely related Hernandiaceae (e.g. Kubitzki, 1970a, b) (see also Table 3.1). In addition, these two families and *Cohongarootonia* share the presence of paired nectariferous appendages, disporangiate anthers, anther dehiscence by flaps, and a single carpel with a more or less apically inserted ovule (Kubitzki, 1970a, b, 1993a, b; Rohwer, 1993; Figs 3.8–3.13).

Notwithstanding these broad similarities, flowers of Hernandiaceae differ from the fossil flower in having an inferior ovary. However, an inferior ovary is also present in many genera of Lauraceae that diverge at an early stage from the main line of diversification in the family. For example an inferior ovary is found in *Hypodaphnis*, which is the sister to all other genera and may indicate that an inferior

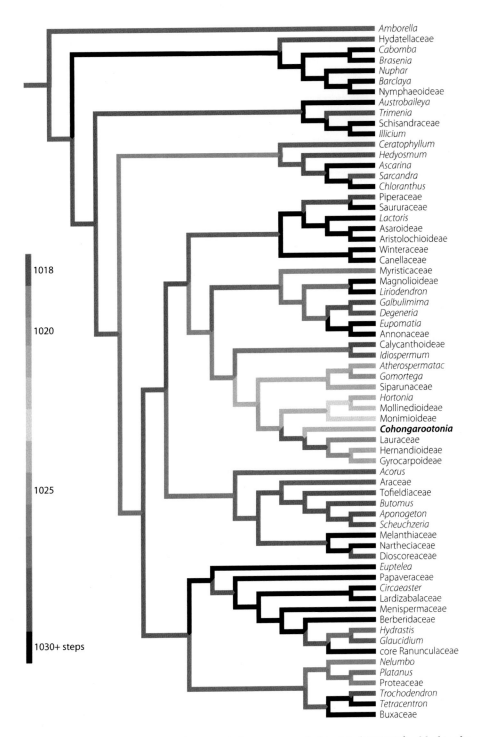

Fig 3.7 Phylogenetic analysis showing *Cohongarootonia hispida* (PP53716) added to the backbone tree of Doyle and Endress (2010). Branch colouring refers to the number of steps under parsimony (morphological character changes) that are required if *Cohongarootonia hispida* is attached to the corresponding branch. For colour illustration see plate section.

(A)

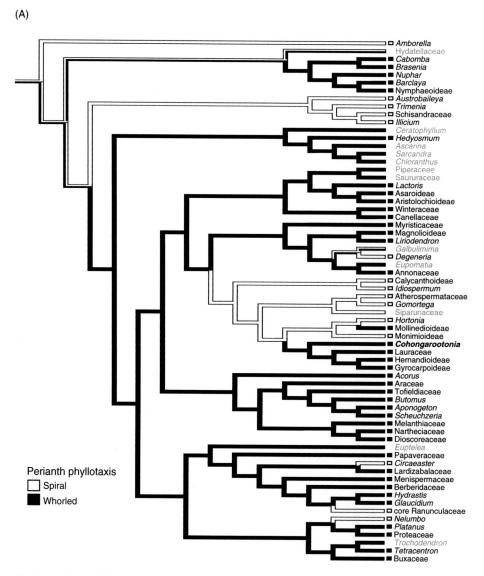

Fig 3.8 Selected floral characters mapped on the backbone tree of Doyle and Endress (2010) with fossil *Cohongarootonia hispida* (PP53716) included. (A) Perianth phyllotaxis. (B) Perianth merism.

(B)

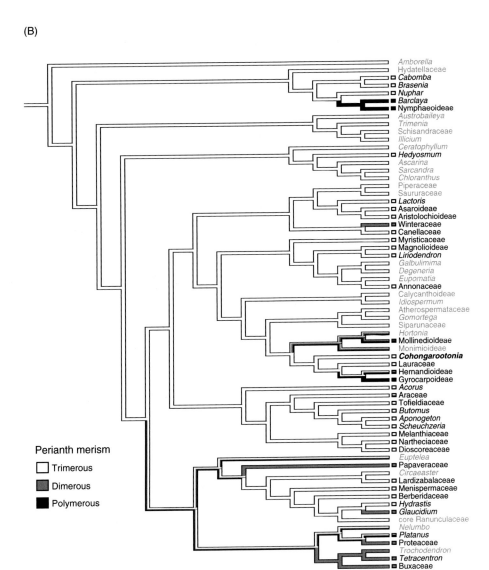

Fig 3.8 (cont.)

(A)

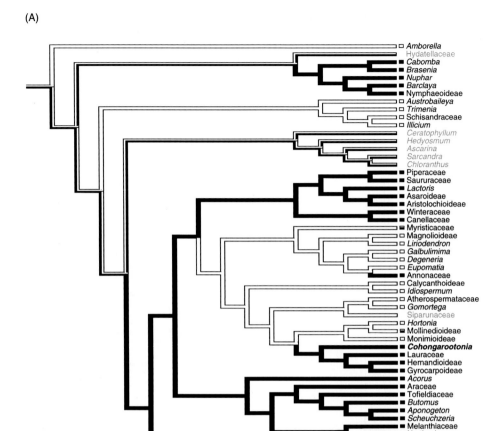

Fig 3.9 Selected floral characters mapped on the backbone tree of Doyle and Endress (2010) with fossil *Cohongarootonia hispida* (PP53716) included. (A) Androecium phyllotaxis. (B) Androecium merism.

(B)

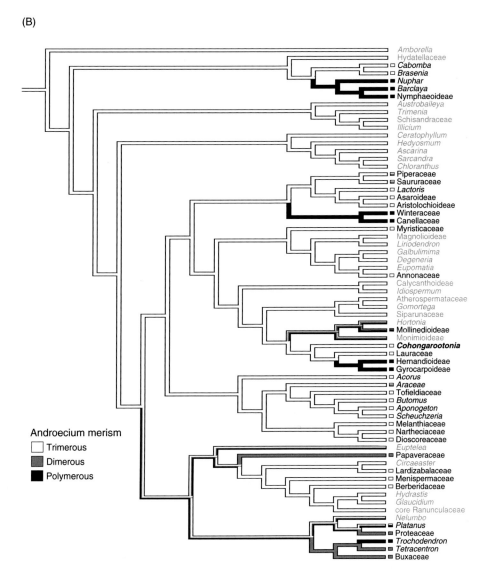

Fig 3.9 (cont.)

(A)

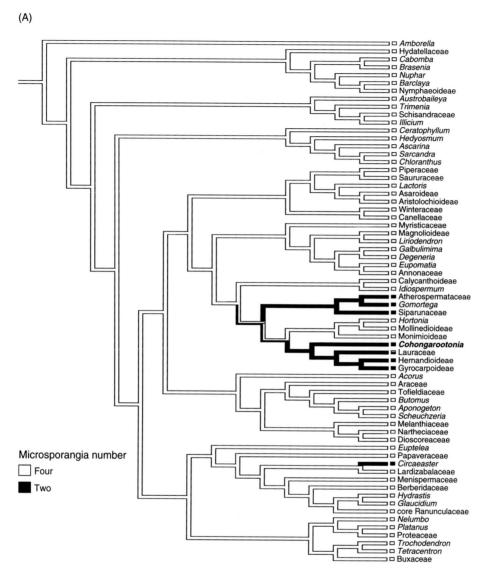

Microsporangia number
- ☐ Four
- ■ Two

Fig 3.10 Selected floral characters mapped on the backbone tree of Doyle and Endress (2010) with fossil *Cohongarootonia hispida* (PP53716) included. (A) Microsporangia number. (B) Anther dehiscence.

(B)

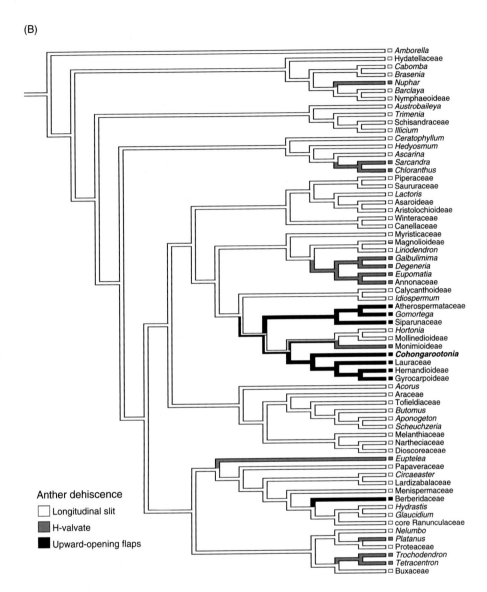

Anther dehiscence

☐ Longitudinal slit

■ H-valvate

■ Upward-opening flaps

Fig 3.10 (cont.)

(A)

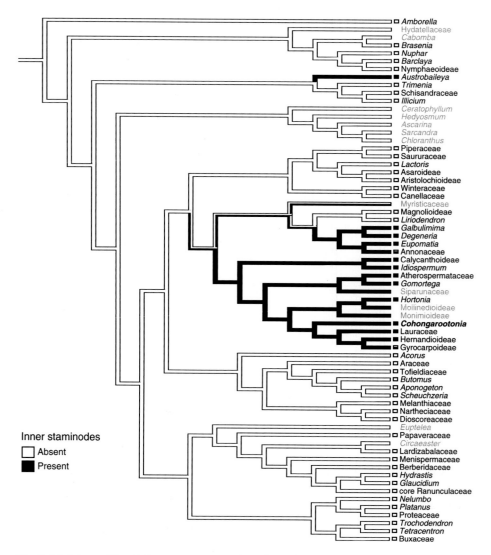

Fig 3.11 Selected floral characters mapped on the backbone tree of Doyle and Endress (2010) with fossil *Cohongarootonia hispida* (PP53716) included. (A) Presence of inner staminodes. (B) Presence of androecial appendages.

(B)

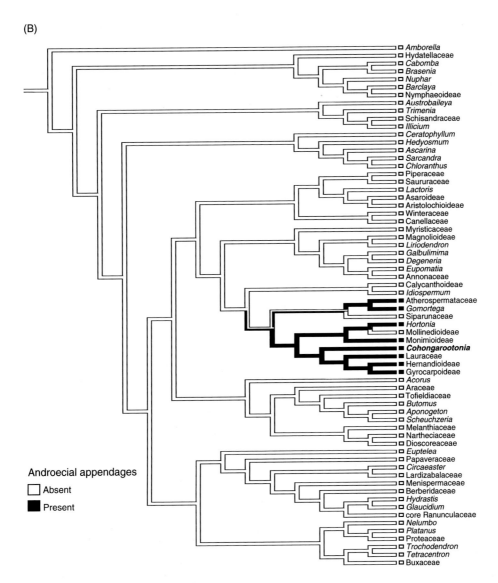

Androecial appendages

☐ Absent

■ Present

Fig 3.11 (cont.)

(A)

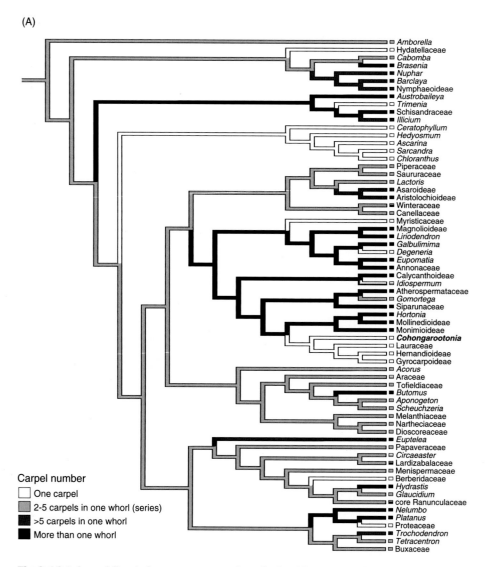

Carpel number
- ☐ One carpel
- ▨ 2-5 carpels in one whorl (series)
- ▨ >5 carpels in one whorl
- ■ More than one whorl

Fig 3.12 Selected floral characters mapped on the backbone tree of Doyle and Endress (2010) with fossil *Cohongarootonia hispida* (PP53716) included. (A) Carpel number. (B) Ovule number.

(B)

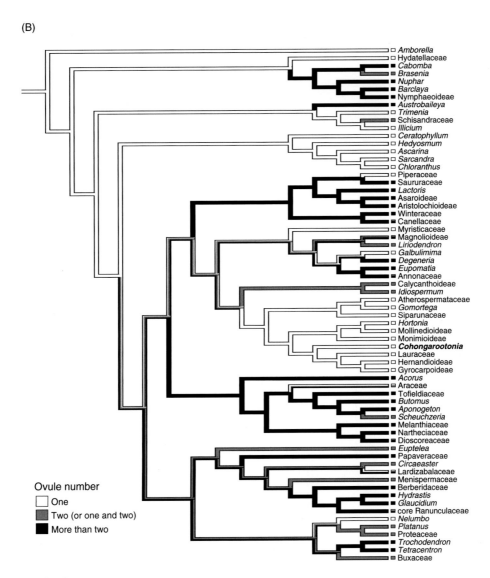

Fig 3.12 (cont.)

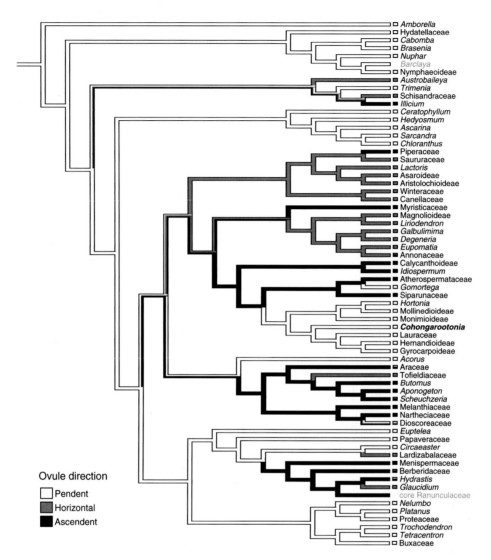

Fig 3.13 Selected floral characters mapped on the backbone tree of Doyle and Endress (2010) with fossil *Cohongarootonia hispida* (PP53716) included. Ovule direction.

ovary is ancestral for the family (Rohwer and Rudolph, 2005). It is also of interest that in Hernandiaceae paired nectariferous appendages occur predominantly in the outer androecial whorl rather than in the inner whorl (Kubitzki, 1970a, b; Endress and Lorence, 2004).

There are also differences between *Cohongarootonia* and the flowers of extant Lauraceae. A stylar canal is present for most of the length of the style of *Cohongarootonia*, but no open canals have been reported for extant representatives of Lauraceae (and Hernandiaceae) (Endress and Igersheim, 1997). However, open stylar canals are known from the closely related modern families Calycanthaceae and Gomortegaceae and from some Monimiaceae (Endress and Igersheim, 1997). In extant Lauraceae, the androecium generally consists of three stamen whorls followed by one whorl of staminodes; nectariferous appendages are present in the third whorl of stamens (e.g. Rohwer, 1993). In a few genera this third whorl may be staminodial and still have appendages (e.g. *Persea* p.p., *Phoebe* p.p., *Dehaasia* p.p., *Cinnamomum* p.p., *Aiouea* p.p., *Dicypellium*, *Aniba* p.p., *Beilschmiedia* p.p., *Potomeia*, *Cassytha* p.p., *Aspidostemon*, *Iteadaphne*; Rohwer, 1993) similar to the situation in *Cohongarootonia*. However, the innermost whorl of staminodes in extant Lauraceae generally has no appendages; these staminodes resemble reduced, almost sessile, anthers with nectariferous tips and are thus morphologically different from third whorl staminodes (e.g. Kasapligil, 1951; Hyland, 1989; Rohwer, 1993; Buzgo et al., 2007).

The presence of nectariferous appendages in more than one whorl, as seen in *Cohongarootonia*, is also rare in extant Lauraceae and is found only in a few genera, such as *Rhodostemonodaphne*, *Urbanodendron*, *Brassiodendron* and *Phyllostemodaphne*, which may have appendages on all stamens, but never on staminodes (Rohwer, 1993). Similarly, in male flowers of unisexual Lauraceae nectariferous appendages occur in more than one whorl, but again only on stamens and not on staminodes (*Litsea*, *Dodecadenia*, *Neolitsea*, *Chlorocardium*, *Lindera*, *Cinnadenia* and *Laurus*; Rohwer, 1993). Interestingly, the situation in *Cohongarootonia* is most similar to that in the female flowers of those genera where there are staminodial whorls with nectariferous appendages (Hyland, 1989; Rohwer, 1993). In particular the organs of these staminodial whorls strongly resemble the shape of the staminodes and appendages observed in the flower of the fossil *Cohongarootonia*. The structures in the fossil are particularly similar to the staminodes in the female flowers of *Litsea* and *Neolitsea* (Hyland, 1989). Among extant taxa with bisexual flowers, the shape of the staminodes and appendages in *Cohongarootonia* also resembles, for example, the third whorl staminodes of some *Aiouea* species (Mez, 1889).

Androecial characters such as presence and position of staminal appendages, as well as number of sporangia are highly variable among extant Lauraceae and may also vary within genera. This relative flexibility in the structure of the androecium

Table 3.1 Comparison of floral characters of *Cohongarootonia* and *Powhatania* with extant families of Laurales.[a]

Taxon Character	*Cohongarootonia*	*Powhatania*	Calycanthaceae	Siparunaceae
Flower	Bisexual	Unknown	Bisexual	Unisexual
Perianth organ number	6	Presumably 6	5–50	4–6(–7)
Perianth phyllotaxis	Trimerous whorls	Trimerous whorls	Spiral	Di- or pentamerous whorls
Stamen number	6	At least 3	5–30	(1–)2–70
Androecium phyllotaxis	Trimerous whorls	Unknown	Spiral	Spiral or whorls (tetramerous)
Sporangia number	2	Unknown	4	2 with one valve
Anther dehiscence	Flaps	Unknown	Longitudinal slits (valves in *Sinocalycanthus*)	Flaps
Staminal appendages	Present	Present	Absent	Absent
Staminodes	Present	Unknown	Present	Absent
Carpel number	1	Unknown	1 to 43	3 to 30
Carpel phyllotaxis	–	Unknown	Spiral	Spiral with transition to irregular (*Siparuna thecaphora*)
Ovary position	Superior	Unknown	Superior	Superior
Ovule number/carpel	1	Unknown	2 (1 sterile)	1
Ovule insertion	Apical	Unknown	Basal	Basal

[a] *Notes*: Data from Schodde (1969), Foreman (1983), Endress (1980a), Endress and Hufford (1989), Endress and Igersheim (1997), Endress and Lorence (2004), Kubitzki (1993a, b, c), Philipson (1993), Renner et al. (1997), Rohwer (1993), Stevens (2001 onwards), Staedler et al. (2007), Staedler and Endress (2009).

Gomortegaceae	Atherospermataceae	Hernandiaceae	Monimiaceae	Lauraceae
Bisexual	Unisexual/bisexual	Unisexual/bisexual	Unisexual (except *Hortonia*)	Unisexual/ bisexual
7–10, intergrading with stamens	6–20	6–8(–12)	3–many	Mostly 6
Spiral	Whorls or spiral (*Daphnandra repandula*); mostly di- or tetramerous whorls, some trimerous or complex whorls	Two whorls of 3–4 (–6) or one whorl of 4–8	Spiral (Monimioideae and *Hortonia*) or dimerous or complex whorls (Mollinedioideae)	Mostly trimerous whorls
7–13	(4–)6–100	3–5(–7)	4–many	Mostly 3–12, some up to 32
Spiral	Mostly di- or tetramerous whorls, otherwise spiral or complex whorls	1 whorl of 3 to 5 organs	Spiral (Monimioideae and *Hortonia*) or dimerous or complex whorls (Mollinedioideae)	Mostly trimerous whorls, some dimerous or irregular
2	2	2	4 (*Monimia* 2)	2 or 4
Flaps	Flaps	Flaps	Longitudinal slits or valves	Flaps
Present	Present	Present (except *Sparattanthelium*)	Present in *Hortonia, Peumus, Monimia*	Mostly present
Present	Present	Absent or present	Present in *Hortonia* and *Peumus* (female)	Mostly present
(2–)3(–5)	3 to many	1	(1–few) to many	1
Spiral	Whorled (di-, tri-, tetra-, hexamerous or complex) or spiral	–	Spiral (Monimioideae and *Hortonia*) Whorls and irregular (Mollinedioideae)	–
Inferior	Superior	Inferior	Superior (inferior in *Tambourissa*)	Mostly superior (some inferior)
1	1	1	1	1
Apical	Basal	Apical	Apical	Apical

has led to conflicting morphology-based classifications of the family in the past (for an overview, see van der Werff and Richter, 1996) and tracing these androecial characters on molecular-based phylogenetic trees demonstrates their high level of homoplasy (e.g. Endress and Hufford, 1989; von Balthazar et al., 2007). Nevertheless, despite the enormous variability in the androecium among extant Hernandiaceae and Lauraceae, the androecium structure of *Cohongarootonia* is not matched in any extant taxon. It seems likely that *Cohongarootonia* belongs to a separate, now extinct, lineage perhaps along the stem lineage of the Lauraceae/Hernandiaceae clade, or along the stem lineage of one of these two families.

3.4.3 Interpretation of the fossil *Powhatania connata* and its systematic affinity

The presence of a basally united perianth composed of two distinct perianth whorls, and an androecium with at least one whorl of stamens associated with stalked, presumably nectariferous, structures in *Powhatania connata* indicates a relationship to extant Laurales. The stalked structures are distinctive in their shape and position. They are free with semi-peltate/peltate heads and occur in a lateral position to the stamens of the first androecial whorl.

Stamens with appendages in the outermost androecial whorl are only present in a few genera of extant Lauraceae (*Rhodostemonodaphne* p.p., *Urbanodendron, Phyllostemodaphne, Brassiodendron, Endiandra, Cryptocarya*; Hyland, 1989; Rohwer, 1993), some Hernandiaceae (Kubitzki, 1970a, b, 1993b), Atherospermataceae (Philipson, 1993), and a few Monimiaceae (*Hortonia, Peumus, Monimia*; Philipson, 1993).

In *Hortonia, Peumus* and *Monimia* (Monimiaceae) the staminal appendages are connected to the base of the filaments, unlike those of *Powhatania*, and their shape is also different: convoluted scale-like in *Hortonia* or club-shaped in *Peumus* and *Monimia* (e.g. Money, 1950; Endress, 1980a, b). In general, the Monimiaceae are characterized by having the reproductive organs on the inner surface of a cup-shaped floral receptacle, which is closed in *Monimia*, but open and shallow in *Peumus* and *Hortonia* (e.g. Endress, 1980a).

In Atherospermataceae, most Hernandiaceae and the genera of Lauraceae listed above, the staminal appendages are attached to the base of the filaments (Money, 1950; Kubitzki, 1970a, b; Philips, 1993; Rohwer, 1993; Endress and Lorence, 2004). Only in a few genera of Lauraceae, for example in *Cryptocarya* p.p. (Hyland, 1989; Rohwer, 1993), are they free from the stamens and positioned laterally, as observed in the fossil *Powhatania*. The appendages in *Cryptocarya* (Hyland, 1989) also resemble those of *Powhatania*, but their position is somewhat unclear. They are reported as occurring in the third androecial whorl, as is normal for flowers of Lauraceae (Kostermans, 1938; Rohwer, 1993), but are also reported to be associated with the outermost stamen whorl (Hyland, 1989).

In summary, *Powhatania* shares the floral cup with Monimiaceae, Hernandiaceae and Lauraceae, and its trimerous organization with Hernandiaceae and Lauraceae. It further shares the shape and position of the staminal appendages with some extant taxa in the Lauraceae. *Powhatania* may thus represent either an extinct taxon along the stem of the clade Monimiaceae/Hernandiaceae/Lauraceae, the clade Hernandiaceae/Lauraceae, or perhaps the stem group of one of the three extant families.

3.4.4 Comparison with other lauralean reproductive structures from the Cretaceous

The distinctiveness of the flowers of core Laurales has made it possible to assign several Cretaceous fossils to the group. Many of these fossils have been assigned to the family Lauraceae, including the unnamed flower fragment (Crane et al., 1994, formally described here as *Powhatania connata*), an isolated stamen illustrated by Crane et al. (1994), *Mauldinia mirabilis* (Drinnan et al., 1990), *Neusenia tetrasporangiata* and two additional unnamed taxa (Eklund, 2000), *Pragocladus lauroides* (Kvaček and Eklund, 2003), *Mauldinia bohemica* (Eklund and Kvaček, 1998), *Mauldinia hirsuta* (Frumin et al., 2004), *Mauldinia* sp. (Herendeen et al., 1994), *Mauldinia angustiloba* (Viehofen et al., 2008), *Prisca reynoldsii* (Retallack and Dilcher, 1981), unnamed flower (Takahashi et al., 1999), *Lauranthus futabensis* (Takahashi et al., 2001) and *Potomacanthus lobatus* (von Balthazar et al., 2007). In all cases where the floral architecture can be interpreted these flowers have a trimerous organization and a single carpel. They are distinguished mainly by differences in the organization of the androecium (Table 3.2).

Most of these fossils (*Mauldinia* spp., *Neusenia tetrasporangiata*, *Pragocladus lauroides, Lauranthus futabensis*) have three whorls of fertile stamens and a fourth inner whorl of staminodes. They all have paired appendages on the stamens of the third whorl. The two new taxa described here also show the same basic construction, but *Cohongarootonia hispida* is distinguished from other Cretaceous fossils by its two innermost androecial whorls being staminodial and bearing paired appendages. *Powhatania connata* is distinguished from all previously described fossil material by the paired stalked structures associated with the outermost androecial whorl (Table 3.2, Fig 3.6). *Potomacanthus lobatus* co-occurs with *Cohongarootonia hispida* and *Powhatania connata* in the Puddledock flora. It is distinguished from the two new taxa by having only two fertile stamen whorls and lacking both staminal appendages and staminodes (von Balthazar et al., 2007). *Potomacanthus* and *Cohongarootonia* are, however, very similar in gynoecium structure. Both taxa have a single carpel with a slender and curved style, an open stylar canal and a single pendent ovule with a median and ventral-apical insertion.

Together *Cohongarootonia*, *Potomacanthus* and *Powhatania* demonstrate that a basically trimerous floral structure, including stamens with or without paired

Table 3.2 Summary of androecium characters from Cretaceous fossil flowers and flower fragments with supposed lauraceous affinity.

Late Cretaceous

Taxon	Number of fertile stamen whorls	Staminode whorl	Staminal appendages	Number of sporangia	Reference
Mauldinia mirabilis	3	Present (4th whorl)	Present (on 3rd whorl)	2-sporangiate	Drinnan et al., 1990
Mauldinia bohemica	3	Present (4th whorl)	Present (on 3rd whorl)	2-sporangiate	Eklund and Kvaček, 1998
Mauldinia sp.	?	?	?	2-sporangiate	Herendeen et al., 1999
Mauldinia hirsuta	?	Present	Present	2-sporangiate	Frumin et al., 2004
Mauldinia angustiloba	3	Present (4th whorl)	Present (on 3rd whorl)	2-sporangiate	Viehofen et al., 2008
Prisca reynoldsii	?	?	?	?	Retallack and Dilcher, 1981
Pragocladus lauroides	3?	Not present	Present (on 3rd whorl)	2-sporangiate	Kvaček and Eklund, 2003
Perseanthus crossmanensis	3	Present (4th whorl)	Present (on 3rd whorl)	?	Herendeen et al., 1994
Neusenia tetrasporangiate	3	Present (4th whorl)	Present (on 3rd whorl)	4-sporangiate	Eklund, 2000
Taxon A	?	?	?	2-sporangiate	Eklund, 2000
Taxon B	?	?	Present	?	Eklund, 2000
Lauranthus futabensis	3	Present (4th whorl)	Present (on 3rd whorl)	4-sporangiate	Takahashi et al., 2001, pers. comm.
Hypogynous Flower Type 2	More than 2	?	?	4-sporangiate	Takahashi et al., 1999

Table 3.2 (cont.)

Early Cretaceous

Taxon	Number of fertile stamen whorls	Staminode whorl	Staminal appendages	Number of sporangia	Reference
Powhatania connata	?	?	Present	?	Crane et al. 1994, this work
Stamen	?	?	?	4-sporangiate	Crane et al. 1994
Potomacanthus lobatus	2	Not present	Not present	2-sporangiate	von Balthazar et al. 2007
Cohongarootonia hispida	2	Present (3rd and 4th whorl)	Present (3rd and 4th whorl)	2-sporangiate	This work

appendages and flap-like dehiscence, together with a single carpel, as seen in some extant core Laurales, was established very early in the evolution of the group by the Early-Middle Albian. They also provide clear evidence that even at this early stage flowers in core Laurales displayed a considerable diversity in the androecium, including combinations of features not seen among extant taxa.

3.5 Conclusions

Cohongarootonia hispida and *Powhatania connata* have several distinctive floral features that among extant angiosperms are common among core Laurales, but that are found only rarely in other taxa. Especially distinctive is the presence of paired nectariferous structures attached to or associated with stamens/staminodes. Based on comparison to similar structures in extant Laurales, these organs are thought to have produced nectar and attracted flower visitors. Today, pollinators of nectar-producing lauralean flowers comprise a relatively broad spectrum of bees, flies and beetles, but also other insects (e.g. Kubitzki and Kurz, 1984; Endress, 1990; Forfang and Olesen, 1998; Endress, 2010).

The character combination observed in the *Powhatania* flower is found mainly in Lauraceae and Hernandiaceae, but also in Monimiaceae. The paired nectariferous structures associated with the first androecial whorl are distinctive and suggest a closer affinity to Hernandiaceae and Lauraceae. The limited number of characters and specimens available preclude more detailed phylogenetic placement. *Cohongarootonia* shares most floral characters with Lauraceae and Hernandiaceae and may represent a separate and now extinct lineage along the stem of Lauraceae/Hernandiaceae clade. Together with other lauraceous fossils from the Puddledock locality these two fossil reproductive taxa expand our knowledge of the diversity within the order Laurales in the Early Cretaceous and demonstrate the extent of early floral diversity, especially in the androecial organization, among early core Laurales.

Acknowledgements

We thank P. K. Endress, J. A. Doyle and J. Schönenberger for helpful discussions; P. K. Endress and an anonymous reviewer for valuable comments on the manuscript; J. A. Doyle for providing the morphological data set of recent taxa and backbone trees; Guido W. Grimm for help with integrating the fossil in the phylogeny of living angiosperms; and M. Stampanoni and F. Marone for help with the SRXTM analyses, which were performed at the Swiss Light Source, Paul Scherrer Institut, Villigen, Switzerland. This work was supported by the Swedish Natural Science Research Council (EMF), the US National Science Foundation (PRC) and a fellowship to the first author by the Swiss National Science Foundation; SRXTM analyses were funded by the Swiss Light Source, European Union FP6, project number 20070197 to P. C. J. Donoghue, S. Bengtson and E. M. Friis.

3.6 References

Brenner, G. J. (1963). The spores and pollen of the Potomac Group of Maryland. *Maryland Department of Geology, Mines and Water Resources Bulletin*, **27**, 1–215.

Buzgo, M., Chanderbali, A., S., Kim, S. et al. (2007). Floral developmental morphology of *Persea americana* (Avocado, Lauraceae): the oddities of male organ identity. *International Journal of Plant Sciences*, **168**, 261–284.

Chanderbali, A. S., van der Werff, H. and Renner, S. S. (2001). Phylogeny and historical biogeography of Lauraceae: evidence from the chloroplast and nuclear genomes. *Annals of the Missouri Botanical Garden*, **88**, 104–134.

Crane, P. R., Friis, E. M. and Pedersen, K. R. (1994). Palaeobotanical evidence on the early radiation of magnoliid angiosperms. *Plant Systematics and Evolution* (Suppl.), **8**, S51–S72.

Crepet, W. L. and Nixon, K. C. (1998). Two new fossil flowers of magnoliid affinity from the Late Cretaceous of New Jersey. *American Journal of Botany*, **85**, 1273–1288.

Crepet, W. L., Nixon, K. C. and Gandolfo, M. A. (2005). An extinct calycanthoid taxon, *Jerseyanthus calycanthoides*, from the Late Cretaceous of New Jersey. *American Journal of Botany*, **92**, 1475–1485.

Dettmann, M. E., Clifford, H. T. and Peters, M. (2009). *Lovellea wintonensis* gen. et sp. nov. – Early Cretaceous (late Albian), anatomically preserved, angiospermous flowers and fruits from the Winton Formation, western Queensland, Australia. *Cretaceous Research*, **30**, 339–355.

Dischinger, J. B. (1987). Late Mesozoic and Cenozoic stratigraphic and structural framework near Hopewell, Virginia. *US Geological Survey Bulletin*, **567**, 1–48.

Donoghue, P. C. J., Bengtson, S., Dong, X.-P. et al. (2006). Synchrotron X-ray tomographic microscopy of fossil embryos. *Nature*, **442**, 680–683.

Doyle, J. A. (1969). Cretaceous angiosperms pollen of the Atlantic Coastal Plain and its evolutionary significance. *Journal of the Arnold Arboretum*, **50**, 1–35.

Doyle, J. A. (1992). Revised palynological correlations of the lower Potomac Group (USA) and the Cocobeach sequence of Gabon (Barremian-Aptian). *Cretaceous Research*, **13**, 337–349.

Doyle, J. A. and Endress P. K. (2000). Morphological phylogenetic analysis of basal angiosperms: comparison and combination with molecular data. *International Journal of Plant Sciences*, **161** (Suppl. 6), S121–S153.

Doyle, J. A., Endress, P. K. and Upchurch, G. R. (2008). Early Cretaceous monocots: A phylogenetic evaluation. *Acta Musei Nationalis Pragae, Series B, Historia Naturalis*, **64**, 59–87.

Doyle, J. A. and Endress P. K. (2010). Integrating Early Cretaceous fossils into the phylogeny of living angiosperms: Magnoliidae and eudicots. *Journal of Systematics and Evolution*, **48**, 1–35.

Doyle, J. A. and Hickey, L. J. (1976). Pollen and leaves from the mid-Cretaceous Potomac Group and their bearing on early angiosperm evolution. pp. 139–206 in Beck, C. B. (ed.), *Origin and Early Evolution of Angiosperms*. New York, NY: Columbia University Press.

Doyle, J. A. and Robbins, E. I. (1977). Angiosperm pollen zonation of the continental Cretaceous of the Atlantic Coastal Plain and its application to deep sea wells in the Salisbury Embayment. *Palynology*, **1**, 43–78.

Drinnan, A. N., Crane, P. R., Friis, E. M. and Pedersen K. R. (1990). Lauraceous flowers from the Potomac group (mid-Cretaceous) of eastern North America. *Botanical Gazette*, **151**, 370–384.

Eklund, H. (2000). Lauraceous flowers from the Late Cretaceous of North Carolina, U.S.A. *Botanical Journal of the Linnean Society*, **132**, 397–428.

Eklund, H. and Kvaček, J. (1998). Lauraceous inflorescences and flowers from the Cenomanian of Bohemia (Czech Republic, Central Europe). *International Journal of Plant Sciences*, **159**, 668–686.

Endress, P. K. (1980a). Ontogeny, function and evolution of extreme floral construction in Monimiaceae. *Plant Systematics and Evolution*, **134**, 79–120.

Endress, P. K. (1980b). Floral structure and relationships of *Hortonia* (Monimiaceae). *Plant Systematics and Evolution*, **133**, 199–221.

Endress, P. K. (1990). Evolution of reproductive structures and function in primitive angiosperms (Magnoliidae). *Memoirs of the New York Botanical Garden*, **55**, 5–34.

Endress, P. K. (2010). The evolution of floral biology in basal angiosperms. *Philosophical Transactions of the Royal Society of London B*, **365**, 411–421.

Endress, P. K. and Doyle, J. A. (2009). Reconstructing the ancestral angiosperm flower and its initial specializations. *American Journal of Botany*, **96**, 22–66.

Endress, P. K. and Hufford, L. (1989). The diversity of stamen structures and dehiscence patterns among Magnoliidae. *Botanical Journal of the Linnean Society*, **100**, 45–85.

Endress, P. K. and Igersheim, A. (1997). Gynoecium diversity and systematics of the Laurales. *Botanical Journal of the Linnean Society* **125**, 93–168.

Endress, P. K. and Lorence, D. H. (2004). Heterodichogamy of a novel type in *Hernandia* (Hernandiaceae) and its structural basis. *International Journal of Plant Sciences*, **165**, 753–763.

Foreman, D. B. (1983). The morphology and phylogeny of the Monimiaceae (sensu lato) in Australia. PhD dissertation, University of New England, Armidale, Australia.

Forfang, A.-S. and Olesen, J. M. (1998). Male-biased sex ratio and promiscuous pollination in the dioecious island tree *Laurus azorica* (Lauraceae). *Plant Systematics and Evolution*, **212**, 143–157.

Friis, E. M., Eklund, H., Pedersen, K. R. and Crane P. R. (1994). *Virginianthus calycanthoides* gen. et sp. nov. – A calycanthaceous flower from the Potomac Groups (Early Cretaceous) of eastern North America. *International Journal of Plant Sciences*, **155**, 772–785.

Friis, E. M., Pedersen K. R. and Crane P. R. (1995). *Appomattoxia ancistrophora* gen. et sp. nov., a new Early Cretaceous plant with similarities to *Circaeaster* and extant Magnoliidae. *American Journal of Botany*, **82**, 933–943.

Friis, E.M., Pedersen, K. R. and Crane, P. R. (1997). *Anacostia*, a new basal angiosperm from the Early Cretaceous of North America and Portugal with monocolpate/trichotomocolpate pollen. *Grana*, **36**, 225–244.

Friis, E. M., Pedersen, K. R. and Crane, P. R. (1999). Early angiosperm diversification: the diversity of pollen associated with angiosperm reproductive structures in Early Cretaceous floras from Portugal. *Annals of the Missouri Botanical Garden*, **86**, 259–296.

Friis, E. M., Pedersen, K. R. and Crane, P. R. (2006). Cretaceous angiosperm flowers: innovation and evolution in plant reproduction. *Palaeography, Palaeoclimate, Palaeoecology*, **232**, 251–293.

Friis, E. M., Pedersen, K. R. and Crane, P. R. (2010). Diversity in obscurity: fossil flowers and the early history of angiosperms. *Philosophical Transaction of the Royal Society B Biological Sciences*, **365**, 369–382.

Frumin, S., Eklund, H. and Friis, E. M. (2004). *Mauldinia hirsuta* sp. nov., a new member of the extinct genus *Mauldinia* (Lauraceae) from the Late Cretaceous (Cenomanian-Turonian) of Kazakhstan. *International Journal of Plant Sciences*, **165**, 883–895.

Herendeen, P. S., Crepet, W. L. and Nixon, K. C. (1994). Fossil flowers and pollen of Lauraceae from the Upper Cretaceous of New Jersey. *Plant Systematics and Evolution*, **189**, 29–40.

Herendeen, P.S., Magallón-Puebla, S., Lupia, R., Crane, P. R. and Kobylinska, J. (1999). A preliminary conspectus of the Allon flora from the Late Cretaceous (late Santonian) of central Georgia, USA. *Annals of the Missouri Botanical Garden*, **86**, 407–471.

Hickey, L. J. and Doyle, J. A. (1977). Early Cretaceous fossil evidence for angiosperm evolution. *Botanical Revue*, **43**, 3–104.

Hufford, L. and Endress, P. K. (1989). The diversity of anther structures and dehiscence patterns among

Hamamelididae. *Botanical Journal of the Linnean Society*, **99**, 301–346.

Hyland, B. P. M. (1989). A revision of Lauraceae in Australia (excluding *Cassytha*). *Australian Systematic Botany*, **2**, 135–367.

Kasapligil, B. (1951). Morphological and ontogenetic studies of *Umbellularia californica* Nutt. and *Laurus nobilis* L. *University of California, Publications in Botany*, **25**, 115–239.

Kostermans, A. J. G. H. (1938). Revision of the Lauraceae V. *Receuil des Travaux Botaniques Néerlandais*, **35**, 834–931.

Kostermans, A. J. G. H. (1939). The African Lauraceae I (Revision of the Lauraceae IV). *Bulletin du Jardin botanique de l'État à Bruxelles*, **15**, 73–108.

Kubitzki, K. (1970a). Monographie der Hernandiaceen. Teil I. *Botanische Jahrbücher für Systematik*, **89**, 78–148.

Kubitzki, K. (1970b). Monographie der Hernandiaceen. Teil II. *Botanische Jahrbücher für Systematik*, **89**, 149–209.

Kubitzki, K. (1993a). Gomortegaceae. pp. 318–320 in Kubitzki, K. (ed.), *The Families and Genera of Vascular Plants*, Vol. 2. Berlin: Springer.

Kubitzki, K. (1993b). Hernandiaceae. pp. 334–338 in Kubitzki, K. (ed.), *The Families and Genera of Vascular Plants*. Vol. 2: Berlin: Springer.

Kubitzki, K. (1993c). Hernandiaceae. pp. 334–338 in Kubitzki, K. (ed.), *The Families and Genera of Vascular Plants*, Vol. 2: Berlin, Springer.

Kubitzki, K. and Kurz, H. (1984). Synchronized dichogamy and dioecy in neotropical Lauraceae. *Plant Systematics and Evolution*, **147**, 253–266.

Kvaček, J. and Eklund, H. (2003). A report on newly recovered reproductive structures from the Cenomanian of Bohemia (Central Europe). *International Journal of Plant Sciences*, **164**, 1021–1039.

Maddison, W. P. and Maddison, D. R. (2007). Mesquite: a modular system for evolutionary analysis. version 2.0. Http://mesquiteproject.org.

Mez, C. (1889). Lauraceae Americanae. *Jahrbuch des Königlichen Botanischen Gartens und des Botanischen Museums zu Berlin*, **5**, 1–556.

Mohr, B. A. R. and Eklund, H. (2003). *Araripia florifera*, a magnoliid angiosperm from the Lower Cretaceous Crato Formation (Brazil). *Review of Palaeobotany and Palynology*, **126**, 279–292.

Money, L. L. (1950). The morphology and relationships of the Monimiaceae. *Journal of the Arnold Arboretum*, **31**, 372–404.

Philipson, W. R. (1993). Monimiaceae. pp. 426–437 in Kubitzki, K. (ed.), *The Families and Genera of Vascular Plants*. Vol. 2. Berlin, Springer,

Qiu, Y.-L., Lee, J., Bernasconi-Quadroni, F. et al. (1999). The earliest angiosperms: evidence from mitochondrial, plastid and nuclear genomes. *Nature*, **402**, 404–407.

Qiu, Y.-L., Li, L., Hendry, T. A. et al. (2006). Reconstructing the basal angiosperm phylogeny: evaluating information content of mitochondrial genes. *Taxon*, **55**, 837–856.

Renner, S. S. (1999). Circumscription and phylogeny of the Laurales: evidence from molecular and morphological data. *American Journal of Botany*, **86**, 1301–1315.

Renner, S. S. (2005). Variation in diversity among Laurales, Early Cretaceous to Present. *Biologiske Skrifter*, **55**, 441–458.

Renner, S. S. and Chanderbali, A. (2000). What is the relationship among Hernandiaceae, Lauraceae, and

Monimiaceae, and why is this question so difficult to answer? *International Journal of Plant Sciences*, **161** (Suppl. 6), S109–S119.

Retallak, G. and Dilcher, D. L. (1981). Early angiosperm reproduction: *Prisca reynoldsii* gen. et sp. nov. from Mid-Cretaceous coastal deposits, Kansas, USA. *Palaeontographica B*, **179**, 103–137.

Rohwer, J. G. (1993). Lauraceae. pp. 366–391 in Kubitzki, K. (ed.), *The Families and Genera of Vascular Plants*. Vol. 2. Berlin, Springer.

Rohwer, J. G. (1994). A note on the evolution of the stamens in the Laurales, with emphasis on the Lauraceae. *Botanica Acta*, **107**, 103–110.

Rohwer, J. G. (2009). The timing of nectar secretion in staminal and staminodial glands in Lauraceae. *Plant Biology*, **11**, 490–492.

Rohwer, J. G. and Rudolph, B. (2005). Jumping genera: the phylogenetic positions of *Cassytha*, *Hypodaphnis*, and *Neocinnamomum* (Lauraceae) based on different analyses of *trnK* intron sequences. *Annals of the Missouri Botanical Garden*, **92**, 153–178.

Schodde, R. (1969). A monograph of the family Atherospermataceae R. Brown. PhD dissertation, University of Adelaide, Adelaide, Australia.

Soltis, D. E., Gitzendanner, M. A. and Soltis, P. S. (2007). A 567-taxon data set for angiosperms: the challenges posed by Bayesian analyses of large data sets. *International Journal of Plant Sciences*, **168**, 137–157.

Staedler, Y. M., Weston, P. H and Endress, P. K. (2007). Floral phyllotaxis and floral architecture in Calycanthaceae (Laurales). *International Journal of Plant Sciences*, **168**, 285–306.

Staedler, Y. M. and Endress, P. K. (2009). Diversity and lability of floral phyllotaxis in the pluricarpellate families of core Laurales (Gomortegaceae, Atherospermataceae, Siparunaceae, Monimiaceae). *International Journal of Plant Sciences*, **170**, 522–550.

Stevens, P. F. (2001 onwards). Angiosperm Phylogeny Website, version 9, June 2008 [and more or less continuously updated since]. Available at http://www.mobot.org/MOBOT/research/APweb/.

Takahashi, M., Crane, P. R. and Ando, H. (1999). Fossil flowers and associated plant fossils from the Kamikitaba locality (Ashizawa Formation, Futaba Group, Lower Conacian, Upper Cretaceous) of northeast Japan. Journal of Plant Research 112: 187–206.

Takahashi, M., Herendeen, P. S. and Crane, P. R. (2001). Lauraceous fossil flower from the Kamikitaba locality (Lower Conacian; Upper Cretaceous) in Northeastern Japan. *Journal of Plant Research*, **114**, 429–434.

van der Werff, H. and Richter, H. G. (1996). Toward an improved classification of Lauraceae. *Annals of the Missouri Botanical Garden*, **83**, 409–418.

Viehofen, A., Hartkopf-Froeder, C. and Friis, E. M. (2008). Inflorescences and flowers of *Mauldinia angustiloba* sp. nov. (Lauraceae) from mid-Cretaceous karst infillings in the Rhenish Massif, Germany. *International Journal of Plant Sciences*, **169**, 871–889.

von Balthazar, M., Pedersen, K. R., Crane, P. R. and Friis, E. M. (2008). *Carpestella lacunata* gen. et sp. nov., a new basal angiosperm flower from the Early Cretaceous (Early to Middle

Albian) of eastern North America. *International Journal of Plant Sciences*, **169**, 890–898.

von Balthazar, M., Pedersen, K. R., Crane, P. R., Stampanoni, M. and Friis, E. M. (2007). *Potomacanthus lobatus* gen. et sp. nov., a new Lauraceae flower from the Early Cretaceous (Early to Middle Albian) of eastern North America. *American Journal of Botany*, **94**, 2041–2053.

4

Tracing the early evolutionary diversification of the angiosperm flower

JAMES A. DOYLE AND PETER K. ENDRESS

4.1 Introduction

The origin of the angiosperm flower and its subsequent evolution have been major topics of discussion and controversy for over a century. Because so many of the distinctive synapomorphies of angiosperms involve the flower, its origin and the homologies of its parts are closely tied to the vexed problem of the origin of angiosperms as a group. From a phylogenetic point of view, the origin of angiosperms involves two related problems: identification of the closest outgroups of angiosperms, which may clarify homologies of their distinctive features with structures seen in other plants, and rooting of the angiosperm phylogenetic tree and identification of its earliest branches, which may allow reconstruction of the flower in the most recent common ancestor of living angiosperms. It is this second topic that we address in this chapter (for the first study, see Frohlich and Chase, 2007; Doyle, 2008). This task has become much easier in the past ten years, thanks to molecular phylogenetics.

Ideas on the ancestral flower have varied greatly since early in the last century. Two extremes were euanthial theories, which postulated that the flower was a simple strobilus that was originally bisexual and had many free parts (Arber and Parkin, 1907), and pseudanthial theories, which assumed that the first angiosperms had unisexual flowers with few parts, as in 'Amentiferae' (now

Flowers on the Tree of Life, ed. Livia Wanntorp and Louis P. Ronse De Craene. Published by Cambridge University Press. © The Systematics Association 2011.

mostly Fagales), which were later grouped to form bisexual flowers (Wettstein, 1907; review in Friis and Endress, 1990). Later variations on the pseudanthial theory proposed that the angiosperms were polyphyletic (Meeuse, 1965, 1975), while recognition of chloranthoid pollen, leaves and flowers in the Early Cretaceous fossil record (Muller, 1981; Upchurch, 1984; Walker and Walker, 1984; Friis et al., 1986; Pedersen et al., 1991; Eklund et al., 2004) contributed to suggestions that Chloranthaceae, which combine putatively primitive wood and monosulcate pollen with extremely simple flowers, often consisting of just one stamen or one carpel, might provide another model for the ancestral flower (Endress, 1986b; Taylor and Hickey, 1992).

The advent of cladistic methods raised hopes of resolving these problems. Cladistic analysis of morphological characters provided a relatively objective method for constructing trees based on as many characters as possible, on which the evolution of individual characters could be traced by using parsimony optimization. However, although morphological cladistic analyses eliminated some alternatives, such as polyphyly of the angiosperms, their implications for rooting of the angiosperms and characters of the first flower were still highly varied. Depending on assumptions on outgroups and character analysis, some studies rooted angiosperms among groups with showy flowers, placing Magnoliales (Donoghue and Doyle, 1989), Calycanthaceae (Loconte and Stevenson, 1991) or Nymphaeales (Doyle, 1996) as the sister group of all other angiosperms, but others rooted them among groups such as Chloranthaceae and/or Piperales with simple flowers (Taylor and Hickey, 1992; Nixon et al., 1994; Hickey and Taylor, 1996).

This picture was clarified dramatically by molecular phylogenetics, which used the vastly greater numbers of characters in nucleotide sequences to generate independent estimates of relationships at the base of angiosperms. The first large molecular studies, on nuclear rRNA (Hamby and Zimmer, 1992) and the chloroplast gene *rbc*L (Chase et al., 1993), suggested that molecular data might also be inconclusive, since the two analyses differed in rooting angiosperms near Nymphaeales and the aquatic genus *Ceratophyllum*, which have multiparted bisexual flowers and extremely simple unisexual flowers, respectively. However, this situation improved with analyses of other genes and concatenated sequences of several genes (Mathews and Donoghue, 1999; Parkinson et al., 1999; Qiu et al., 1999, 2006; Soltis et al., 1999, 2000, 2005; Antonov et al., 2000; Barkman et al., 2000; Zanis et al., 2002). Despite variations in outgroup relationships, all of these studies rooted angiosperms among the so-called ANITA lines, namely *Amborella*, Nymphaeales and Austrobaileyales, while confirming many clades within the remaining angiosperms (*Mesangiospermae* of Cantino et al., 2007) that had been inferred from rRNA and *rbc*L. The main variations concern whether *Amborella* and Nymphaeales form two successive branches or a clade (e.g. Barkman et al., 2000; Qiu et al., 2006) and different arrangements of basal lines in the mesangiosperms. In Doyle and Endress (2000), we evaluated conflicts between morphological and molecular results by combining a morphological data set with sequences of three genes (18S rDNA, *rbc*L, *atp*B), with the angiosperm tree rooted on *Amborella*. As expected from the great number of molecular characters, this combined analysis generally confirmed the molecular results (for example, Nymphaeales were in the basal grade rather than linked with monocots), but there were a few exceptions. Most notably, in Laurales the sister group of Lauraceae was Hernandiaceae rather than Monimiaceae or Monimiaceae + Hernandiaceae.

Several studies have used molecular and/or combined trees as a framework for parsimony reconstruction of the evolution of floral characters (Doyle and Endress, 2000; Ronse De Craene et al., 2003; Zanis et al., 2003; Endress and Doyle, 2009), thus avoiding the circular reasoning that plagued earlier discussions. For these purposes, it is fortunate that the ANITA groups form a series of low-diversity lines that diverge sequentially below the vast majority of angiosperms; as a result, the many character states that are shared by these lines can be interpreted as ancestral in angiosperms. To some extent this circumvents the problem of identification of angiosperm outgroups, which remains one of the most intractable problems in plant evolution. Whereas morphological analyses associated living Gnetales in various ways with angiosperms, molecular analyses appear to be converging on trees with Gnetales nested within conifers (summarized in Burleigh and Mathews, 2004; Soltis et al., 2005). Morphological analyses that constrained Gnetales to a position in conifers (Doyle, 2006, 2008) identified fossil glossopterids, *Pentoxylon*, Bennettitales and *Caytonia* as extinct outgroups of the angiosperms, but there is no consensus that any of these taxa are related to angiosperms (cf. Taylor and Taylor, 2009).

The ancestral flower reconstructed by optimizing characters on trees of living taxa is the flower in the most recent common ancestor of all living angiosperms, or the crown group node. This is not necessarily the first flower, as flowers could have originated much earlier on the stem lineage leading to angiosperms. If the flower originated earlier there is no way to reconstruct its characters without fossil evidence, except perhaps to some extent by studies of the evolution of genes involved in development (Frohlich and Chase, 2007). Numerous fossil flowers are now known from the Early Cretaceous (Friis et al., 2006), but so far they have provided no clear evidence on this question, because none have been convincingly placed on the angiosperm stem lineage (Doyle, 2008; Endress and Doyle, 2009).

In this study we explore early floral evolution using the morphological data set of Endress and Doyle (2009), with a few changes made in Doyle and Endress (2010), where we used this data set to integrate Cretaceous fossils into the tree of living angiosperms. In Endress and Doyle (2009), we examined the implications of eight alternative trees, designed to represent the spectrum of currently viable hypotheses. These included trees with the two arrangements of *Amborella* and Nymphaeales and with two arrangements within mesangiosperms – relationships among major mesangiosperm lines are still poorly resolved, presumably because they radiated in a geologically short period of time. In one mesangiosperm arrangement (called J/M), *Ceratophyllum* was the sister group of eudicots and Chloranthaceae were sister to the magnoliid clade (*Magnoliidae* of Cantino et al., 2007: Magnoliales, Laurales, Canellales, Piperales), as in analyses of nearly complete chloroplast genomes (Jansen et al., 2007; Moore et al., 2007). In the other (D&E), an updated version of the combined tree of Doyle and Endress (2000), *Ceratophyllum* (not included in Doyle and Endress, 2000) was linked with Chloranthaceae, as indicated by morphology and some molecular analyses (Antonov et al., 2000; Duvall et al., 2006, 2008; Qiu et al., 2006), and the resulting clade was sister to the remaining mesangiosperms, as were Chloranthaceae alone in Doyle and Endress (2000). However, we moved Piperales from a position linked with monocots in Doyle and Endress (2000) into the magnoliids, as the sister group of Canellales based on accumulating molecular data (Soltis et al., 2005). Relationships within major clades in the D&E and J/M trees are the same, with Lauraceae and Hernandiaceae linked based on Doyle and Endress (2000), and with *Euptelea* moved to the base of Ranunculales (following Kim et al., 2004). New taxa were inserted in positions based on molecular data. We also considered trees with and without *Archaefructus*, an Early Cretaceous aquatic plant with reproductive axes variously interpreted as flowers with numerous stamens and carpels but no perianth (Sun et al., 1998, 2002) or as inflorescences of flowers consisting of one or two stamens or carpels (Friis et al., 2003). An analysis by Sun et al. (2002) identified *Archaefructus* as the sister group of all living angiosperms, but the seed-plant analysis of Doyle (2008) placed it within the crown group, linked with the aquatic family Hydatellaceae, which were formerly considered highly reduced monocots, but have been recently shown to be basal Nymphaeales (Saarela et al., 2007). The most important changes in Doyle and Endress (2010) were re-scoring of androecial characters in Piperales in accordance with developmental data and interpretations of Liang and Tucker (1995), Hufford (1996) and Tucker and Douglas (1996), and the phylogenetic results of Wanke et al. (2007); re-scoring of floral phyllotaxis and merism in some Laurales based on Staedler et al. (2007) and Staedler and Endress (2009); and increasing the number of states recognized in the carpel number character.

In Endress and Doyle (2009) we presented inferences on the evolution of all the floral characters in our data set, emphasizing implications for the morphology of the ancestral flower and for suggestions that the simple flowers of living and fossil aquatic taxa might be ancestral. Here we take a complementary, more taxon-oriented approach, in which we concentrate more on evolution of the flower as a whole, working upward from the base of the tree, stressing general aspects of floral organization such as phyllotaxis and number of parts. Besides reviewing our reconstruction of the flower at the basal node of extant angiosperms, we extend this approach to several important higher nodes, such as mesangiosperms, magnoliids, monocots and eudicots, noting important trends within these clades. Instead of considering all eight trees, we concentrate on the D&E tree, where *Amborella* and Nymphaeales form two successive branches and Chloranthaceae and *Ceratophyllum* are sister to the remaining mesangiosperms. The differences among the eight trees had relatively little impact on scenarios for floral evolution, presumably because few changes occurred between the initial splitting events in mesangiosperms. Perhaps most significantly, in the D&E tree the shift from the ancestral barrel-shaped ascidiate carpel to the leaf-like plicate carpel occurred once just above the base of the mesangiosperms, after divergence of the ascidiate Chloranthaceae–*Ceratophyllum* line, but it may have occurred anywhere between one and four times with the J/M chloroplast tree. In general, evolutionary scenarios for several characters are more ambiguous with the J/M tree.

For ease of discussion and economy of space, we have combined the four characters for perianth and androecium phyllotaxis and merism recognized in Endress and Doyle (2009) and Doyle and Endress (2010) into two characters (Figs 4.1A, 4.2A), in which spiral phyllotaxis is treated as a state coordinate with the trimerous, dimerous and polymerous whorled states. Phyllotaxis and merism were kept separate in Endress and Doyle (2009) and Doyle and Endress (2010), with spiral taxa scored as unknown (inapplicable) for merism, on the grounds that the contrast between whorled and spiral phyllotaxis may be a phylogenetically informative distinction, independent of merism, that would be masked by treating spiral as a state of an unordered multistate character. Similarly, presence or absence of a perianth and number of perianth whorls (series in spiral taxa) were treated as separate characters in Endress and Doyle (2009) and Doyle and Endress (2010), on the assumption that origin or loss of a perianth is a phylogenetically significant event, independent of the number of whorls, but in Fig 4.1B we treat absence of a perianth as a state of the character for number of whorls (as in Doyle and Endress, 2000). Number of stamen whorls and presence of one versus more than one stamens were also treated as separate characters in Endress and Doyle (2009) and Doyle and Endress (2010), on the assumption that flowers with one stamen deserve special recognition and cannot be assumed to be a result of reduction from one whorl, but here one stamen is treated as a state in the character for number of stamen whorls (Fig 4.2B). It may be theoretically preferable to separate characters for the presence versus absence of structures from characters for their different forms (Sereno, 2007), and the same may be true of characters such as spiral versus whorled phyllotaxis and merism of whorls, although this coding may introduce a risk of 'long distance' effects (Maddison, 1993) that bias toward the same ancestral state in a character when it only exists in widely separated clades. However, in practice there are only a few cases in which the two approaches give different results, as is discussed below.

Sources of data on characters and taxa, and discussion of problems in character analysis can be found in Endress and Doyle (2009) and Doyle and Endress (2010). Characters were optimized on the tree using MacClade (Maddison and Maddison, 2003).

4.2 From the base of the angiosperms to mesangiosperms

We begin by discussing inferences concerning morphology of the flower at the basal node of angiosperms and near the basal node of mesangiosperms, considering first organization of the perianth (Fig 4.1), then the androecium (Fig 4.2) and finally the gynoecium (Fig 4.3). Because the basal branch in mesangiosperms in the D&E tree is the Chloranthaceae-*Ceratophyllum* clade, whose members have extremely simple flowers, the exact point of origin of several features that characterize the rest of the mesangiosperms is ambiguous. These inferences will serve as a foundation for discussion of floral evolution in major clades within the mesangiosperms. Throughout this study, it should be recognized that such parsimony-based statements are only the most economical explanations of the data; in reality, there may have been more fluctuations along evolutionary lines, as assumed by likelihood-based methods.

Because there is an alternation of lines with spiral and whorled perianth at the base of the angiosperm tree (Fig 4.1A), the inferred ancestral perianth phyllotaxis is equivocal – either spiral or whorled and trimerous. If the ancestral state was spiral, it became whorled and trimerous either once, with a reversal in Austrobaileyales, or twice, in Nymphaeales and in mesangiosperms. If it was whorled and trimerous, it became spiral independently in *Amborella* and in Austrobaileyales. This and analyses of other characters presented below indicate that many aspects of floral organization were highly labile early in angiosperm evolution, as emphasized by Endress (1987a) and Ronse De Craene et al. (2003). In Nymphaeales (setting aside Hydatellaceae) the perianth was originally trimerous, as in Cabombaceae and *Nuphar*, but it became polymerous (specifically tetramerous) within Nymphaeaceae (*Barclaya* + Nymphaeoideae = *Nymphaea*, *Euryale* and *Victoria*). Perianth phyllotaxis remained spiral within Austrobaileyales. However, the perianth is unambiguously reconstructed as whorled and trimerous at the basal node of mesangiosperms. The same results are obtained if perianth

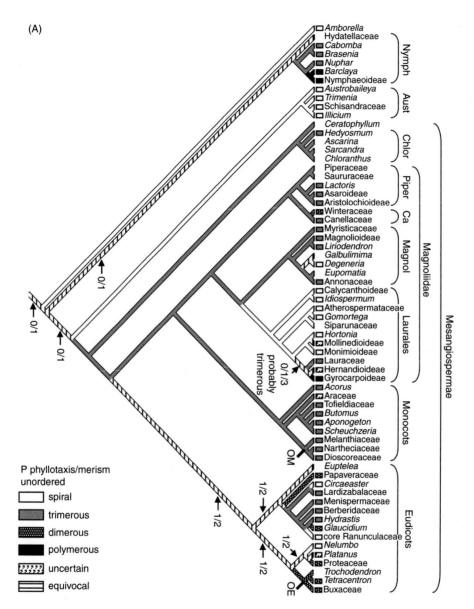

(A)

Fig 4.1 D&E tree of Endress and Doyle (2009), based on molecular and morphological data, with shading of branches showing the most parsimonious course of evolution of perianth characters (Endress and Doyle, 2009; Doyle and Endress, 2010), reconstructed with MacClade (Maddison and Maddison, 2003). (A) Character combining Endress and Doyle (2009) characters for perianth phyllotaxis (32) and merism (33). (B) Character combining Endress and Doyle (2009) characters for perianth presence (31) and number of whorls or series (34). Arrows indicate possible states on branches where the inferred character state is equivocal (e.g. in A, 0/1 = either spiral or trimerous). OM and OE indicate the probable positions of other monocots (*Petrosaviidae*) and other eudicots (*Gunneridae*, including *Pentapetalae*). Abbreviations: Nymph = Nymphaeales, Aust = Austrobaileyales, Chlor = Chloranthaceae, Piper = Piperales, Ca = Canellales, Magnol = Magnoliales.

(B)

Fig 4.1 (cont.)

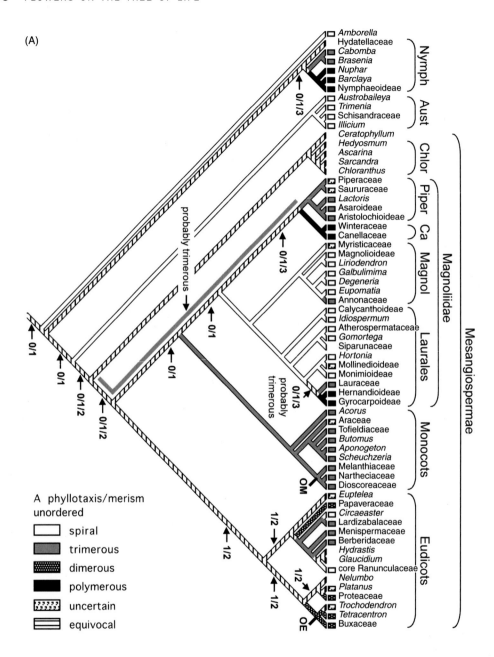

Fig 4.2 Same tree as in Fig 4.1, showing the most parsimonious course of evolution of androecium characters (Endress and Doyle, 2009; Doyle and Endress, 2010). (A) Character combining Endress and Doyle (2009) characters for androecium phyllotaxis (41) and merism (42). Grey bar indicates where use of separate characters for phyllotaxis and merism (as in Endress and Doyle, 2009) implies that the androecium was trimerous. (B) Character combining Endress and Doyle (2009) characters for one versus more stamens (40) and number of stamen whorls or series (43). Abbreviations as in Fig 4.1.

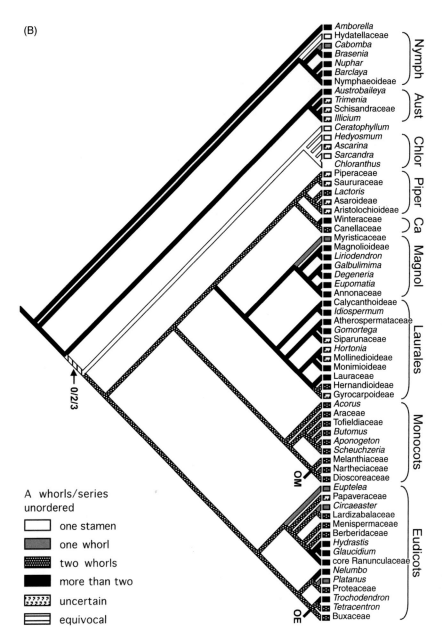

Fig 4.2 (cont.)

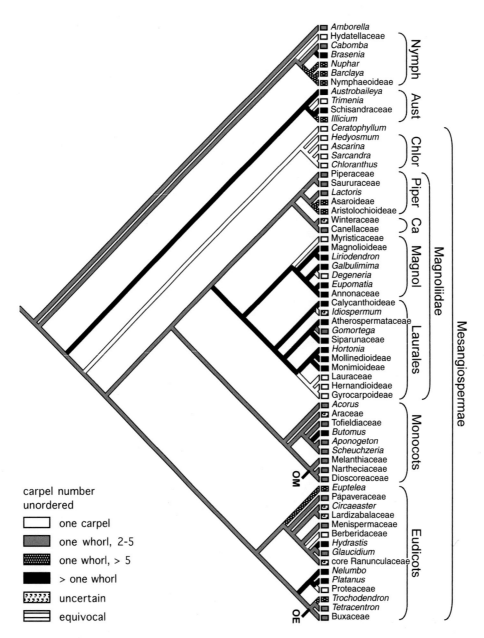

Fig 4.3 Same tree as in Fig 4.1, showing the most parsimonious course of evolution of the carpel number character (96) of Doyle and Endress (2010). Abbreviations as in Fig 4.1.

phyllotaxis and merism are treated as separate characters (Fig 4 in Endress and Doyle, 2009).

The character for number of perianth whorls (Fig 4.1B) is not strictly applicable in flowers with spiral phyllotaxis, but we have treated the number of so-called series of tepals that roughly fill the circumference of such flowers as equivalent to the number of whorls (Endress and Doyle, 2007; Staedler et al., 2007). The ancestral state of this character is more than two whorls or series, as in *Amborella*, Nymphaeaceae and Austrobaileyales, with reduction to two whorls in Cabombaceae and complete loss of the perianth in Hydatellaceae. This state persists to the base of the mesangiosperms; because the perianth was whorled and trimerous by this point, this implies that the common ancestor of mesangiosperms had three or more whorls of three tepals. The inference that the perianth was lost in Hydatellaceae assumes that the superficially flower-like reproductive units of this group are inflorescences with basal bracts and unisexual flowers consisting of one stamen or one carpel (Endress and Doyle, 2009), rather than incompletely organized 'pre-flowers' (one possibility considered by Rudall et al., 2007) or 'non-flowers' (Rudall et al., 2009). But even if the inflorescence interpretation is incorrect, it would be most parsimonious to assume that the lack of typical floral organization in Hydatellaceae was derived rather than primitive.

The case of Chloranthaceae and *Ceratophyllum*, which are basal in mesangiosperms in the D&E tree, is more complex. In Chloranthaceae, the male flowers of *Hedyosmum* and all flowers of *Ascarina*, *Sarcandra* and *Chloranthus* lack a perianth, but female flowers of *Hedyosmum* bear three small appendages on top of the ovary of the single carpel that are usually interpreted as tepals (Endress, 1987b). Assuming that *Ceratophyllum* has a single carpel (Endress, 1994) rather than a pseudomonomerous gynoecium consisting of one fertile and one sterile carpel (Shamrov, 2009), its female flowers consist of one carpel surrounded by appendages that have been considered tepals, but are more likely bracts, since they sometimes have carpels in their axils (Aboy, 1936; Iwamoto et al., 2003). The male structures have been interpreted as flowers with tepals and numerous stamens, but they are more likely spicate inflorescences with basal bracts and male flowers consisting of one stamen (Endress, 2004), because phyllotaxis of the stamens is highly labile and their maturation is markedly delayed toward the apex (Endress, 1994), which would be anomalous for a flower, but is typical for an inflorescence. This is a case where different treatments of relevant characters give different results. In Doyle and Endress (2000), where lack of a perianth was treated as one state of the character for number of whorls and *Ceratophyllum* was not included, it was equivocal whether the perianth of *Hedyosmum* was retained from lower in the tree or secondarily derived from no perianth. In Doyle et al. (2003) and Eklund et al. (2004), where the character for presence versus absence of perianth was introduced, the presence of a perianth in *Hedyosmum* was inferred to be primitive in Chloranthaceae. However, the situation is more confused if *Ceratophyllum* is linked with Chloranthaceae. If presence versus absence of a perianth and number of whorls are treated as separate characters, as in Endress and Doyle (2009), the perianth of *Hedyosmum* may be either primitive or secondarily derived. However, when the two characters are combined (Fig 4.1B), it appears that the perianth was lost on the line to Chloranthaceae and *Ceratophyllum*, and reappeared in *Hedyosmum*. Since there are theoretical reasons to separate presence versus absence of a structure from its different forms (Sereno, 2007), this question should be considered unresolved. In either case, as with Hydatellaceae, it would be unparsimonious to suggest that the reproductive structures of Chloranthaceae and *Ceratophyllum* are primitive.

In the ANITA lines (except *Nuphar*, which has both tepals and petals), the perianth parts are all tepals, which may be sepaloid (*Amborella*, *Trimenia*), petaloid (Cabombaceae) or differentiated into sepaloid outer organs and petaloid inner organs (Endress, 2008). With *Amborella* at the base of the angiosperms, the ancestral state is equivocal, either all sepaloid or both sepaloid and petaloid (Fig 5B in Endress and Doyle, 2009). However, the perianth can be reconstructed as sepaloid and petaloid in the common ancestor of all angiosperms except *Amborella*, and this condition persisted well into the mesangiosperms. The petaloid perianth of Cabombaceae and the sepaloid perianth of *Trimenia* were apparently derived from this state. *Nuphar* has not only sepaloid and petaloid tepals, but also inner perianth parts that fit the anatomical and developmental definition of petals (Endress and Doyle, 2009); this is clearly a convergence with the typical petals of Ranunculales in the eudicots.

Parsimony optimization indicates that the ancestral flower may have been either bisexual (the dominant traditional view) or unisexual, with both the D&E and J/M trees. This is because *Amborella* is unisexual, as are Hydatellaceae, Schisandraceae, *Ceratophyllum* and the chloranthaceous genera *Hedyosmum* and *Ascarina*. However, after the divergence of Chloranthaceae and *Ceratophyllum*, the flower in the common ancestor of the remaining mesangiosperms (magnoliids, monocots and eudicots) can be reconstructed as bisexual. As discussed in Endress and Doyle (2009), an argument in favour of the bisexual hypothesis is the fact that female flowers of *Amborella* have one or two sterile stamens – they are structurally bisexual. In any case, sex expression appears to have been remarkably labile early in the radiation of angiosperms, a conclusion supported by the mixture of unisexual and bisexual flowers in Early Cretaceous fossil floras (Friis et al., 2006).

Interestingly, the fact that *Sarcandra* and *Chloranthus* are nested among groups with unisexual flowers (*Ceratophyllum*, *Hedyosmum* and *Ascarina*) implies that their curious bisexual flowers, which consist of one carpel and either one stamen (*Sarcandra*) or a trilobed structure variously interpreted as one subdivided stamen or three fused stamens (*Chloranthus*), were derived from unisexual flowers (scenario 2 of Doyle et al., 2003). This is the only case in our data set where phylogenetic analysis implies that bisexual flowers were derived from unisexual. Cases elsewhere in angiosperms may be *Lacandonia*, nested within the otherwise unisexual monocot family Triuridaceae (Rudall and Bateman, 2006; Rudall et al.,

2009); Centrolepidaceae, if these are nested within Restionaceae (Sokoloff et al., 2009) and *Rhoiptelea*, nested within Fagales (although the occurrence of bisexual flowers in Late Cretaceous Fagales could affect this picture: Schönenberger et al., 2001). An alternative view (considered less likely by Endress, 1987b) is that the supposed bisexual flowers of *Sarcandra* and *Chloranthus* are actually pseudanthia.

As with the perianth, the ancestral state for androecium phyllotaxis (Fig 4.2A) is equivocal: either spiral or whorled and trimerous. If the ancestral androecium was spiral, it became whorled and initially either trimerous or polymerous in Nymphaeales, whereas if it was originally whorled and trimerous it became spiral in both *Amborella* and Austrobaileyales. Above this point, in constrast to the situation for the perianth, combining androecium phyllotaxis and merism has a significant effect on the results. When the two characters were kept separate (Endress and Doyle, 2009), the ancestral state in the mesangiosperms was equivocal, because Chloranthaceae and *Ceratophyllum* (both with basically one stamen) were scored as unknown, but in the common ancestor of magnoliids, monocots and eudicots the androecium was unambiguously reconstructed as whorled and trimerous. However, when the two characters are combined (Fig 4.2A), the situation becomes more ambiguous: the state both at the base of the mesangiosperms and after divergence of the Chloranthaceae–*Ceratophyllum* line, in the common ancestor of magnoliids, monocots and eudicots, may be either spiral, trimerous or even dimerous (as in many basal eudicots). This is because treating phyllotaxis and merism as an unordered multistate character obscures the fact that androecia that are trimerous (monocots, Piperales), polymerous (Canellales) and dimerous (eudicots) are all similar in being whorled. Because this is a potentially serious loss of information, it seems most likely that the ancestral state in the magnoliid-monocot–eudicot clade was whorled and trimerous, as inferred when phyllotaxis and merism were kept separate. However, this issue is not settled, because with the J/M tree the androecium may be either spiral or trimerous from the base of the mesangiosperms up to the magnoliids, independent of whether the two characters are kept separate or combined.

The scenario for number of stamen whorls (series) is less ambiguous (Fig 4.2B). The ancestral state was more than two whorls or series of stamens, which were reduced to one stamen in Chloranthaceae and *Ceratophyllum* and two whorls or series of stamens in the common ancestor of other mesangiosperms. Within Nymphaeales, the androecium was reduced to one stamen in Hydatellaceae and one whorl of six stamens in double positions in *Cabomba*. The same scenario was inferred in Endress and Doyle (2009), where the contrast between one and more than one stamen and the number of stamen whorls (series) were treated as two separate characters. Within Chloranthaceae, some *Ascarina* species have more than one stamen, but this condition was apparently secondarily derived from one stamen, and the same is true of the three-lobed androecium of *Chloranthus*, irrespective of whether this is one subdivided stamen or three fused stamens (Endress, 1987b; Doyle et al., 2003). These results imply that the flower in the common ancestor of magnoliids, monocots and eudicots (and possibly of mesangiosperms as a whole) had three or more whorls of three tepals, but two whorls/series of stamens. Curiously, this floral diagram is not retained in any of the derivative clades, all of which undergo their own further modifications, although it may have re-appeared later within some groups (e.g. some species of *Orophea* in the Annonaceae: Buchheim, 1964; Tofieldiaceae, if the calyculus is interpreted as a perianth whorl: Remizova and Sokoloff, 2003). The closest approach is in some trimerous Ranunculales, but as discussed below, it is uncertain whether their trimerous condition is a direct retention, and they differ in having true petals.

Perhaps the most interesting characters of the stamens themselves concern overall shape and orientation of dehiscence (position of the microsporangia). A distinction is often made between laminar (leaf-like) stamens, traditionally considered primitive (e.g. Canright, 1952; Takhtajan, 1969), and filamentous stamens (with a long, narrow base), but we find it more useful to distinguish three types of stamen base (short, long and wide, long and narrow; for numerical limits see Endress and Doyle, 2009) and connective apex (extended, truncated, 'peltate'). Doyle and Endress (2000) inferred that the ancestral stamen was introrse (with adaxial microsporangia) and had a long, wide base and an extended apex, as in *Amborella*, Nymphaeoideae and (except for a truncate apex) *Illicium*. However, because Hydatellaceae, which have a typical long, narrow filament and latrose dehiscence (lateral microsporangia), have more recently been associated with Nymphaeales, the situation is now more confused (Fig 7 in Endress and Doyle, 2009). Thus, the ancestral stamen may have had either a long and wide or a long and narrow base and either introrse or latrorse dehiscence, although it does appear that the connective apex was extended. Within Nymphaeales the stamen base shows an intriguing trend from long and narrow (Hydatellaceae, Cabombaceae) through short (*Nuphar, Barclaya*) to long and wide (Nymphaeoideae), while it was shortened in Chloranthaceae and *Ceratophyllum*. Stamen characters at the base of mesangiosperms are highly ambiguous (an extended apex but any type of base and orientation). However, we can infer that the common ancestor of magnoliids, monocots and eudicots had a long, narrow filament. These characters are among the most homoplastic in our data set, perhaps because they are highly sensitive to changes in pollination biology.

In Endress and Doyle (2009) we recognized fewer characters for organization of the gynoecium, on the grounds that its phyllotaxis and merism are too closely correlated with those of the androecium, so treating them as independent would over-weight changes in the two sets of organs. An obvious exception is presence of a single carpel versus more than one, a character used in Endress and Doyle (2009). In part to aid in placement of fossils, we refined this scheme in Doyle and Endress (2010) by breaking more than one carpel into three states (Fig 4.3): two to five in one whorl or series; more than five in one whorl or series (the 'star-shaped' state of von Balthazar et al., 2008), as in Nymphaeaceae and *Illicium*; more than one whorl or series, as in Schisandraceae and Magnoliaceae. With this character, the inferred ancestral state in angiosperms is one whorl or series of two to five carpels, which is retained well into the mesangiosperms. Carpels were independently multiplied by increasing their number in a single whorl/series in Nymphaeaceae (correlated with carpel fusion) and *Illicium*, and by increasing the number of whorls/series in *Brasenia* and Austrobaileyales. Reduction to one carpel occurred many times, from two to five carpels in Hydatellaceae and the Chloranthaceae–*Ceratophyllum* line, and from more than one series in *Trimenia*, as well as other times in other mesangiosperms.

As discussed in detail in Doyle and Endress (2000), Endress and Igersheim (2000) and Endress and Doyle (2009), the molecular rooting implies that the carpel was originally ascidiate, growing up as a tube from a ring-shaped primordium and sealed by secretion. This is in sharp contrast to older views that it was originally plicate (conduplicate), like a leaf folded down the middle and fused along its margins (Bailey and Swamy, 1951). If the D&E tree is correct, origin of the plicate carpel and sealing of its margins by postgenital fusion occurred in the mesangiosperms after divergence of the Chloranthaceae–*Ceratophyllum* line, in the common ancestor of magnoliids, monocots and eudicots, with several reversals (in which carpel form and sealing were less closely correlated) within these groups (Fig 4.4) and partial convergences in Nymphaeaceae and *Illicium* (Endress and Doyle, 2009). A related ancestral feature is formation of an extragynoecial compitum, where pollen tubes from one stigma can grow to ovules in another carpel through surface secretion (Endress and Igersheim, 2000; Williams, 2009). This was lost in the magnoliid–monocot–eudicot clade, but it re-appeared once or twice in Magnoliales and Laurales. Parsimony optimization indicates that the ancestral carpel contained one pendent ovule, a condition retained in *Amborella*, Hydatellaceae, *Trimenia*, *Ceratophyllum* and Chloranthaceae, which may be significant in the search for homologues of the carpel in fossil seed-plants (Doyle, 2008). The number of ovules increased several times: in Nymphaeales other than Hydatellaceae, *Austrobaileya*, Schisandraceae and one or more lines in the mesangiosperms. Among basal groups, carpel fusion occurred in Nymphaeaceae, followed by formation of an inferior ovary in *Barclaya* and Nymphaeoideae.

These results are summarized in Fig 4.4, which shows the D&E tree with the carpel form character and reconstructed floral diagrams for key nodes, assuming that the ancestral flower was bisexual. At the base are two equally parsimonious

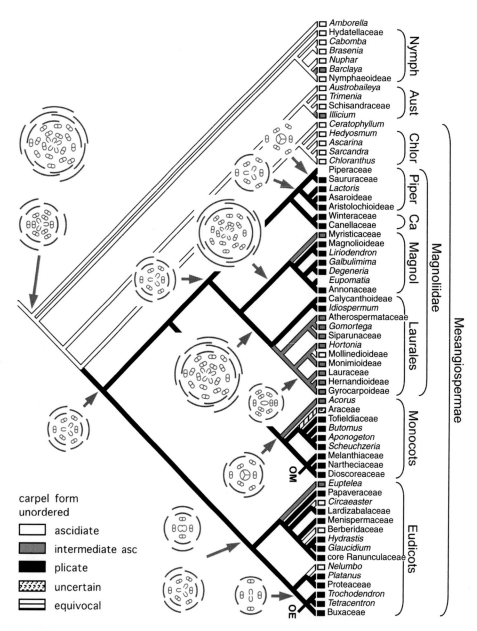

Fig 4.4 Same tree as in Fig 4.1, showing the most parsimonious course of evolution of the carpel form character (75) of Endress and Doyle (2009), with reconstructed floral diagrams for key nodes. Abbreviations as in Fig 4.1.

floral diagrams for the first angiosperms, with either spiral tepals and stamens in more than two series each, or more than two whorls of three tepals and stamens (combinations of spiral and whorled parts are also theoretically possible, but perhaps less plausible). Interestingly, there are no unequivocal changes in floral organization along the 'backbone' from the basal node to within the mesangiosperms, which may reflect both the fairly consistent floral morphology of most ANITA groups and homoplasy in the characters that do vary among them. If the ancestral flower had spiral parts, there was a shift from spiral to whorled phyllotaxis, but its location is uncertain; if the ancestral flower was trimerous, there may have been no change until reduction in the number of stamen whorls in the mesangiosperms, either before or after divergence of Chloranthaceae and *Ceratophyllum*. The reconstructed ancestral flower in the remaining mesangiosperms is trimerous, with more than two whorls of tepals, but only two whorls of stamens; as noted above, this precise architecture is not retained in any of the derived groups. Considering characters of individual floral parts, the only unequivocal change below the mesangiosperms is loss of curved hairs on the carpel before the divergence of Austrobaileyales, followed by the shift from ascidiate to plicate carpels and postgenital fusion of the carpel margins after divergence of Chloranthaceae and *Ceratophyllum*.

4.2.1 Magnoliidae

The magnoliid clade (*Magnoliidae* of Cantino et al., 2007), which consists of the APG II (2003) orders Magnoliales, Laurales, Canellales (including Winteraceae) and Piperales, includes many of the taxa that were thought to be primitive before recognition of the ANITA rooting (e.g. Takhtajan, 1966, 1969, 1980; Cronquist, 1968, 1981; Thorne, 1974). This monophyletic group should not be confused with Magnoliidae in the older sense of Takhtajan (1966, 1969, 1980) and Cronquist (1968, 1981), which was a paraphyletic grade taxon that included not only Magnoliidae in the present sense, but also the ANITA lines, Chloranthaceae and *Ceratophyllum* (as well as basal eudicot groups such as Ranunculales, removed by Takhtajan, 1969, 1980). Magnoliids are located well above the base of the angiosperm tree, but in addition to non-floral features such as monosulcate pollen and pinnately veined leaves (Doyle, 2005, 2007) some of their supposedly ancestral floral features, such as more than two whorls or series of tepals (Fig 4.1B), may have indeed been retained from the first crown group angiosperms.

Based on our data set and the D&E topology, the only unequivocal changes in floral morphology (either general organization or characters of individual organs) between the base of mesangiosperms and the base of magnoliids were the shift to plicate carpels and carpel sealing by postgenital fusion after divergence of the Chloranthaceae–*Ceratophyllum* line. The reconstructed flower in the common ancestor of magnoliids (Fig 4.4) therefore has the same floral diagram as the common ancestor of magnoliids, monocots and eudicots: more than two whorls of three tepals, sepaloid below and petaloid above; two whorls or series of stamens with a long, narrow filament and two to five plicate carpels sealed by postgenital fusion.

Besides several whorls or series of tepals, many members of the clade consisting of Magnoliales and Laurales have other floral features that have been traditionally interpreted as primitive, but are here inferred to be derived, such as spiral perianth phyllotaxis and more than two series of spiral stamens. Given the uncertainty over the ancestral floral phyllotaxis (Figs 4.1A, 4.2A), if these cases of spiral phyllotaxis are derived, it is unclear whether they are reversals to the original angiosperm state or convergences with parallel shifts from whorled to spiral phyllotaxis in the ANITA grade, specifically in *Amborella* and Austrobaileyales. Intriguingly, phyllotactic patterns in the perianth and the androecium have been closely correlated in the groups seen so far, but this correlation breaks down in Magnoliales and Laurales (Endress and Doyle, 2007).

The first inferred change is a shift to spirally arranged stamens in more than two series in the common ancestor of Magnoliales and Laurales (Figs 4.2A, 4.2B). Under both the character definition scheme of Endress and Doyle (2009) and that shown in Fig 4.2B, the increase in number of whorls/series is a reversal to the ancestral angiosperm condition, seen in the ANITA grade, not a direct retention of the ancestral state. The status of spiral stamen phyllotaxis is more ambiguous. With the character in Fig 4.2A, where androecium phyllotaxis and merism are combined, spiral may be either secondarily derived or a retention from the first angiosperms, but under the arguably preferable scheme of Endress and Doyle (2009), where the two characters were kept separate, it is clearly derived, either as a reversal to the original condition in angiosperms (if spiral was ancestral) or a convergence with *Amborella* and Austrobaileyales (if trimerous was ancestral). However, as noted above, the scenario is also ambiguous with the J/M chloroplast tree when phyllotaxis and merism are kept separate, where spiral stamens may have extended from the first angiosperms into mesangiosperms.

Laurales show an additional shift from whorled and trimerous to spiral phyllotaxis of the perianth (Fig 4.1A), which is either a reversal to the original angiosperm condition or a convergence with *Amborella* and Austrobaileyales. With the data set of Endress and Doyle (2009), it was equivocal whether a spiral perianth originated once at the base of Laurales or more than once within the order, but with the changes in scoring of some taxa in Laurales by Doyle and Endress (2010), based on Staedler et al. (2007) and Staedler and Endress (2009), it is now an unequivocal synapomorphy of the order (Fig 4.1A). This shift coincides with origin of a hypanthium (floral cup), a conspicuous synapomorphy of Laurales. Inner staminodes are another derived feature that occurred at the base of Laurales, but as discussed below it is unclear whether this is a synapomorphy of Laurales or of both Laurales and Magnoliales. The reconstructed ancestral flower for Laurales would be generally similar to that of living Calycanthaceae.

The peculiar stamens of most Laurales, with two basal glands and anther dehiscence by upward-opening flaps, may have originated after divergence of Calycanthaceae, but this is equivocal, because Siparunaceae and Mollinedioideae lack glands (possibly as a consequence of packing in the hypanthium) and typical lauralean flaps are absent in Monimiaceae. A partial reversal from plicate to intermediate ascidiate (and uniovulate) carpels also occurred above Calycanthaceae. A marked departure from spiral phyllotaxis and numerous series of parts occurred in Lauraceae and Hernandiaceae, with two trimerous whorls of tepals and more than two trimerous whorls of stamens in Lauraceae, and with one to three whorls of two, three or more tepals and usually two polymerous whorls of stamens in Hernandiaceae. In Figs 4.1A and 4.2A, where spiral phyllotaxis and the three different whorled conditions are states of one character, it is equivocal whether the change from spiral to whorled phyllotaxis occurred once, in the common ancestor of the two families, or independently in both of them. However, a single shift is favoured when phyllotaxis and merism are treated as separate characters (Endress and Doyle, 2009), and this scenario is further supported by the mid-Cretaceous fossil *Mauldinia* (Drinnan et al., 1990), which had a typical lauraceous floral diagram. In our analysis of the phylogenetic positions of fossils (Doyle and Endress, 2010), *Mauldinia* was attached to the stem lineage of both Lauraceae and Hernandiaceae, implying that their common ancestor had similar flowers. Other changes at this point are reduction to the characteristic single carpel of the two families and adnation of the hypanthium to produce an inferior ovary; the basal lines in Lauraceae have an inferior ovary, implying that the superior ovary of most Lauraceae is a reversal (Rohwer and Rudolph, 2005).

Decoupling of perianth and androecium phyllotaxis is most obvious in Magnoliales, where the reconstructed ancestor had more than two whorls of three tepals, but numerous spiral stamens, as in most Magnoliaceae. Correlation between the two sets of organs was restored by shifts to spiral tepals in *Degeneria* (and some derived Magnolioideae) and to whorled outer stamens in Annonaceae, becoming chaotic inwards (Endress, 1987a). Myristicaceae underwent several major modifications: a shift to unisexual flowers, reduction to one perianth whorl, reduction in number and connation of the stamens and reduction to one carpel. In the current phylogenetic context, the elongate receptacle of Magnoliaceae, often considered primitive, is instead derived. The clade consisting of *Galbulimima* (= Himantandraceae), *Degeneria*, *Eupomatia* and Annonaceae differs from Magnoliaceae in having inner staminodes (retained in the basal genus *Anaxagorea*, but lost in other Annonaceae). Since this feature also occurs in most Laurales, but not in Myristicaceae and Magnoliaceae, it may have originated independently within Magnoliales and in Laurales. However, the optimization of this character is ambiguous with our data set, because we scored Myristicaceae as unknown, on the grounds that the absence of inner staminodes could be a side-effect of the shift to unisexual flowers, reduction in stamen number and union of stamens. As a result, it is equally parsimonious to assume that inner staminodes arose in the common ancestor of Magnoliales and Laurales, and were lost in Magnoliaceae. The fact that the staminodes of Magnoliales have distinctive food bodies might be taken as evidence for an independent origin. *Galbulimima* and *Eupomatia* show bizarre and independent departures from the basic floral type, linked with loss of the perianth and modification of the inner staminodes (and outer staminodes in *Galbulimima*) into petaloid organs, as discussed in Endress (1984, 2003) and Kim et al. (2005).

Another supposed primitive feature of many Magnoliales and Laurales is laminar stamens, partly expressed by the stamen base character (either long and wide or short, versus long and narrow). In angiosperms as a whole, it is equivocal whether the ancestral stamen base was long and wide or long and narrow, but in either case the laminar stamens of magnoliids appear to be derived: the reconstructed ancestral stamen in mesangiosperms had a long, narrow filament, which was shortened in the common ancestor of Magnoliales and Laurales, where the stamens also became more numerous and spiral (Fig 7A in Endress and Doyle, 2009). This laminar stamen was later modified again to a filamentous type in more derived Laurales. Presumably these changes are related to the well-known beetle pollination syndrome of these plants. These inferences shed light on a contrast in the orientation of dehiscence in laminar stamens in different taxa, between extrorse in Magnoliales and introrse in ANITA groups such as *Austrobaileya*. Takhtajan (1969) interpreted this difference as evidence that the ancestral stamen had lateral sporangia, which shifted to the adaxial side in some lines, but the abaxial side in others. Whereas the inferred ancestral state in angiosperms was either introrse or latrorse, and the state at the base of mesangiosperms is unresolved, magnoliids are basically extrorse, with abaxial sporangia. This is consistent with a scenario in which stamens were originally laminar and introrse, became filamentous and latrorse near the base of the mesangiosperms, shifted to extrorse at the base of the magnoliids, and then became laminar again in the Magnoliales–Laurales clade, with the sporangia now located on the abaxial side. When the stamens became filamentous again within Laurales, they shifted from extrorse to introrse, often with variation among whorls in Lauraceae.

Based on our data set, there are no unequivocal changes in floral characters on the line from the base of the mesangiosperms to the common ancestor of Canellales and Piperales. If mesangiosperms originally had a trimerous androecium, as inferred when phyllotaxis and merism are treated as separate characters, there was an increase in the number of stamens per whorl in Canellales (i.e. the androecium became polymerous), followed by connation of the stamens into a peculiar tubular androecium in Canellaceae. In Winteraceae there was a shift from a trimerous to a dimerous perianth and an increase in the number of stamen whorls, resulting in a convergence with Magnoliales and Laurales.

Whereas Magnoliales and Laurales show floral elaboration, Piperales show a marked opposite trend for floral simplification, with reduction to one whorl of three tepals in *Lactoris* and Aristolochiaceae and complete loss of the perianth in Saururaceae and Piperaceae. With the character definitions in Fig 4.1B and in Endress and Doyle (2009), it is equivocal whether the perianth was reduced twice or in a stepwise fashion, from two whorls to one and then none. The genus *Saruma* in the Asaroideae has three petals, as well as three sepaloid tepals, which was considered a primitive feature by Thorne (1974), but given the phylogenetic position of Asaroideae, these petals are most parsimoniously interpreted as derived. In Doyle and Endress (2010) we changed the scoring of androecium merism and number of whorls in Saururaceae and Piperaceae to take into account developmental evidence that some taxa with six stamens are dimerous, with lateral stamens in double positions, rather than trimerous (Liang and Tucker, 1995; Hufford, 1996; Tucker and Douglas, 1996), and the discovery that *Verhuellia*, which has only two stamens, is basal in Piperaceae (Wanke et al., 2007; Samain et al., 2010). Nevertheless, it is still most parsimonious to reconstruct the common ancestors of both Piperales as a whole and the Saururaceae–Piperaceae clade as having two whorls of three stamens, as in the first mesangiosperms (Figs 4.2B, 4.4).

An important derived feature of many Canellales and Piperales is syncarpy, seen in Canellaceae, the bicarpellate basal genus *Takhtajania* in the Winteraceae, Aristolochiaceae (with an increase in the number of carpels) and the Saururaceae-Piperaceae clade. The type of syncarpy varies between eusyncarpous (carpels fused at the centre of the gynoecium) in Aristolochiaceae and paracarpous (carpels fused into a unilocular gynoecium with parietal placentation) in Canellaceae, *Takhtajania*, Saururaceae and Piperaceae. With apocarpous, paracarpous and eusyncarpous treated as three unordered states, optimization of this character is highly ambiguous (Fig 10B in Endress and Doyle, 2009), allowing scenarios ranging from separate origins of syncarpy in Canellaceae, *Takhtajania*, Aristolochiaceae and the Saururaceae-Piperaceae clade, to origin of paracarpous syncarpy at the base of the Canellales-Piperales clade and secondary reversals to free carpels in *Lactoris* and within Winteraceae.

4.2.2 Monocots

Assuming that the common ancestor of mesangiosperms had three or more whorls of three tepals and two whorls of three stamens, the main floral change in the origin of monocots was reduction to two whorls of tepals (Fig 4.4). This resulted in the familiar floral formula retained through monocots until the origin of highly derived groups such as Iridaceae, orchids, sedges and grasses. Another inferred change on the monocot stem lineage was from both sepaloid and petaloid tepals to all sepaloid tepals. This condition was apparently later modified to all petaloid tepals in the common ancestor of Melanthiaceae, Nartheciaceae and Dioscoreaceae, which are the three representatives in our data set of the 'core' monocots (*Petrosaviidae* of Cantino et al., 2007).

A possibly more surprising conclusion is that the three carpels were fused into a syncarpous gynoecium in the common ancestor of monocots, and the free carpels of aquatic Alismatales, often considered primitive, are instead a reversal, as inferred by Chen et al. (2004). Remizowa et al. (2006) questioned this conclusion on the grounds that fusion of the carpels is congenital in some basal monocots (*Acorus*, Araceae, *Narthecium*), but postgenital in others (*Tofieldia*), which they suggested was evidence for multiple origins of syncarpy. However, under our organizational definition, the gynoecium of Tofieldiaceae is apocarpous with postgenital carpel connection (pseudosyncarpous in the sense of some authors) rather than syncarpous (Igersheim et al., 2001), and in any case their nested position, between syncarpous Araceae and apocarpous aquatic Alismatales, implies that their condition was derived from congenital syncarpy.

4.2.3 Eudicots

There are other major changes and problems of floral evolution in eudicots, the clade united by tricolpate (and tricolpate-derived) pollen, which includes some 75% of angiosperm species. Many members of the basal order Ranunculales have trimerous flowers (e.g. Menispermaceae, Berberidaceae), and it might be thought that these represent a retention of the trimerous state reconstructed in the perianth and probably the androecium of the common ancestor of mesangiosperms (Figs 4.1A, 4.2A, 4.4). However, Drinnan et al. (1994) suggested that eudicots originally had dimerous flowers, like most Papaveraceae, near the base of Ranunculales, and Proteaceae, *Tetracentron* and Buxaceae in the other line, which also includes the remaining or 'core' eudicots (*Gunneridae*, consisting of Gunnerales and *Pentapetalae*, Cantino et al., 2007). Often these flowers appear to be tetramerous, but a dimerous interpretation is supported by the fact that the sets of four organs arise as two successive decussate pairs and/or the stamens are seemingly opposite the perianth parts, as expected if two alternating pairs of perianth parts are followed by two pairs of stamens (Fig 4.4). The same relation of perianth parts and stamens is seen in the trimerous flowers of monocots and Ranunculales, which are also often misleadingly described as having stamens opposite the perianth parts. With a few autapomorphic exceptions, stamens in basal eudicots have a long, narrow filament, apparently retained from the base of mesangiosperms.

Our analysis indicates that the common ancestor of eudicots had more than two whorls of tepals and two whorls of stamens, as in the reconstructed common ancestor of mesangiosperms (Figs 4.1B, 4.2B, 4.4), but their merism is equivocal: either dimerous or trimerous. In Ranunculales, if eudicots were originally dimerous, Papaveraceae retain the ancestral state, and a reversal to trimery occurred in the remaining groups; if trimery was ancestral, the dimerous condition in Papaveraceae is convergent with dimery in the other eudicot branch. Ranunculales also retain more than two whorls of perianth parts and two whorls of stamens from the common ancestor of mesangiosperms, but the inner perianth parts differ in being true petals as defined anatomically and developmentally, an important morphological innovation and a convergence with Pentapetalae. Within Ranunculales, Ranunculaceae show a shift to spiral phyllotaxis in both the perianth and androecium, with a reduction to two series of perianth parts and an increase to more than two series of stamens.

The ancestral merism is also ambiguous in the main eudicot line, as is the number of perianth whorls/series, because *Nelumbo* has more than two series of spiral tepals. However, the perianth was reduced to two whorls of tepals and became entirely sepaloid in the main eudicot clade, either once (with reversals in *Nelumbo*) or twice. Fossil evidence may help resolve some of these ambiguities. When Early Cretaceous relatives of *Platanus* are added to the analysis (Doyle and Endress, 2010), it can be inferred that both the perianth and the androecium were originally dimerous in the main eudicot line, and on the line leading to *Platanus* there was first a shift from two dimerous whorls of stamens to one whorl of five stamens (as in fossil 'platanoids'), followed by reduction to three or four stamens in *Platanus*. With or without fossil evidence, two dimerous whorls of both tepals and stamens can be reconstructed in the common ancestor of Trochodendraceae and Buxaceae, which are united by another origin of syncarpy and nectaries on the abaxial side of the carpels. Given the dramatically different flowers of *Nelumbo*, *Platanus* and Proteaceae, it may be hard to accept that they constitute a clade (Proteales), but these differences are not really a problem, because they are all a function of autapomorphies: floral gigantism and increase in the number of floral organs in *Nelumbo*, unisexuality and crowding of flowers into heads in *Platanus* and reduction of the gynoecium to one carpel in Proteaceae.

A key taxon for the question of the original merism in eudicots is *Euptelea*, which lacks a perianth. Hoot et al. (1999) suggested that *Euptelea* is fundamentally dimerous, because its floral primordium is bilateral (Endress, 1986a), but because the organs do not develop in a dimerous pattern and the shape of the primordium may be a result of space constraint by the subtending bracts (Ren et al., 2007) we scored *Euptelea* as unknown. If *Euptelea* could be shown to be basically dimerous, this would strengthen the view that dimery was ancestral in eudicots. In general, our inferences on floral evolution in eudicots are somewhat more tentative than those in other groups because we have not included other basal eudicot taxa such as Sabiaceae (Wanntorp and Ronse De Craene, 2007) and *Didymeles* (von Balthazar et al., 2003), to say nothing of potentially relevant basal members of the remaining eudicots.

These results are of broader significance for floral evolution, because the clade that includes Trochodendraceae and Buxaceae also appears to contain the remaining eudicots, or Gunneridae, in which Gunnerales (*Gunnera*, *Myrothamnus*) have simple, apetalous flowers (clearly dimerous in *Gunnera*: Wanntorp and Ronse De Craene, 2005), whereas Pentapetalae have basically pentamerous flowers. This implies that the typical flowers of Pentapetalae, with alternating pentamerous whorls of sepals, petals and stamens, were derived from much simpler flowers with four sepaloid tepals and four stamens, whether by multiplication of whorls, incorporation of bracts into the perianth, increase in number of parts per whorl, change in identity of floral organs or some combination of these processes (cf. Soltis et al., 2003; Ronse De Craene, 2007). Wanntorp and Ronse De Craene (2005) argued that the simple flowers of *Gunnera* are reduced as an adaptation to wind pollination and therefore not significant for origin of the flowers of Pentapetalae. However, adaptive explanations and phylogenetic significance need not be mutually exclusive. As recognized by Ronse De Craene (2007), phylogenetic relationships imply that the ancestors of Pentapetalae had simple, apetalous flowers, which could have been an adaptation to wind pollination at an earlier stage that was maintained (and perhaps intensified) in Gunnerales. The resulting picture recalls scenarios for floral evolution proposed by Walker and Walker (1984) and Ehrendorfer (1989), although these were based on different sets of taxa, many of which no longer appear to be phylogenetically relevant.

4.3 Conclusions

These results can be summarized with reference to Fig 4.4, which shows reconstructed floral diagrams for key nodes. Our results indicate that the ancestral flower had more than two whorls or series of tepals and stamens, and several ascidiate carpels containing a single pendent ovule, but floral phyllotaxis appears to have been labile at first, so it is equivocal whether the floral parts were originally spiral or whorled and trimerous. Extreme floral reduction occurred in Hydatellaceae and the clade including Chloranthaceae and *Ceratophyllum*. However, a trimerous flower with more than two whorls of tepals and two whorls of stamens appears to have been established near the base of the mesangiosperms. Near this point the carpel became plicate. Within the magnoliid clade, the perianth was reduced to one whorl of tepals and lost in the Piperales, but on the line to Magnoliales and Laurales the stamens became more numerous and spirally arranged, correlated with a strong tendency for beetle pollination, and in Laurales the perianth became spiral as well, until a reversal to trimerous whorls of both tepals and stamens, and reduction to one carpel occurred in the Lauraceae–Hernandiaceae line. Monocots lost one whorl of tepals and underwent carpel fusion, with a reversal of the latter in Alismatales. In eudicots there was a shift from trimerous to dimerous flowers, either once on the stem lineage or two or three times within the clade, followed by reduction to two pairs of tepals and stamens on the line that gave rise to pentamerous core eudicots (Pentapetalae), presumably as an adaptation to wind pollination. Cretaceous fossils have considerable potential for resolving the ambiguities in this scheme, as exemplified by the cases of *Mauldinia* in the Laurales and fossil 'platanoids' in the Proteales. It is our hope that this improved picture of patterns of early floral evolution will provide a more robust framework for process-oriented investigations of functional and developmental factors involved in the early angiosperm radiation.

Acknowledgements

We wish to thank Louis Ronse De Craene and Livia Wanntorp for inviting us to the symposium on which this book is based, Alex Bernhard for help with illustrations, Paula Rudall for useful discussions, Richard Bateman and an anonymous reviewer for valuable comments on the manuscript and the NSF Deep Time Research Collaboration Network (RCN0090283) for facilitating collaboration.

4.4 References

Aboy, H. E. (1936). A study of the anatomy and morphology of *Ceratophyllum demersum*. Unpublished MS thesis, Ithaca, NY: Cornell University.

Antonov, A. S., Troitsky, A. V., Samigullin, T. K. et al. (2000). Early events in the evolution of angiosperms deduced from cp rDNA ITS 2-4 sequence comparisons. pp. 210–214 in Liu, Y.-H., Fan, H.-M., Chen, Z.-Y., Wu, Q.-G. and Zeng, Q.-W. (eds.), *Proceedings of the International Symposium on the Family Magnoliaceae*. Beijing: Science Press.

APG [Angiosperm Phylogeny Group] II (2003). An update of the Angiosperm Phylogeny Group classification for the orders and families of flowering plants: APG II. *Botanical Journal of the Linnean Society*, **141**, 399–436.

Arber, E. A. N. and Parkin, J. (1907). On the origin of angiosperms. *Journal of the Linnean Society, Botany*, **38**, 29–80.

Bailey, I. W. and Swamy, B. G. L. (1951). The conduplicate carpel of dicotyledons and its initial trends of specialization. *American Journal of Botany*, **38**, 373–379.

Barkman, T. J., Chenery, G., McNeal, J. R. et al. (2000). Independent and combined analyses of sequences from all three genomic compartments converge on the root of flowering plant phylogeny. *Proceedings of the National Academy of Sciences USA*, **97**, 13166–13171.

Buchheim, G. (1964). Magnoliales. pp. 108–131 in Melchior, H. (ed.), *A. Engler's Syllabus der Pflanzenfamilien*, Vol. 2, 12th edn. Berlin: Borntraeger.

Burleigh, J. G. and Mathews, S. (2004). Phylogenetic signal in nucleotide data from seed plants: implications for resolving the seed plant tree of life. *American Journal of Botany*, **91**, 1599–1613.

Canright, J. E. (1952). The comparative morphology and relationships of the Magnoliaceae. I. Trends of specialization in the stamens. *American Journal of Botany*, **39**, 484–497.

Cantino, P. D., Doyle, J. A., Graham, S. W. et al. (2007). Towards a phylogenetic nomenclature of *Tracheophyta*. *Taxon*, **56**, 822–846.

Chase, M. W., Soltis, D. E., Olmstead, R. G. et al. (1993). Phylogenetics of seed plants: an analysis of nucleotide sequences from the plastid gene *rbc*L. *Annals of the Missouri Botanical Garden*, **80**, 526–580.

Chen, J.-M., Chen, D., Gituru, W. R., Wang, Q.-F. and Guo, Y.-H. (2004). Evolution of apocarpy in Alismatidae using phylogenetic evidence from chloroplast *rbc*L gene sequence data. *Botanical Bulletin of Academia Sinica*, **45**, 33–40.

Cronquist, A. (1968). *The Evolution and Classification of Flowering Plants*. Boston: Houghton Mifflin.

Cronquist, A. (1981). *An Integrated System of Classification of Flowering Plants*. New York: Columbia University Press.

Donoghue, M. J. and Doyle, J. A. (1989). Phylogenetic analysis of angiosperms and the relationships of

Hamamelidae. pp. 17–45 in Crane, P. R. and Blackmore, S. (eds.), *Evolution, Systematics, and Fossil History of the Hamamelidae*, Vol. 1. Oxford: Clarendon Press.

Doyle, J. A. (1996). Seed plant phylogeny and the relationships of Gnetales. *International Journal of Plant Sciences*, **157** (Supplement), S3–S39.

Doyle, J. A. (2005). Early evolution of angiosperm pollen as inferred from molecular and morphological phylogenetic analyses. *Grana*, **44**, 227–251.

Doyle, J. A. (2006). Seed ferns and the origin of angiosperms. *Journal of the Torrey Botanical Society*, **133**, 169–209.

Doyle, J. A. (2007). Systematic value and evolution of leaf architecture across the angiosperms in light of molecular phylogenetic analyses. *Courier Forschungsinstitut Senckenberg*, **258**, 21–37.

Doyle, J. A. (2008). Integrating molecular phylogenetic and paleobotanical evidence on origin of the flower. *International Journal of Plant Sciences*, **169**, 816–843.

Doyle, J. A. and Endress, P. K. (2000). Morphological phylogenetic analysis of basal angiosperms: comparison and combination with molecular data. *International Journal of Plant Sciences*, **161** (Supplement), S121–S153.

Doyle, J. A. and Endress, P. K. (2010). Integrating Early Cretaceous fossils into the phylogeny of living angiosperms: Magnoliidae and eudicots. *Journal of Systematics and Evolution*, **48**, 1–35.

Doyle, J. A., Eklund, H. and Herendeen, P. S. (2003). Floral evolution in Chloranthaceae: implications of a morphological phylogenetic analysis. *International Journal of Plant Sciences*, **164** (Suppl.), S365–S382.

Drinnan, A. N., Crane, P. R., Friis, E. M. and Pedersen, K. R. (1990). Lauraceous flowers from the Potomac Group (mid-Cretaceous) of eastern North America. *Botanical Gazette*, **151**, 370–384.

Drinnan, A. N., Crane, P. R. and Hoot, S. B. (1994). Patterns of floral evolution in the early diversification of non-magnoliid dicotyledons (eudicots). *Plant Systematics and Evolution Supplement*, **8**, 93–122.

Duvall, M. R., Mathews, S., Mohammad, N. and Russell, T. (2006). Placing the monocots: conflicting signal from trigenomic analyses. *Aliso*, **22**, 79–90.

Duvall, M. R., Robinson, J. W., Mattson, J. G. and Moore, A. (2008). Phylogenetic analyses of two mitochondrial metabolic genes sampled in parallel from angiosperms find fundamental interlocus incongruence. *American Journal of Botany*, **95**, 871–884.

Ehrendorfer, F. (1989). The phylogenetic position of the Hamamelidae. pp. 1–7 in Crane, P. R. and Blackmore, S. (eds.), *Evolution, Systematics, and Fossil History of the Hamamelidae*, Vol. 1. Oxford: Clarendon Press.

Eklund, H., Doyle, J. A. and Herendeen, P. S. (2004). Morphological phylogenetic analysis of living and fossil Chloranthaceae. *International Journal of Plant Sciences*, **165**, 107–151.

Endress, P. K. (1984). The role of inner staminodes in the floral display of some relic *Magnoliales*. *Plant Systematics and Evolution*, **146**, 269–282.

Endress, P. K. (1986a). Floral structure, systematics, and phylogeny in Trochodendrales. *Annals of the Missouri Botanical Garden*, **73**, 297–324.

Endress, P. K. (1986b). Reproductive structures and phylogenetic significance of extant primitive angiosperms. *Plant Systematics and Evolution*, **152**, 1–28.

Endress, P. K. (1987a). Floral phyllotaxis and floral evolution. *Botanische Jahrbücher für Systematik*, **108**, 417–438.

Endress, P. K. (1987b). The Chloranthaceae: reproductive structures and phylogenetic position. *Botanische Jahrbücher für Systematik*, **109**, 153–226.

Endress, P. K. (1994). Evolutionary aspects of the floral structure in *Ceratophyllum*. *Plant Systematics and Evolution Supplement*, **8**, 175–183.

Endress, P. K. (2003). Early floral development and nature of the calyptra in Eupomatiaceae (Magnoliales). *International Journal of Plant Sciences*, **164**, 489–503.

Endress, P. K. (2004). Structure and relationships of basal relictual angiosperms. *Australian Systematic Botany*, **17**, 343–366.

Endress, P. K. (2008). Perianth biology in the basal grade of extant angiosperms. *International Journal of Plant Sciences*, **169**, 844–862.

Endress, P. K. and Doyle, J. A. (2007). Floral phyllotaxis in basal angiosperms: development and evolution. *Current Opinion in Plant Biology*, **10**, 52–57.

Endress, P. K. and Doyle, J. A. (2009). Reconstructing the ancestral angiosperm flower and its initial specializations. *American Journal of Botany*, **96**, 22–66.

Endress, P. K. and Igersheim, A. (2000). Gynoecium structure and evolution in basal angiosperms. *International Journal of Plant Sciences*, **161** (Suppl.), S211–S223.

Friis, E. M. and Endress, P. K. (1990). Origin and evolution of angiosperm flowers. *Advances in Botanical Research*, **17**, 99–162.

Friis, E. M., Crane, P. R. and Pedersen, K. R. (1986). Floral evidence for Cretaceous chloranthoid angiosperms. *Nature*, **320**, 163–164.

Friis, E. M., Doyle, J. A., Endress, P. K. and Leng, Q. (2003). *Archaefructus* – angiosperm precursor or specialized early angiosperm? *Trends in Plant Science*, **8**, 369–373.

Friis, E. M., Pedersen, K. R. and Crane, P. R. (2006). Cretaceous angiosperm flowers: innovation and evolution in plant reproduction. *Palaeogeography Palaeoclimatology Palaeoecology*, **232**, 251–293.

Frohlich, M. W. and Chase, M. W. (2007). After a dozen years of progress the origin of angiosperms is still a great mystery. *Nature*, **450**, 1184–1189.

Hamby, R. K. and Zimmer, E. A. (1992). Ribosomal RNA as a phylogenetic tool in plant systematics. pp. 50–91 in Soltis, P. S., Soltis, D. E. and Doyle, J. J. (eds.), *Molecular Systematics of Plants*. New York: Chapman and Hall.

Hickey, L. J. and Taylor, D. W. (1996). Origin of the angiosperm flower. pp. 176–231 in Taylor, D.W. and Hickey, L. J. (eds.), *Flowering Plant Origin, Evolution and Phylogeny*. New York: Chapman and Hall.

Hoot, S. B., Magallón, S. and Crane, P. R. (1999). Phylogeny of basal eudicots based on three molecular data sets: *atp*B, *rbc*L, and 18S nuclear ribosomal DNA sequences. *Annals of the Missouri Botanical Garden*, **86**, 1–32.

Hufford, L. (1996). Ontogenetic evolution, clade diversification, and homoplasy. pp. 271–301 in Sanderson, M. J. and Hufford, L. (eds.), *Homoplasy – The Recurrence of Similarity in Evolution*. San Diego, CA: Academic Press.

Igersheim, A., Buzgo, M. and Endress, P. K. (2001). Gynoecium diversity and systematics in basal monocots. *Botanical Journal of the Linnean Society*, **136**, 1–65.

Iwamoto, A., Shimizu, A. and Ohba, H. (2003). Floral development and phyllotactic variation in *Ceratophyllum demersum* (Ceratophyllaceae). *American Journal of Botany*, **90**, 1124–1130.

Jansen, R. K., Cai, Z., Raubeson, L. A. et al. (2007). Analysis of 81 genes from 64 plastid genomes resolves relationships in angiosperms and identifies genome-scale evolutionary patterns. *Proceedings of the National Academy of Sciences USA*, **104**, 19369–19374.

Kim, S., Koh, J., Ma, H. et al. (2005). Sequence and expression studies of A-, B-, and E-class MADS-box homologues in *Eupomatia* (Eupomatiaceae): support for the bracteate origin of the calyptra. *International Journal of Plant Sciences*, **166**, 185–198.

Kim, S., Soltis, D. E., Soltis, P. S., Zanis, M. J. and Suh, Y. (2004). Phylogenetic relationships among early-diverging eudicots based on four genes: were the eudicots ancestrally woody? *Molecular Phylogenetics and Evolution*, **31**, 16–30.

Liang, H.-X. and Tucker, S. C. (1995). Floral ontogeny of *Zippelia begoniaefolia* and its familial affinity: Saururaceae or Piperaceae? *American Journal of Botany*, **82**, 681–689.

Loconte, H. and Stevenson, D. W. (1991). Cladistics of the Magnoliidae. *Cladistics*, **7**, 267–296.

Maddison, D. R. and Maddison, W. P. (2003). *MacClade 4: Analysis of Phylogeny and Character Evolution*, version 4.06. Sunderland, MA: Sinauer Associates.

Maddison, W. P. (1993). Missing data versus missing characters in phylogenetic analysis. *Systematic Biology*, **42**, 576–581.

Mathews, S. and Donoghue, M. J. (1999). The root of angiosperm phylogeny inferred from duplicate phytochrome genes. *Science*, **286**, 947–950.

Meeuse, A. D. J. (1965). *Angiosperms – Past and Present*. New Delhi: Institute for the Advancement of Science and Culture.

Meeuse, A. D. J. (1975). Changing floral concepts: anthocorms, flowers, and anthoids. *Acta Botanica Neerlandica*, **24**, 23–36.

Moore, M. J., Bell, C. D., Soltis, P. S. and Soltis, D. E. (2007). Using plastid genome-scale data to resolve enigmatic relationships among basal angiosperms. *Proceedings of the National Academy of Sciences USA*, **104**, 19363–19368.

Muller, J. (1981). Fossil pollen records of extant angiosperms. *Botanical Review*, **47**, 1–142.

Nixon, K. C., Crepet, W. L., Stevenson, D. and Friis, E. M. (1994). A reevaluation of seed plant phylogeny. *Annals of the Missouri Botanical Garden*, **81**, 484–533.

Parkinson, C. L., Adams, K. L. and Palmer, J. D. (1999). Multigene analyses identify the three earliest lineages of extant flowering plants. *Current Biology*, **9**, 1485–1488.

Pedersen, K. R., Crane, P. R., Drinnan, A. N. and Friis, E. M. (1991). Fruits from the mid-Cretaceous of North America with pollen grains of the *Clavatipollenites* type. *Grana*, **30**, 577–590.

Qiu, Y.-L., Lee, J., Bernasconi-Quadroni, F. et al. (1999). The earliest angiosperms: evidence from mitochondrial, plastid and nuclear genomes. *Nature*, **402**, 404–407.

Qiu, Y.-L., Li, L., Hendry, T. A. et al. (2006). Reconstructing the basal angiosperm phylogeny: evaluating information content of mitochondrial genes. *Taxon*, **55**, 837–856.

Remizova, M. and Sokoloff, D. (2003). Inflorescence and floral morphology in *Tofieldia* (Tofieldiaceae) compared with Araceae, Acoraceae and Alismatales s.str. *Botanische Jahrbücher für Systematik*, **124**, 255–271.

Remizowa, M., Sokoloff, D. and Rudall, P. J. (2006). Evolution of the monocot gynoecium: evidence from comparative morphology and development in *Tofieldia, Japonolirion, Petrosavia*, and *Narthecium*. *Plant Systematics and Evolution*, **258**, 183–209.

Ren, Y., Li, H.-F., Zhao, L. and Endress, P. K. (2007). Floral morphogenesis in *Euptelea* (Eupteleaceae, Ranunculales). *Annals of Botany*, **100**, 185–193.

Rohwer, J. G. and Rudolph, B. (2005). Jumping genera: the phylogenetic positions of *Cassytha, Hypodaphnis*, and *Neocinnamomum* (Lauraceae) based on different analyses of *trn*K intron sequences. *Annals of the Missouri Botanical Garden*, **92**, 153–178.

Ronse De Craene, L. P. (2007). Are petals sterile stamens or bracts? The origin and evolution of petals in the core eudicots. *Annals of Botany*, **100**, 621–630.

Ronse De Craene, L. P., Soltis, P. S. and Soltis, D. E. (2003). Evolution of floral structures in basal angiosperms. *International Journal of Plant Sciences*, **164** (Supplement), S329–S363.

Rudall, P. J. and Bateman, R. M. (2006). Morphological phylogenetic analysis of Pandanales: testing contrasting hypotheses of floral evolution. *Systematic Botany*, **31**, 223–238.

Rudall, P. J., Remizowa, M. V., Prenner, G. et al. (2009). Nonflowers near the base of extant angiosperms? Spatiotemporal arrangement of organs in reproductive units of Hydatellaceae and its bearing on the origin of the flower. *American Journal of Botany*, **96**, 67–82.

Rudall, P. J., Sokoloff, D. D., Remizowa, M. V. et al. (2007). Morphology of Hydatellaceae, an anomalous aquatic family recently recognized as an early-divergent angiosperm lineage. *American Journal of Botany*, **94**, 1073–1092.

Saarela, J. M., Rai, H. S., Doyle, J. A. et al. (2007). Hydatellaceae identified as a new branch near the base of the angiosperm phylogenetic tree. *Nature*, **446**, 312–315.

Samain, M. S., Vrijdaghs, A., Hesse, M. et al. (2010). *Verhuellia* is a segregate lineage in Piperaceae: more evidence from flower, fruit and pollen morphology, anatomy and development). *Annals of Botany*, **105**, 677–688.

Schönenberger, J., Pedersen, K. R. and Friis, E. M. (2001). Normapolles flowers of fagalean affinity from the Late Cretaceous of Portugal. *Plant Systematics and Evolution*, **226**, 205–230.

Sereno, P. C. (2007). Logical basis for morphological characters in phylogenetics. *Cladistics*, **23**, 565–587.

Shamrov, I. I. (2009). Morfologicheskaya priroda ginetseya i ploda u *Ceratophyllum* (*Ceratophyllaceae*). *Botanicheskiy Zhurnal*, **94**, 938–961.

Sokoloff, D. D., Remizowa, M. V., Linder, H. P. and Rudall, P. J. (2009). Morphology and development of the gynoecium in Centrolepidaceae: the most remarkable range of variation in Poales. *American Journal of Botany*, **96**, 1925–1940.

Soltis, D. E., Senters, A. E., Zanis, M. J. et al. (2003). Gunnerales are sister to other core eudicots: implications for the evolution of pentamery. *American Journal of Botany*, **90**, 461–470.

Soltis, D. E., Soltis, P. S., Chase, M. W. et al. (2000). Angiosperm phylogeny inferred from 18S rDNA, *rbcL*, and *atpB* sequences. *Botanical Journal of the Linnean Society*, **133**, 381–461.

Soltis, D. E., Soltis, P. S., Endress, P. K. and Chase, M. W. (2005). *Phylogeny and Evolution of Angiosperms*. Sunderland, MA: Sinauer Associates.

Soltis P. S., Soltis, D. E. and Chase, M. W. (1999). Angiosperm phylogeny inferred from multiple genes as a tool for comparative biology. *Nature*, **402**, 402–404.

Staedler, Y. M. and Endress, P. K. (2009). Diversity and lability of floral phyllotaxis in the pluricarpellate families of core Laurales (Gomortegaceae, Atherospermataceae, Siparunaceae, Monimiaceae). *International Journal of Plant Sciences*, **170**, 522–550.

Staedler, Y. M., Weston, P. H. and Endress, P. K. (2007). Floral phyllotaxis and floral architecture in Calycanthaceae (Laurales). *International Journal of Plant Sciences*, **168**, 285–306.

Sun, G., Dilcher, D. L., Zheng, S. and Zhou, Z. (1998). In search of the first flower: a Jurassic angiosperm, *Archaefructus*, from northeast China. *Science*, **282**, 1692–1695.

Sun, G., Ji, Q., Dilcher, D. L. et al. (2002). Archaefructaceae, a new basal angiosperm family. *Science*, **296**, 899–904.

Takhtajan, A. L. (1966). *Sistema i Filogeniya Tsvetkovykh Rasteniy*. Moscow: Nauka.

Takhtajan, A. L. (1969). *Flowering Plants: Origin and Dispersal.* Washington, DC: Smithsonian Institution.

Takhtajan, A. L. (1980). Outline of the classification of flowering plants (Magnoliophyta). *Botanical Review*, **46**, 225–359.

Taylor, D. W. and Hickey, L. J. (1992). Phylogenetic evidence for the herbaceous origin of angiosperms. *Plant Systematics and Evolution*, **180**, 137–156.

Taylor, E. L. and Taylor, T. N. (2009). Seed ferns from the late Paleozoic and Mesozoic: any angiosperm ancestors lurking there? *American Journal of Botany*, **96**, 237–251.

Thorne, R. F. (1974). A phylogenetic classification of the Annoniflorae. *Aliso*, **8**, 147–209.

Tucker, S. C. and Douglas, A. W. (1996). Floral structure, development, and relationships of paleoherbs: *Saruma*, *Cabomba*, *Lactoris*, and selected Piperales. pp. 141–175 in Taylor, D.W. and Hickey, L. J. (eds.), *Flowering Plant Origin, Evolution and Phylogeny*. New York: Chapman and Hall.

Upchurch, G. R. (1984). Cuticular anatomy of angiosperm leaves from the Lower Cretaceous Potomac Group. I. Zone I leaves. *American Journal of Botany*, **71**, 192–202.

von Balthazar, M., Pedersen, K. R., Crane, P. R. and Friis, E. M. (2008). *Carpestella lacunata* gen. et sp. nov., a new basal angiosperm flower from the Early Cretaceous (Early to Middle Albian) of eastern North America. *International Journal of Plant Sciences*, **169**, 890–898.

von Balthazar, M., Schatz, G. E. and Endress, P. K. (2003). Female flowers and inflorescences of Didymelaceae. *Plant Systematics and Evolution*, **237**, 199–208.

Walker, J. W. and Walker, A. G. (1984). Ultrastructure of Lower Cretaceous angiosperm pollen and the origin and early evolution of flowering plants. *Annals of the Missouri Botanical Garden*, **71**, 464–521.

Wanke, S., Vanderschaeve, L., Mathieu, G. et al. (2007). From forgotten taxon to a missing link? The position of the genus *Verhuellia* (Piperaceae) revealed by molecules. *Annals of Botany*, **99**, 1231–1238.

Wanntorp, L. and Ronse De Craene, L. P. (2005). The *Gunnera* flower: key to eudicot diversification or response to pollination mode? *International Journal of Plant Sciences*, **166**, 945–953.

Wanntorp, L. and Ronse De Craene, L. P. (2007). Flower development of *Meliosma* (Sabiaceae): evidence for multiple origins of pentamery in the eudicots. *American Journal of Botany*, **94**, 1828–1836.

Wettstein, R. R. von (1907). *Handbuch der Systematischen Botanik*, Vol. 2. Vienna: Franz Deuticke.

Williams, J. H. (2009). *Amborella trichopoda* (Amborellaceae) and the evolutionary developmental origins of the angiosperm progamic phase. *American Journal of Botany*, **96**, 144–165.

Zanis, M. J., Soltis, D. E., Soltis, P. S., Mathews, S. and Donoghue, M. J. (2002). The root of the angiosperms revisited. *Proceedings of the National Academy of Sciences USA*, **99**, 6848–6853.

Zanis, M. J., Soltis, P. S., Qiu, Y. L., Zimmer, E. and Soltis, D. E. (2003). Phylogenetic analyses and perianth evolution in basal angiosperms. *Annals of the Missouri Botanical Garden*, **90**, 129–150.

5

Changing views of flower evolution and new questions

Peter K. Endress

5.1 Flowers in phylogenetic and evolutionary studies

The role of flowers in evolutionary biology has changed in the past 20 years, as the major foci are constantly changing with new approaches and better understanding of evolutionary processes. The revolution of molecular phylogenetics and molecular developmental genetics produced a trend in flower studies away from phylogenetics and towards evolution. In turn, the discovery of many well-preserved Cretaceous fossil flowers led to a new trend in flower studies towards phylogenetics, because fossil flowers do not provide DNA. The following three current fields of flower structural studies may be distinguished:

(1) *Comparative morphological analysis of flowers* – Many new major angiosperm clades have been recognized by molecular phylogenetic studies since Chase et al. (1993), as surveyed in APG (1998, 2009), Stevens (2001 onwards) and Soltis et al. (2005). These new clades need now to be critically studied comparatively in their structure and biology as they are largely unknown (e.g. Endress and Matthews, 2006; Endress, 2010a).

(2) *Morphology for phylogenetic studies* – Flowers were generally used for phylogenetic studies in the era before the molecular revolution. In the past 20 years,

Flowers on the Tree of Life, ed. Livia Wanntorp and Louis P. Ronse De Craene. Published by Cambridge University Press. © The Systematics Association 2011.

phylogenetics has concentrated on molecular approaches, which yield more results in a shorter time than morphology. However, morphological phylogenetic analyses are still performed and yield interesting results, either alone or in combination with molecular analyses (at higher systematic levels, e.g. Nandi et al., 1998; Doyle and Endress, 2000, or lower levels, e.g. Carillo-Reyes et al., 2008; Sweeney, 2008). There has been a pessimistic attitude towards the use of morphological features in phylogenetics because of too much homoplasy (e.g. Givinish and Sytsma, 1997; Patterson and Givnish, 2002; Givnish, 2003; Scotland et al., 2003) and difficulties in scoring structural characters (Stevens, 2000). This is true if superficial structural features that are easy to spot are used (e.g. tepals large and showy versus small and inconspicuous, or fruits capsules versus berries, or storage organs rhizomes versus bulbs). However, morphology encompasses much more than such features. It can be expected that as our knowledge of flowers increases, there will be a resurgence in morphological phylogenetic analyses. In addition, the more fossil flowers become available, the more important morphological phylogenetic analyses will become (e.g. Friis et al., 2009; Doyle and Endress, 2010). There are not only many more fossil flowers available than 20 years ago, but there are also new techniques to reconstruct their morphology: the use of microtome section series (Schönenberger, 2005) and tomography (Friis et al., 2009). The search for and the detection of new structural patterns of interest is a continuing challenge. Characters and character states 'cannot be defined but need to be discussed,' as Wagner (2005) put it, meaning that definitions need to be constantly evaluated and updated to fit the current knowledge with each change in the phylogenetic framework. New knowledge on phylogeny (and evolution) continuously creates a new basis for discussion. Of course, if morphological characters are used for phylogenetic studies, this also means the necessity of repeated reciprocal illumination (see also Kelly and Stevenson, 2005). 'Tree-thinking' has been encouraged in evolutionary studies (O'Hara, 1988; Donoghue and Sanderson, 1992). This is of course also relevant for the focus on structural features, including the construction of morphological matrices for phylogenetic studies. The more detailed a tree under reconstruction already is and the more detailed our knowledge about the distribution of traits on this tree is, the better we can judge the quality of characters and character states to be scored.

(3) *Morphology for evolutionary studies* – The new phylogenetic results can now be used to study the evolution of flowers on a much more solid basis than was possible before. A general result is that many features are more evolutionarily flexible than previously assumed. A number of examples are surveyed in this study. Rarely is a character more stable than previously assumed at macro-systematic level; such an exception are features of ovules (Endress, 2003, 2005a, 2010). However, such flexibility is not randomly distributed through

the larger clades. Given features are more concentrated (but not universal) in a certain clade than in another one. Why is this so? Answers can be expected from better knowledge of the genetic systems that operate in the development of such features (e.g. Borowsky, 2008; Melzer et al., 2008). Thus, homoplasy in structure is pervasive, much more common than earlier imagined and is a fascinating aspect of flower evolution (e.g. Cantino, 1985; Endress, 1996). For more evolutionary aspects of flower morphology, see Endress (1994, 2003, 2005b, 2006).

5.2 Homology

The terms homology and homologous were used originally to express that two parts that superficially look very different are in fact more similar and are evolutionarily derived from the same ancestral structure. Later, overapplication of the term in trying to find homologies for every little part has led to an inflation of the term. The attempt to work with 'partial homologies' in trying to determine at which percentage two parts are homologous (Sattler, 1992) is especially not helpful in understanding the evolution of structures. That there are continuities in the evolution of organisms in many ways is trivial. However, organisms are also characterized by a stratified complexity, and evolution proceeds with more or less hierarchical modules and patterns (e.g. Simon, 1962; Endress, 2005b). These patterns are the big issues to tackle in evolutionary biology. Thus a sensible evolutionary question in the detailed comparison of two parts is not by what percentage they are homologous, but in which respects they are homologous. Thus, one would like to know which aspects (submodules or subpatterns) they share with each other over their common ancestor.

How to proceed in the study of homology? Earlier recipes to assess homologies by morphologists such as Remane (1956), Eckardt (1964) and Kaplan (1984) are too strict and too restricted from a current perspective. However, this was reasonable at the time they were proposed. We are now in a different position. Earlier biologists used homologies to find systematic relationships and to understand phylogeny. Today it goes the other way around. We use homologies to understand evolution (and not phylogeny/relationships). Work with homologies gives a guideline. Thus we may proceed in the following way:

■ Detailed study of the structure in question, if possible with living material from entire plants available

■ Study of the development of the structure

■ Comparison of the structure with the corresponding structure in the closest relatives (which may be less difficult to interpret)

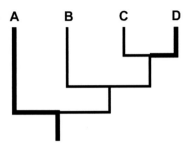

Fig 5.1 Re-appearance or re-institution in clade D of a feature (bold) present in clade A, which is missing in clades B and C.

- Comparison with other groups with the same ecology (if an extreme ecology is connected with the special structure, for instance water plants) (e.g. Endress and Doyle, 2009).

In the era of cladistics, homology has sometimes been equated with synapomorphy. This is, however, only one side of the coin and thus of limited use in understanding the evolution of structures. A more differentiated view is proposed by Wagner (1989, 2007) who distinguishes between biological versus cladistic (phylogenetic) homology. Cladistic homology is used for synapomorphy, whereas biological homogy is used for morphological structures that have the same underlying genetic structure (see also Collin and Miglietta, 2008). The evolution of the genetic structure is an ever more fascinating field to tackle. A number of possible mechanisms for such evolutionary changes are currently being discussed, such as gene duplication and co-option of conserved modules (e.g. Causier et al., 2005; Irish and Litt, 2005; Rosin and Kramer, 2009) or frame-shifts (e.g. Lamb and Irish, 2003; Vargas and Wagner, 2009; Wagner, 2009).

In the example illustrated in Fig 5.1, the genetic machinery for the bold feature of clade A has not disappeared in clades B and C, which do not exhibit the feature (the feature is repressed or not turned on), and is re-activated or re-instituted so that the feature reappears in clade D. Thus the bold feature shared by clades A and D is not homologous in a cladistic sense, but is homologous in a biological sense. Both aspects, cladistic homology (synapomorphy) and biological homology, are important in evolutionary biology.

5.3 Alpha- and omega-morphology

Morphology of plants has many different applications. *Alpha-morphology* (term coined here) is in morphology what alpha-taxonomy (Turrill, 1938) is in taxonomy. It is what many non-morphologists understand by morphology. It focuses on simple descriptions of plants as they are used for species descriptions, for floras or taxonomic revisions, mainly at the infrageneric level, or for the characterization of individuals in population biology. It deals with features that can rapidly and easily be seen, such as number and size of organs. It is one-dimensional, linear and measures simple parameters. In contrast, *omega-morphology* is an attempt to understand the structure of plants in evolutionary biology. It is multidimensional and includes all that can be attempted at a given time (comparative and molecular developmental morphology, floral biology, of extant and, if possible, fossil plants, evo-devo, evo-devo-eco) and it can always be improved by incorporating new aspects from various directions as they become available. A general goal of omega-morphology is an integrated understanding of form and function, and development and evolution. Morphological traits are shaped by history, likewise by organizational and ecological constraints, which both need to be considered to gain a deepened understanding of flower evolution (see also Endress, 2003; Givnish, 2003; Friedman et al., 2008).

Alpha-morphology can also be seen as a starting point for omega-morphology. Morphometrics (Klingenberg, 2009) and automated measurements of simple shapes and their variation (Lexer et al., 2009) are mostly used at the level of alpha-morphology. Omega-morphology is indispensable for evolutionary studies at higher levels, because changes in the organization of structures or structural complexes cannot commonly be represented by simple numbers and distances between reference points in structural units. Patterns of interest may be complex three-dimensional structures (especially the gynoecium) or complex branching systems of inflorescences with specific patterns of concatenation of branches, both of which may need developmental studies, or evaluation of a sequence of leaves with changing morphology on a shoot and patterns in this sequence, or patterns in the sequence of changing lateral branch differentiation along a gradient of subtending leaves in an inflorescence.

	Easy to assemble	Interesting for (macro-) phylogeny
Simple features, Size, number: (e.g. distance between two defined points, number of specific organs)	+	– +
Complex features, Shape, pattern: (e.g. position of parts in a system, developmental sequence of specific events)	–	

5.4 Detailed comparative morphological analysis for the elucidation of structural conundra

An example of the different focuses in alpha- and omega-morphology are two publications on the family Oliniaceae (Myrtales). The study by Sebola and Balkwill (2009) deals with alpha-morphology, with the aim of an infrafamilial classification (low systematic level). This approach enabled a better understanding of the infrafamilial taxonomic differentiation. In contrast, the work by Schönenberger and Conti (2003) has a more omega-morphological approach, including developmental studies and comparison with the most closely related families Penaeaceae, Rhynchocalycaceae and Alzateaceae, and aims to understand the morphological and evolutionary relationships between Oliniaceae and related families. This approach enabled an understanding of the morphology of the puzzling perianth of Oliniaceae and, because the study included a phylogenetic analysis, an evaluation of its evolutionary relationships with the perianth of the other families.

Another example that goes beyond the level of alpha-morphology is the study of the morphological structure of cupules, which may be complex structures with a protective function for flowers and especially fruits. They are mainly known from several subclades of rosids. It has been shown by comparative developmental studies that, although they may look similar, not all cupules are homologous. In Fagales (Nothofagaceae, Fagaceae) and Sapindales (Anacardiaceae) they are complex, consisting of two or more sterile cymose branching systems, which are condensed, forming coenosomes (Fey and Endress, 1983; Rozefelds and Drinnan, 2002; Bachelier and Endress, 2007). In contrast, in Malpighiales (Balanopaceae), they are simple, consisting of a uniaxial rosette of bracts (Merino Sutter and Endress, 2003).

A third example is the morphological interpretation of reproductive units in Cyperaceae, in which a distinction between flowers and spikelets sometimes poses problems. A comparative developmental study in *Exocarya* (Cyperaceae) paved the way for a delimitation of flowers (Richards et al., 2006).

5.5 Morphology of syncarpous gynoecia

In flowers, syncarpous gynoecia are in general the most complex morphological structures. To describe syncarpous gynoecia there are two different classical approaches, that by Troll (1928) and that by Leinfellner (1950). Troll's approach has been used more often in morphological studies than Leinfellner's, because it is easier to apply. However, in view of developmental biology, Leinfellner's approach is more informative. Troll's approach with the distinction of (eu)syncarpous versus paracarpous focuses on the architecture in distinguishing septate versus non-septate ovaries. The weakness of this approach is that it does not consider that a septate ovary can arise by two different developmental processes: the septa can be congenitally uniform or they can arise by postgenital fusion. In contrast, Leinfellner focuses on this difference between congenital and postgenital organization. He recognizes that syncarpous gynoecia have two zones: at the base a zone in which there is congenital intracarpellary fusion, and above a zone with postgenital or no intracarpellary fusion. Thus, for Leinfellner it is not primarily important whether or not the ovary is septate, but whether intracarpellary fusion is congenital or post-genital. His focus is primarily on organization, and not architecture. It includes the reconstruction of the entire primary morphological surface, i.e. the surface that is derived from the floral apex. Thus the information content is greater than in Troll's approach. Concomitantly, Leinfellner's approach requires a more detailed morphological analysis with transverse microtome section series, and sometimes also study of younger developmental stages.

In general, if we don't know what happens with the primary morphological surface, we don't know a crucial aspect of development. Unfortunately, in many publications on gynoecium and other floral structures this aspect is not studied.

It is also useful to distinguish between morphology, anatomy and histology. Morphology is related to the development of the primary morphological surface. Histology deals with all the kinds of tissues in the organs. Anatomy is related to the patterns of distribution of the tissues in the organs and organ complexes, especially the architecture of the vasculature. For the analysis of all three levels, histology, anatomy and morphology, anatomical techniques are necessary.

In the studies on rosids in my lab, Merran Matthews and Julien Bachelier made detailed analyses of the gynoecium structure in larger clades and showed that all these aspects just mentioned, the topography of the inner morphological surface, anatomy and histology, are of systematic interest (Matthews et al., 2001; Matthews and Endress, 2002, 2004, 2005a, b, 2006, 2008; Bachelier and Endress, 2007, 2008, 2009). For related studies on other angiosperms, see, e.g. Remizowa et al. (2006).

5.6 Concept of sepals and petals

The concept of sepals and petals in angiosperms has been tackled from different perspectives, and new aspects have been discussed, but it is still not convincingly resolved how sepals and petals should be distinguished and where in the phylogenetic tree petals originated. It is an example of how difficult it is for the diversity perspective (Endress, 1994, 2005a, 2008a, 2010; Erbar et al., 1998; Ronse De Craene, 2007, 2008; Wanntorp and Ronse De Craene, 2009) and the model organism perspective to meet and be reconciled (Kramer and Irish, 1999; Davies et al., 2006;

Litt, 2007; Irish, 2009; Rasmussen et al., 2009; Soltis et al., 2009; Warner et al., 2009; Kramer and Hodges, 2010; Yoo et al., 2010). A critical problem here is that for the shaping of organs of a given species, not only organ identity genes, but also genes for further differentiation are required. This 'fourth category' (following meristem identity genes, cadastral genes and organ identity genes) is not considered in the original ABC model of flower development by Coen and Meyerowitz (1991) and is commonly neglected in discussions. Within the framework of a single species (a 'model organism') this doesn't pop up as a problem, but if the entire diversity of angiosperms is considered, it does.

5.7 Initiation of organs

Unfortunately the terms initiation and primordium are often used in a sloppy and incorrect way in the literature, which may lead to misunderstandings of early developmental patterns. Organ initiation takes place within the floral apex and cannot be seen from the outside. An organ primordium is the very first developmental stage following initiation. It is at first not visible at the surface and then becomes visible as a shallow bump. An organ in a later stage is no longer a primordium. If primordial and slightly older stages are not distinguished, but lumped together, developmental patterns may be misinterpreted. The initiation of floral organs in the floral apex commonly proceeds in the centripetal (acropetal) direction. However, in multi-parted, especially polystemonous, flowers, there is centrifugal (basipetal) direction in some clades (e.g. Leins, 1964; Endress, 1997; Rudall, 2010). Caution is especially needed because in some clades with completely centripetal stamen initiation patterns, the pattern of later development (maturation) becomes reversed, and is thus centrifugal. This is the case in some Winteraceae (Tucker, 1959; Doust, 2001; Doust and Drinnan, 2004) and some Ranunculaceae (Anemoneae, *Aquilegia*, and probably *Glaucidium*) (Tepfer, 1953; Ren et al., 2010). A functional aspect of centrifugal stamen initiation is that in polystemonous androecia the development of stamens becomes decoupled from that of the gynoecium, and in extreme cases the last initiated (outermost) stamens may have a different shape from the older ones. This pattern is used for heteranthery, which evolved especially in various pollen flowers (Endress, 2006). The significance in pollination biology of centrifugal stamen maturation has, to my knowledge, not been explored.

A more subtle problem is whether in highly integrated flowers with stable organ number and syncarpous gynoecium the gynoecium is initiated slightly before the stamens, or at least whether the prospective gynoecium area is determined earlier than the individual stamens, as suggested by molecular developmental genetic studies in *Antirrhinum majus* (Zachgo et al., 2000; Z. Schwarz-Sommer, pers. comm.) and by comparative structural studies in *Arabidopsis thaliana* (Choob and Penin, 2004). In

Duparquetia (Leguminosae) the single carpel appears before the stamens (Prenner and Klitgaard, 2008). In polystemonous flowers, such as in *Dillenia* (Dilleniaceae) or *Couroupita* (Lecythidaceae) the carpels appear earlier than the individual stamens, which are formed on primary ring primordia (e.g. Endress, 1994, 1997).

5.8 Current perception of evolutionary patterns

The course and direction of evolution is more flexible than previously thought. This is brought clearly to light by the ever more detailed phylogenetic studies on flowering plants and elsewhere. Previously, for example, evolution of a superior ovary from an inferior one was thought to be so unusual that an obvious example was published in *Science* (*Tetraplasandra*, Eyde and Tseng, 1969). One of the most cited features of flexible evolution is the multiple and complex evolution of C4 photosynthesis (e.g. Besnard et al., 2009). There are also many examples from flowers, which are shown here. Thus we have to be much more differentiated as to what we call apomorphies or synapomorphies.

5.8.1 Inferior ⟷ superior ovary

The above-mentioned example of *Tetraplasandra* species with secondarily superior ovaries in Araliaceae has been further corroborated. It has also been shown, however, that development begins as for an inferior ovary (Costello and Motley, 2001, 2004, 2007). Thus the superior state is both developmentally and evolutionarily secondary. In addition, species with a 'superior' ovary in *Tetraplasandra* form a single clade (Costello and Motley, 2007). Another genus in Araliaceae with a secondarily superior ovary is *Dipanax* (Wen et al., 2001). Also in Rubiaceae several genera were found to have secondarily superior ovaries (*Gaertnera*, Igersheim et al., 1994; *Mitrasacmopsis*, Groeninckx et al., 2007; overview, Igersheim et al., 1994; Endress, 2002). Families in rosids with secondarily superior ovaries are Vochysiaceae (Litt and Stevenson, 2003a, b), and Saxifragales are especially flexible in the transitions between superior and inferior ovaries at various systematic levels (e.g. level of family Saxifragaceae, level of genus *Lithophragma*, Hufford and McMahon, 2003; Soltis et al., 2005). Examples of families in monocots containing clades with secondarily superior ovaries include Hemerocallidaceae and Xanthorrhoeaceae (Rudall, 2002), Haemodoraceae (Simpson, 1998) and Bromeliaceae (Sajo et al., 2004).

5.8.2 Decrease ⟷ increase in floral organ number

Earlier, evolution was mainly seen in terms of reduction from many to a few parts. We now know that this is not the case at many levels. In basal angiosperms evolution in number of all floral parts, including ovules, is flexible (Doyle and Endress,

2000, this volume; Endress and Doyle, 2009). In Rubiaceae (Psychotrieae alliance) ovule number per carpel is increased from one to many (Razafimandimbison et al., 2008).

5.8.3 Whorled ⟷ spiral floral phyllotaxis

Contrary to earlier beliefs, floral phyllotaxis patterns are very flexible. The direction is not only from spiral to whorled, but there was repeated evolution of spiral phyllotaxis from whorled in the perianth in basal angiosperms (Endress and Doyle, 2007, 2009; Doyle and Endress, 2011) and there is much evolutionary flexibility of patterns within families, genera or even species (Endress, 1987; Staedler and Endress, 2009).

5.8.4 Repeated evolution of a double perianth

That a double perianth was reduced during evolution in many groups of eudicots has long been known. However, in some clades there was also repeated evolution of a double perianth with sepals and petals from a simple perianth with only one kind of organ or a fluctuation between both traits. An example is Caryophyllales sensu lato (APG, 2009), which originally had a simple perianth and a double perianth evolved independently in several families (Brockington et al., 2009), with petals at least in part derived from the androecium (Ronse De Craene, 2008).

5.8.5 Floral monosymmetry ⟷ floral polysymmetry

There are evolutionary trends from polysymmetric to monosymmetric flowers in various angiosperm groups. However, the opposite direction can also be found (Endress, 1999). The best-known pathway is polysymmetry by pelorization of monosymmetric flowers (e.g. *Cadia*, Leguminosae, Citerne et al., 2006). A more complex case is the trend from pronounced monosymmetric to more or less polysymmetric flowers in some Lecythidaceae-Lecythidoideae (*Allantoma/Cariniana decandra*) (Tsou and Mori, 2007).

5.8.6 Repeated evolution of floral asymmetry from monosymmetry at several levels within a larger clade

In Fabales both large families, Fabaceae and Polygalaceae, have groups with conspicuously asymmetric flowers (e.g. Prenner, 2004). At the level of the family Fabaceae, asymmetric flowers evolved in several subclades, especially in Cassiinae of caesalpinioids, and in Phaseoleae and Vicieae of papilionoids. At the level of the subtribe Cassiinae, species of *Chamaecrista* and *Senna* have asymmetric flowers. At genus level, in *Senna* floral asymmetry evolved independently in several subclades (Marazzi et al., 2006; Marazzi and Endress, 2008). For distribution of asymmetry in angiosperms, see Endress (accepted).

5.8.7 Complex flowers ⟷ simple flowers

The evolution of complex flowers is based on repeated synorganization of structural units (e.g. Endress, 2006). However, complexity may also be lost in evolution, as shown in Amorpheae (Leguminosae), in which the architecture of complex keel flowers has been lost (McMahon and Hufford, 2005), or in *Besseya* (Plantaginaceae), in which the corolla tube disappeared (Hufford, 1995).

5.8.8 Shortening and secondary elongation of sepals with lost primary function

In Thunbergioideae (Acanthaceae), flower evolution went through a conspicuous transference of function. The two floral prophylls became the protective organs up to anthesis, whereas the sepals that previously had this function became reduced (Schönenberger and Endress, 1998; Schönenberger, 1999). This reduction is expressed in shortening and narrowing, and sometimes complete loss of the sepals. Concomitant with this reduction, if the sepals are not completely lost, is an increase in number of the small sepals. In some moth-pollinated species with extremely elongate corolla, the reduced sepals are secondarily elongated as well, however, without re-gaining the lost protective function, and this secondary elongation seems to be merely a passive by-product of the elongation of the corolla (Endress, 2008b). Phylogenetic analysis shows that such secondary elongation evolved more than once (Borg et al., 2008).

5.8.9 Inflation of calyx

In some Solanaceae, the synsepalous calyx becomes inflated and balloon-like during fruit development. This 'inflated-calyx syndrome' evolved several times (or became lost several times) within the family (Hu and Saedler, 2007).

5.8.10 Centripetal versus centrifugal stamen initiation in polystemonous androecia

The subclass Dilleniidae was introduced into angiosperm macrosystematics by Takhtajan (1964) largely influenced by the occurrence of centrifugal stamen initiation in polymerous androecia, which was supposed to be a fundamentally important pattern in macrosystematics. However, the subclass was later dismantled, first by structural cladistic studies (Hufford, 1992) and then also by molecular studies (Chase et al., 1993) (see also Endress et al., 2000). From the present perspective, the feature is not stable at very high systematic levels, however, often it is still at family level.

5.8.11 Rapid evolution and diversification of floral traits

In some groups with a relatively recent diversification, evolutionary changes in floral size or floral structure are unusually massive. The most striking example is perhaps *Rafflesia*, which surprisingly is related to Euphorbiaceae (Davis et al., 2007), a generally small-flowered group, but which has flowers that reach almost one metre in diameter in some species; this gigantism evolved relatively recently, in less than

50 my (Barkman et al., 2008; Davis, 2008). Another example is the genus *Impatiens*, which became very diverse, with currently more than 1000 species, in a short time (mostly in less than 5 my) (Janssens et al., 2009). In what ways inner conditions enabled these rapid changes is unknown.

5.8.12 Flexibility of pollination system transitions

Flexibility of pollination system transitions has been studied in numerous groups of angiosperms. Some conspicuous examples are be mentioned here. Among Phyllanthaceae, 'in Phyllantheae, specialization to pollination by *Epicephala* moths evolved at least five times, involving more than 500 Phyllantheae species in this obligate association' (Kawakita and Kato, 2009).

Floral architecture in *Mitella* (including *Tolmiea, Lithophragma, Heuchera, Bensoniella*) (Saxifragaceae), many of which are pollinated by fungus gnats, fluctuates in evolution, including 'saucer-shaped' flowers, flowers with 'pollination organs projected', and flowers with 'pollination organs enclosed' (Okuyama et al., 2008).

Fluctuation between bee and hummingbird pollination has found special attention (Cronk and Ojeda, 2008). Case studies of such flexibility are those of *Penstemon* and *Keckiella* (Veronicaceae) (Wilson et al., 2007), Sinningieae (Gesneriaceae) (Perret et al., 2007) and *Ruellia* (Acanthaceae) (Tripp and Manos, 2008). Fluctuation between different pollination syndromes occurs in Iochrominae (Solanaceae) (Smith and Baum, 2006). For *Pedicularis* (Orobanchaceae) it has been shown which functional floral traits are more homoplastic than others and also that floral tube tube length is especially plastic (Ree, 2005). Renner and Schaefer (2010) calculated that oil flowers in angiosperms evolved at least 28 times and floral oil was lost at least 36–40 times.

5.8.13 Stability of patterns

However, there are a few features that now appear to be more stable than previously thought, especially in ovule structure, such as nucellus thickness and number of integuments. New studies have continuously reinforced this picture (Endress et al., 2000; Endress, 2003, 2010; Endress and Matthews, 2006). Additional ovule structural features have also recently been found to be of macrosystematic significance (relative thickness of inner and outer integument, Endress and Matthews, 2006).

5.9 Summary and outlook

The combination of ever more fine-grained phylogenetic analyses and mapping of morphological features in such cladograms has shown that evolutionary transitions of all kinds in floral structure are much easier than previously believed. There are almost no limitations for evolutionary directions in floral features. Nevertheless, certain directions are clearly favoured compared with others. Specific trends can be found in certain clades, i.e. the concentrated and often several times repeated appearance of a feature. This suggests 'biological homology' in the sense of Wagner

(1989, 2007). It will be important to know in more detail in which clades and at what evolutionary level particular features are stable or labile. This will have repercussions on the coding of floral morphological characters for phylogenetic analyses.

Comparative floral morphology needs to include a developmental component for the reconstruction of the primary morphological surface (which is enclosed in the gynoecium and potentially in other organ complexes and then does not form the topographical surface of the organ or organ complex). The primary morphological surface is sometimes no longer apparent at anthesis in cases in which postgenital fusion takes place.

Patterns in the sequence of origin of floral organs are often constant at lower or higher phylogenetic levels. Initiation takes place before the organ becomes visible at the surface. There are errors in the literature with regard to centripetal and centrifugal initiation, because the term 'primordium' is sometimes used in a sloppy way, and initiation and post-initiation development are not correctly distinguished. But when is an organ initiated and when is it a 'primordium'? This needs more profound critical study.

New conceptional and methodical developments need to be constantly integrated into morphological research. A lot is to be expected from evo-devo research, from the study of the wealth of floral fossils available, and from the comparative study of newly found clades.

Acknowledgements

Louis Ronse De Craene and Livia Wanntorp are thanked for the invitation to participate in the symposium on 'Flowers on the Tree of Life', held in Leiden, Netherlands, in August 2009. I especially thank James A. Doyle, Merran L. Matthews, Julien B. Bachelier and Yannick M. Staedler for discussions. Zsuzsanna Schwarz-Sommer is thanked for information on floral development in *Antirrhinum majus*. Alex Bernhard is acknowledged for graphic work. I also thank the Swiss National Science Foundation. Although this publication is not directly part of a project supported by the Foundation, it greatly profited from two earlier projects that were (3100–040327.94 and 3100–059149.99/1).

5.10 References

APG (1998). An ordinal classification for the families of flowering plants. *Annals of the Missouri Botanical Garden*, **85**, 531–553.

APG (2009). An update of the Angiosperm Phylogeny Group classification for the orders and families of flowering plants: APG III. *Botanical Journal of the Linnean Society*, **161**, 105–121.

Bachelier, J. B. and Endress, P. K. (2007). Development of inflorescences, cupules, and flowers in

Amphipterygium, and comparison with *Pistacia* (Anacardiaceae). *International Journal of Plant Sciences*, **168**, 1237–1253.

Bachelier, J. B. and Endress, P. K. (2008). Floral structure of *Kirkia* (Kirkiaceae) and its position in Sapindales. *Annals of Botany*, **102**, 539–550.

Bachelier, J. B. and Endress, P. K. (2009). Comparative floral morphology and anatomy of Anacardiaceae and Burseraceae (Sapindales), with a special focus on gynoecium structure and evolution. *Botanical Journal of the Linnean Society*, **159**, 499–571.

Barkman, T. J., Bendiksby, M., Lim, S.-H. et al. (2008). Accelerated rates of floral evolution at the upper size limit for flowers. *Current Biology*, **18**, 1508–1513.

Besnard, G., Muasya, A. M., Russier, F. et al. (2009). Phylogenomics of C4 photosynthesis in sedges (Cyperaceae): multiple appearances and genetic convergence. *Molecular Biology and Evolution*, **26**, 1909–1919.

Borg, A. J., McDade, L. A. and Schönenberger, J. (2008). Molecular phylogenetics and morphological evolution of Thunbergioideae (Acanthaceae). *Taxon*, **57**, 811–822.

Borowsky, R. (2008). Restoring sight in blind cavefish. *Current Biology*, **18**, R23–R24.

Brockington, S. F., Alexandre, R., Ramdial, J. et al. (2009). Phylogeny of the Caryophyllales sensu lato: Revisiting hypotheses on pollination biology and perianth differentiation in the core Caryophyllales. *International Journal of Plant Sciences*, **170**, 627–643.

Cantino, P. (1985). Phylogenetic inference from nonuniversal derived character states. *Systematic Botany*, **10**, 119–122.

Carrillo-Reyes, P., Sosa, V. and Mort, M. E. (2008). *Thompsonella* and the 'Echeveria group' (Crassulaceae): Phylogenetic relationships based on molecular and morphological characters. *Taxon*, **57**, 863–874.

Causier, B., Castillo, R., Zhou, J. L. et al. (2005). Evolution in action: Following function in duplicated floral homeotic genes. *Current Biology*, **15**, 1508–1512.

Chase, M. W., Soltis, D. E., Olmstead, R. G. et al. (1993). Phylogenetics of seed plants: An analysis of nucleotide sequences from the plastid gene *rbcL*. *Annals of the Missouri Botanical Garden*, **80**, 528–580.

Choob, V. V. and Penin, A. A. (2004). Structure of flower in *Arabidopsis thaliana*: Spatial pattern formation. *Russian Journal of Developmental Biology*, **35**, 224–227.

Citerne, H. L., Pennington, R. T. and Cronk, Q. C. B. (2006). An apparent reversal in floral symmetry in the legume *Cadia* is a homeotic transformation. *Proceedings of the National Academy of Sciences USA*, **103**, 12017–12020.

Coen, E. S. and Meyerowitz, E. M. (1991). The war of the whorls: Genetic interactions controlling flower development. *Nature*, **353**, 31–37.

Collin, R. and Miglietta, M. P. (2008). Reversing opinions on Dollo's law. *Trends in Ecology and Evolution*, **23**, 602–609.

Costello, A. and Motley, T. J. (2001). Molecular systematics of *Tetraplasandra*, *Munroidendron* and *Reynoldsia sandwicensis* (Araliaceae) and the evolution of superior ovaries in *Tetraplasandra*. *Edinburgh Journal of Botany*, **58**, 229–242.

Costello, A. and Motley, T. J. (2004). The development of the superior ovaries in

Tetraplasandra (Araliaceae). *American Journal of Botany*, **91**, 644–655.

Costello, A. and Motley, T. J. (2007). Phylogenetics of the *Tetraplasandra* group (Araliaceae) inferred from ITS, 5S-NTS, and morphology. *Systematic Botany*, **32**, 464–477.

Cronk, Q. and Ojeda, I. (2008). Bird-pollinated flowers in an evolutionary and molecular context. *Journal of Experimental Botany*, **59**, 715–727.

Davies, B., Cartolano, M. and Schwarz-Sommer, Z. (2006). Flower development: The *Antirrhinum* perspective. *Advances in Botanical Research*, **44**, 279–319.

Davis, C. C. (2008). Floral evolution: Dramatic size change was recent and rapid in the world's largest flowers. *Current Biology*, **18**, R1102-R1104.

Davis, C. C., Latvis, M., Nickrent, D. L., Wurdack, K. J. and Baum, D. A. (2007). Floral gigantism in Rafflesiaceae. *Science*, **315**, 1812.

Donoghue, M. J. and Sanderson, M. J. (1992). The suitability of molecular and morphological evidence in reconstructing plant phylogeny. pp. 340–368 in Soltis, P. S., Soltis, D. E. and Doyle, J. J. (eds.), *Molecular Systematics of Plants*. New York: Chapman and Hall.

Doust, A. N. (2001). The developmental basis of floral variation in *Drimys winteri* (Winteraceae). *International Journal of Plant Sciences*, **162**, 697–717.

Doust, A. N. and Drinnan, A. N. (2004). Floral development and molecular phylogeny support the generic status of *Tasmannia* (Winteraceae). *American Journal of Botany*, **91**, 321–331.

Doyle, J. A. and Endress, P. K. (2000). Morphological phylogenetic analysis of basal angiosperms: Comparison and combination with molecular data. *International Journal of Plant Sciences*, **161**, S121-S153.

Doyle, J. A. and Endress, P. K. (2010). Integrating Early Cretaceous fossils into the phylogeny of living angiosperms: Magnoliidae and eudicots. *Journal of Systematics and Evolution*, **48**, 1–35.

Doyle, J. A. and Endress, P. K. (2011). Tracing the early evolutionary diversification of the angiosperm flower. pp. 85–117 in Wanntrop, L., Ronse De Craene, (eds.), *Flowers on the Tree of Life*. Cambridge: Cambridge University Press.

Eckardt, T. (1964). Das Homologieproblem und Fälle strittiger Homologien. *Phytomorphology*, **14**, 79–92.

Endress, P. K. (1987). Floral phyllotaxis and floral evolution. *Botanische Jahrbücher für Systematik*, **108**, 417–438.

Endress, P. K. (1994). *Diversity and Evolutionary Biology of Tropical Flowers*. Cambridge: Cambridge University Press.

Endress, P. K. (1996). Homoplasy in angiosperm flowers. pp. 301–323 in Sanderson, M. J. and Hufford, L. (eds.), *Homoplasy and the Evolutionary Process*. Orlando: Academic Press.

Endress, P. K. (1997). Relationships between floral organization, architecture, and pollination mode in *Dillenia* (Dilleniaceae). *Plant Systematics and Evolution*, **206**, 99–118.

Endress, P. K. (1999). Symmetry in flowers – diversity and evolution. *International Journal of Plant Sciences*, **160**, S3–S23.

Endress, P. K. (2002). Morphology and angiosperm systematics in the molecular era. *Botanical Review*, **68**, 545–570.

Endress, P. K. (2003). What should a 'complete' morphological phylogenetic analysis entail? *Regnum Vegetabile*, **141**, 131–164.

Endress, P. K. (2005a). Links between embryology and evolutionary floral morphology. *Current Science*, **89**, 749–754.

Endress, P. K. (2005b). The role of morphology in angiosperm evolutionary studies. *Nova Acta Leopoldina*, **92** (342), 221–238.

Endress, P. K. (2006). Angiosperm floral evolution: Morphological and developmental framework. *Advances in Botanical Research*, **44**, 1–61.

Endress, P. K. (2008a). Perianth biology in the basal grade of extant angiosperms. *International Journal of Plant Sciences*, **169**, 844–862.

Endress, P. K. (2008b). The whole and the parts: relationships between floral architecture and floral organ shape, and their repercussions on the interpretation of fragmentary floral fossils. *Annals of the Missouri Botanical Garden*, **95**, 101–120.

Endress, P. K. (2010). Floral structure and trends of evolution in eudicots and their major subclades. *Annals of the Missouri Botanical Garden*, **97**, 541–583.

Endress, P. K. (accepted). The immense diversity of floral monosymmetry and asymmetry.

Endress, P. K. and Doyle, J. A. (2007). Floral phyllotaxis in basal angiosperms – development and evolution. *Current Opinion in Plant Biology*, **10**, 52–57.

Endress, P. K. and Doyle, J. A. (2009). Reconstructing the ancestral flower and its initial specializations. *American Journal of Botany*, **96**, 22–66.

Endress, P. K. and Matthews, M. L. (2006). First steps towards a floral structural characterization of the major rosid subclades. *Plant Systematics and Evolution*, **260**, 223–251.

Endress, P. K., Baas, P. and Gregory, M. (2000). Systematic morphology and anatomy: 50 years of progress. *Taxon*, **49**, 401–434.

Erbar, C., Kusma, S. and Leins, P. (1998). Development and interpretation of nectary organs in Ranunculaceae. *Flora*, **194**, 317–332.

Eyde, R. H. and Tseng, C. C. (1969). Flower of *Tetraplasandra gymnocarpa*. Hypogyny with epigynous ancestry. *Science*, **166**, 506–508.

Fey, B. S. and Endress, P. K. (1983). Development and morphological interpretation of the cupule in Fagaceae. *Flora*, **173**, 451–468.

Friedman, W. E., Barrett, S. C. H., Diggle, P. K., Irish, V. F. and Hufford, L. (2008). Whither plant evo-devo? *New Phytologist*, **178**, 468–471.

Friis, E. M., Pedersen, K. R., von Balthazar, M., Grimm, G. W. and Crane, P. R. (2009). *Monetianthus mirus* gen. et sp. nov., a nymphaealean flower from the Early Cretaceous of Portugal. *International Journal of Plant Sciences*, **170**, 1086–1101.

Givnish, T. J. (2003). How a better understanding of adaptations can yield better use of morphology in plant systematics: Toward eco-evo-devo. *Regnum Vegetabile*, **141**, 273–295.

Givnish, T. J. and Sytsma, K. J. (1997). Homoplasy in molecular vs. morphological data: The likelihood of correct phylogenetic inference. pp. 55–101 in Givnish, T. J. and Sytsma, K. J. (eds.), *Molecular Evolution and Adaptive Radiation*. Cambridge: Cambridge University Press.

Groeninckx, I., Vrijdaghs, A., Huysmans, S., Smets, E. and Dessein, S. (2007). Floral ontogeny of the Afro-Madagascan genus *Mitrasacmopsis* with comments on the development of superior ovaries in Rubiaceae. *Annals of Botany*, **100**, 41–49.

Hu, J.-Y. and Saedler, H. (2007). Evolution of the inflated calyx syndrome in Solanaceae. *Molecular Biology and Evolution*, **24**, 2443–2453.

Hufford, L. (1992). Rosidae and their relationships to other nonmagnoliid dicotyledons: A phylogenetic analysis using morphological and chemical data. *Annals of the Missouri Botanical Garden*, **79**, 218–248.

Hufford, L. (1995). Patterns of ontogenetic evolution in perianth diversification of *Besseya* (Scrophulariaceae). *American Journal of Botany*, **82**, 655–680.

Hufford, L. and McMahon, M. (2003). Beyond morphoclines and trends: The elements of diversity and the phylogenetic patterning of morphology. *Regnum Vegetabile*, **141**, 165–186.

Igersheim, A., Puff, C., Leins, P. and Erbar, C. (1994). Gynoecial development of *Gaertnera* Lam. and of presumably allied taxa of the Psychotrieae (Rubiaceae): secondarily 'superior' vs. inferior ovaries. *Botanische Jahrbücher für Systematik*, **116**, 401–414.

Irish, V. F. (2009). Evolution of petal identity. *Journal of Experimental Botany*, **60**, 2517–2527.

Irish, V. F. and Litt, A. (2005). Flower development and evolution: Gene duplication, diversification and redeployment. *Current Opinion in Genetics and Development*, **15**, 454–460.

Janssens, S. B., Knox, E. B., Huysmans, S., Smets, E. F. and Merckx, V. S. F. T. (2009). Rapid radiation of *Impatiens* (Balsaminaceae) during Pliocene and Pleistocene: Result of a global climate change. *Molecular Phylogenetics and Evolution*, **52**, 806–824.

Kaplan, D. R. (1984). The concept of homology and its central role in the elucidation of plant systematic relationships. pp. 51–70 in Duncan, T. and Stuessy, T. F. (eds.), *Cladistics. Perspectives on the Reconstruction of Evolutionary History*. New York: Columbia University Press.

Kawakita, A. and Kato, M. (2009). Repeated independent evolution of obligate pollination mutualism in the Phyllantheae-*Epicephala* association. *Proceedings of the Royal Society B*, **276**, 417–426.

Kelly, L. M. and Stevenson, D. W. (2005). Floral morphological character coding and the use of trees. XVII International Botanical Congress, Vienna. Abstract 12.3.1.

Klingenberg, C. P. (2009). Morphometric integration and modularity in configurations of landmarks: tools for evaluating a priori hypotheses. *Evolution and Development*, **11**, 405–421.

Kramer, E. M. and Hodges, S. A. (2010). *Aquilegia* as a model system for the evolution and ecology of petals. *Philosophical Transactions of the Royal Society B*, **365**, 477–490.

Kramer, E. M. and Irish, V. F. (1999). Evolution of genetic mechanisms controlling petal development. *Nature*, **399**, 144–148.

Lamb, R. S. and Irish, V. F. (2003). Functional divergence within the *APETALA3/PISTILLATA* floral homeotic gene lineages. *Proceedings of the National Academy of Sciences USA*, **100**, 6558–6563.

Leinfellner, W. (1950). Der Bauplan des synkarpen Gynoeceums. *Österreichische Botanische Zeitschrift*, **97**, 403–436.

Leins, P. (1964). Die frühe Blütenentwicklung von *Hypericum hookerianum* Wight et Arn. und *H. aegypticum* L. *Berichte der Deutschen Botanischen Gesellschaft*, **77**, 112–123.

Lexer, C., Joseph, J., van Loo, M. et al. (2009). The use of digital image-based morphometrics to study the phenotypic mosaic in taxa with porous genomes. *Taxon*, **58**, 349–364.

Litt, A. (2007). An evaluation of A-function: evidence from the *APETALA1* and *APETALA2* gene lineages. *International Journal of Plant Sciences*, **168**, 73–91.

Litt, A. and Stevenson, D. W. (2003a). Floral development and morphology of Vochysiaceae. I. The structure of the gynoecium. *American Journal of Botany*, **90**, 1533–1547.

Litt, A. and Stevenson, D. W. (2003b). Floral development and morphology of Vochysiaceae. II. The position of the single fertile stamen. *American Journal of Botany*, **90**, 1548–1559.

Marazzi, B. and Endress, P. K. (2008). Patterns and development of floral asymmetry in *Senna* (Leguminosae, Cassiinae). *American Journal of Botany*, **95**, 22–40.

Marazzi, B., Endress, P. K., Paganucci de Queiroz, L. and Conti, E. (2006). Phylogenetic relationships within *Senna* (Leguminosae, Cassiinae) based on three chloroplast DNA regions: Patterns in the evolution of floral symmetry and extrafloral nectaries. *American Journal of Botany*, **93**, 288–303.

Matthews, M. L. and Endress, P. K. (2002). Comparative floral structure and systematics in Oxalidales (Oxalidaceae, Connaraceae, Cephalotaceae, Brunelliaceae, Cunoniaceae, Elaeocarpaceae, Tremandraceae). *Botanical Journal of the Linnean Society*, **140**, 321–381.

Matthews, M. L. and Endress, P. K. (2004). Comparative floral structure and systematics in Cucurbitales (Corynocarpaceae, Coriariaceae, Datiscaceae, Tetramelaceae, Begoniaceae, Cucurbitaceae, Anisophylleaceae). *Botanical Journal of the Linnean Society*, **145**, 129–185.

Matthews, M. L. and Endress, P. K. (2005a). Comparative floral structure and systematics in Celastrales (Celastraceae, Parnassiaceae, Lepidobotryaceae). *Botanical Journal of the Linnean Society*, **149**, 129–194.

Matthews, M. L. and Endress, P. K. (2005b). Comparative floral structure and systematics in Crossosomatales (Crossosomataceae, Stachyuraceae, Staphyleaceae, Aphloiaceae, Geissolomataceae, Ixerbaceae, Strasburgeriaceae). *Botanical Journal of the Linnean Society*, **147**, 1–46.

Matthews, M. L. and Endress, P. K. (2006). Floral structure and systematics in four orders of rosids, including a broad survey of floral mucilage cells. *Plant Systematics and Evolution*, **260**, 199–221.

Matthews, M. L. and Endress, P. K. (2008). Comparative floral structure and systematics in Chrysobalanaceae s.l. (Chrysobalanaceae, Dichapetalaceae, Euphroniaceae, and Trigoniaceae; Malpighiales). *Botanical Journal of the Linnean Society*, **157**, 249–309.

Matthews, M. L., Endress, P. K., Schönenberger, J. and Friis, E. M. (2001). A comparison of floral structures of Anisophylleaceae and Cunoniaceae and the problem of their systematic position. *Annals of Botany*, **88**, 439–455.

McMahon, M. M. and Hufford, L. (2005). Evolution and development in the amorphoid clade (Amorpheae: Papilionoideae: Leguminosae): Petal loss and dedifferentiation. *International Journal of Plant Sciences*, **166**, 383–396.

Melzer, S., Lens, F., Geman, J. et al. (2008). Flowering-time genes modulate meristem determinacy and growth form in *Arabidopsis thaliana*. *Nature Genetics*, **40**, 1489–1492.

Merino Sutter, D. and Endress, P. K. (2003). Structure of female flowers an cupules in Balanopaceae, an enigmatic rosid family. *Annals of Botany*, **92**, 459–469.

Nandi, O. I., Chase, M. W. and Endress, P. K. (1998). A combined cladistic analysis of angiosperms using *rbcL* and nonmolecular data sets. *Annals of the Missouri Botanical Garden*, **85**, 137–212.

O'Hara, R. J. (1988). Homage to Clio, or, toward an historical philosophy for evolutionary biology. *Systematic Zoology*, **37**, 142–155.

Okuyama, Y., Pellmyr, O. and Kato, M. (2008). Parallel floral adaptations to pollination by fungus gnats within the genus *Mitella* (Saxifragaceae). *Molecular Phylogenetics and Evolution*, **46**, 560–575.

Patterson, T. B. and Givnish, T. J. (2002). Phylogeny, concerted convergence, and phylogenetic niche conservatism in the core Liliales: Insights from *rbcL* and *ndh*F sequence data. *Evolution*, **56**, 233–252.

Perret, M., Chautems, A., Spichiger, R., Barraclough, T. G. and Savolainen, V. (2007). The geographical pattern of speciation and floral diversification in the Neotropics: The tribe Sinningieae (Gesneriaceae) as a case study. *Evolution*, **61**, 1641–1660.

Prenner, G. (2004). Floral development in *Polygala myrtifolia* (Polygalaceae) and its similarities with Leguminosae. *Plant Systematics and Evolution*, **249**, 67–76.

Prenner, G. and Klitgaard, B. B. (2008). Towards unlocking the deep nodes of Leguminosae: Floral development and morphology of the enigmatic *Duparquetia orchidacea* (Leguminosae, Caesalpinioideae). *American Journal of Botany*, **95**, 1349–1365.

Rasmussen, D. A., Kramer, E. M. and Zimmer, E. A. (2009). One size fits all? Molecular evidence for a commonly inherited petal identity program in Ranunculales. *American Journal of Botany*, **96**, 96–109.

Razafimandimbison, S. G., Rydin, C. and Bremer, B. (2008). Evolution and trends in the Psychotrieae alliance (Rubiaceae) – A rarely reported evolutionary change of many-seeded carpels from one-seeded carpels. *Molecular Phylogenetics and Evolution*, **48**, 207–223.

Ree, R. H. (2005). Phylogeny and the evolution of floral diversity in *Pedicularis* (Orobanchaceae). *International Journal of Plant Sciences*, **166**, 595–613.

Remane, A. (1956). *Die Grundlagen des natürlichen Systems, der vergleichenden Anatomie und der Phylogenetik*. Leipzig: Akademische Verlagsanstalt.

Remizowa, M. V., Sokoloff, D. and Rudall, P. J. (2006). Evolution of the monocot gynoecium: Evidence from comparative morphology and development in *Tofieldia, Japonolirion, Petrosavia* and *Narthecium*. *Plant Systematics and Evolution*, **258**, 183–209.

Ren, Y., Chang, H.-L. and Endress, P. K. (2010). Floral development in Anemoneae (Ranunculaceae). *Botanical Journal of the Linnean Society*, **162**, 77–100.

Renner, S. S. and Schaefer, H. (2010). The evolution and loss of oil-offering flowers: New insights from dated phylogenies for angiosperms and bees. *Philosophical Transactions of the Royal Society B*, **365**, 423–435.

Richards, J. H., Bruhl, J. J. and Wilson, K. L. (2006). Flower or spikelet? Understanding the morphology and development of reproductive structures in *Exocarya* (Cyperaceae, Mapanioideae, Chrysotricheae). *American Journal of Botany*, **93**, 1241-1250.

Ronse De Craene, L. P. (2007). Are petals sterile stamens or bracts? The origin and evolution of petals in the core eudicots. *Annals of Botany*, **100**, 621-630.

Ronse De Craene, L. P. (2008). Homology and evolution of petals in the core eudicots. *Systematic Botany*, **33**, 301-325.

Rosin, F. M. and Kramer, E. M. (2009). Old dogs, new tricks: Regulatory evolution in conserved genetic modules leads to novel morphologies in plants. *Developmental Biology*, **332**, 25-35.

Rozefelds, A. C. and Drinnan, A. N. (2002). Ontogeny of pistillate flowers and inflorescences in *Nothofagus* subgenus *Lophozonia* (Nothofagaceae). *Plant Systematics and Evolution*, **233**, 105-126.

Rudall, P. J. (2002). Homologies of inferior ovaries and septal nectaries in monocotyledons. *International Journal of Plant Sciences*, **163**, 261-276.

Rudall, P. J. (2010). All in a spin: Centrifugal organ formation and floral patterning. *Current Opinion in Plant Biology*, **13**, 108-114.

Sajo, M. G., Rudall, P. J. and Prychid, C. J. (2004). Floral anatomy of Bromeliaceae, with particular reference to the evolution of epigyny and septal nectaries in commelinid monocots. *Plant Systematics and Evolution*, **247**, 215-231.

Sattler, R. (1992). Process morphology: Structural dynamics in development

and evolution. *Canadian Journal of Botany*, **70**, 708-714.

Schönenberger, J. (1999). Floral structure, development, and diversity in *Thunbergia* (Acanthaceae). *Botanical Journal of the Linnean Society*, **130**, 1-36.

Schönenberger, J. (2005). Rise from the ashes – reconstruction of charcoal fossil flowers. *Trends in Plant Science*, **10**, 436-443.

Schönenberger, J. and Conti, E. (2003). Molecular phylogeny and floral evolution of Penaeaceae, Oliniaceae, Rhynchocalycaceae, and Alzateaceae (Myrtales). *American Journal of Botany*, **90**, 292-309.

Schönenberger, J. and Endress, P. K. (1998). Structure and development of the flowers in *Mendoncia, Pseudocalyx*, and *Thunbergia* (Acanthaceae) and their systematic implications. *International Journal of Plant Sciences*, **159**, 446-465.

Scotland, R. W., Olmstead, R. G. and Bennett, J. R. (2003). Phylogeny reconstruction: The role of morphology. *Systematic Biology*, **52**, 539-548.

Sebola, R. J. and Balkwill, K. (2009). Numerical phenetic analysis of *Olinia rochetiana* sensu lato (Oliniaceae). *Kew Bulletin*, **64**, 95-121.

Simon, H. A. (1962). The architecture of complexity. *Proceedings of the American Philosophical Society*, **106**, 467-482.

Simpson, M. G. (1998). Reversal in ovary position from inferior to superior in the Haemodoraceae: Evidence from floral ontogeny. *International Journal of Plant Sciences*, **159**, 466-479.

Smith, S. D. and Baum, D. A. (2006). Phylogenetics of the florally diverse Andean clade Iochrominae (Solanaceae). *American Journal of Botany*, **93**, 1140-1153.

Soltis, D. E., Soltis, P. S., Endress, P. K. and Chase, M. W. (2005). *Phylogeny and Evolution of Angiosperms*. Sunderland, MA: Sinauer.

Soltis, P. S., Brockington, S. F., Yo, M.-J. et al. (2009). Floral variation and floral genetics in basal angiosperms. *American Journal of Botany*, **96**, 110–128.

Staedler, Y. M. and Endress, P. K. (2009). Diversity and lability of floral phyllotaxis in the pluricarpellate families of core Laurales (Gomortegaceae, Atherospermataceae, Siparunaceae, Monimiaceae). *International Journal of Plant Sciences*, **170**, 522–550.

Stevens, P. F. (2000). On characters and character states: Do overlapping and non-overlapping variation, morphology and molecules all yield data of the same value? pp. 81–105 in Scotland, R. W. and Pennington, T. (eds.), *Homology and Systematics. Coding Characters for Phylogenetic Analysis*. London: Taylor and Francis.

Stevens, P. F. (2001 onwards). Angiosperm Phylogeny Website. Available at http://www.mobot.org/MOBOT/research/APweb/.

Sweeney, P. W. (2008). Phylogeny and floral diversity in the genus *Garcinia* (Clusiaceae) and relatives. *International Journal of Plant Sciences*, **169**, 1288–1303.

Takhtajan, A. (1964). The taxa of the higher plants above the rank of order. *Taxon*, **13**, 160–164.

Tepfer, S. S. (1953). Floral anatomy and ontogeny in *Aquilegia formosa* var. *truncata* and *Ranunculus repens*. *University of California Publications in Botany*, **25**, 513–648.

Tripp, E. A. and Manos, P. S. (2008). Is floral specialization an evolutionary dead-end? Pollination system transitions in *Ruellia* (Acanthaceae). *Evolution*, **62**, 1712–1737.

Troll, W. (1928). Zur Auffassung des parakarpen Gynaeceums und des coenocarpen Gynaeceums überhaupt. *Planta*, **6**, 255–276.

Tsou, C.-H. and Mori, S. A. (2007). Floral organogenesis and floral evolution of the Lecythidoideae (Lecythidaceae). *American Journal of Botany*, **94**, 716–736.

Tucker, S. C. (1959). Ontogeny of the inflorescence and the flower in *Drimys winteri* var. *chilensis*. *University of California Publications in Botany*, **30**, 257–336.

Turrill, W. B. (1938). The expansion of taxonomy with special reference to Spermatophyta. *Biological Review*, **13**, 342–373.

Vargas, A. O. and Wagner, G. P. (2009). Frame-shifts of digit identity in bird evolution and cyclopamine-treated wings. *Evolution and Development*, **11**, 163–169.

Wagner, G. P. (1989). The biological homology concept. *Annual Reviews in Ecology and Systematics*, **20**, 51–69.

Wagner, G. P. (2005). Concepts matter: characters as units of evolutionary change. XVII International Botanical Congress, Vienna. Abstracts, p. 215.

Wagner, G. P. (2007). The developmental genetics of homology. *Nature Reviews Genetics*, **8**, 473–479.

Wagner, G. P. (2009). BIO. *Evolution and Development*, **11**, 139–141.

Wanntorp, L. and Ronse De Craene, L. P. (2009). Perianth evolution in the sandalwood order Santalales. *American Journal of Botany*, **96**, 1361–1371.

Warner, K. A., Rudall, P. J. and Frohlich, M. W. (2009). Environmental control of sepalness and petalness in perianth organs of waterlilies: A new mosaic theory for the evolutionary origin of a differentiated perianth. *Journal of Experimental Botany*, **60**, 3559–3574.

Wen, J., Plunckett, G. M., Mitchell, A. D. and Wagstaff, S. J. (2001). The evolution of Araliaceae: A phylogenetic analysis based on ITS sequences of nuclear ribosomal DNA. *Systematic Botany*, **26**, 144–167.

Wilson, P., Wolfe, A. D., Armbruster, W. S. and Thomson, J. D. (2007). Constrained lability in floral evolution: Counting convergent origins of hummingbird pollination in *Penstemon* and *Keckiella*. *New Phytologist*, **176**, 883–890.

Yoo, M.-J., Soltis, P. S. and Soltis, D. E. (2010). Expression of floral MADS-box genes in two divergent 'water lilies': Nymphaeales and *Nelumbo*. *International Journal of Plant Sciences*, **171**, 121–146.

Zachgo, S., Perbal, M.-C., Saedler, H. and Schwarz-Sommer, Z. (2000). In situ analysis of RNA and protein expression in whole mounts facilitates detection of floral gene expression dynamics. *Plant Journal*, **23**, 697–702.

6

Centrifugal stamens in a modern phylogenetic context: was Corner right?

PAULA J. RUDALL

6.1 Introduction

The widespread occurrence of centrifugal stamen formation in some members of the group that we now term eudicots (tricolpates) led the influential tropical botanist E. J. H. Corner to suggest that 'so profound a disturbance in floral development as the reversal of the androecium' (Corner, 1946, p. 435) must be of considerable phylogenetic significance. Has Corner's prediction proved correct, when viewed in a modern phylogenetic context? One reason that this question remains relevant today is that there is a considerable shortage of useful morphological synapomorphies that are taxonomically applicable at deep nodes within angiosperms. Most major groupings above the family level are defined by molecular, rather than morphological, differences, making them difficult to teach to students or to recognize in the field. In angiosperm taxonomy, the majority of characters relating to the flower are traditionally employed primarily at the species and genus level, or at best the family level. For example, differences in the degree of fusion between floral organs, both within and between floral whorls (especially petals and stamens), are often used to delimit species and genera. With the exception of syncarpy, characters relating to floral organ fusion are mostly too homoplastic to be taxonomically applicable at deeper nodes in angiosperms.

Flowers on the Tree of Life, ed. Livia Wanntorp and Louis P. Ronse De Craene. Published by Cambridge University Press. © The Systematics Association 2011.

The character state 'centrifugal stamens', as defined by Corner (1946) and others, represents one aspect of centrifugal organ growth. Centrifugal growth in flowers was first documented by early morphologists (Payer, 1857; Eichler, 1875, 1878) and has been reviewed by several authors (e.g. Leins, 1964; Sattler, 1972b; Tucker, 1972, 1984; Rudall, 2010). It represents a complex set of morphogenetic phenomena, rather than a single developmental process (Rudall, 2010). Consistent with the *Tree of Life* theme of this volume, this chapter explores the taxonomic significance of this feature in the context of recent angiosperm classifications based on molecular data (e.g. Angiosperm Phylogeny Group III, 2009).

6.2 Deep-node characters

Examples of deep-node taxonomic characters that have apparently survived the molecular revolution relatively intact include cotyledon number, pollen aperture number, number of integuments in the ovule and flower merism. Some of these characters have been recognized for a long time. John Ray's *Historia Plantarum* (1686–1704) first cited cotyledon number as a useful means of subdividing flowering plants into monocotyledons and dicotyledons. In the 'postmolecular' era, we can re-evaluate this distinction in a phylogenetic context: a dicotylar embryo occurs in most early-divergent angiosperms (except Nymphaeales) and most extant gymnosperms, and probably represents the plesiomorphic (primitive) angiosperm condition. However, monocotyly remains a defining feature of monocots, one of the major subgroups of flowering plants (Sokoloff et al., 2008).

Among reproductive structures, some pollen and ovule characters are also taxonomically significant at deep nodes in angiosperms. Gunnar Erdtman's (1960) palynological distinction between two major groups of flowering plants, 'monosulcates' and 'tricolpates' – possessing a single (polar) pollen aperture or three (equatorial) pollen apertures, respectively – has at least partly survived the molecular revolution, as the tricolpates (now generally termed eudicots) are an important feature of modern classifications (e.g. Angiosperm Phylogeny Group III, 2009). Monosulcate pollen is the plesiomorphic condition, occurring in monocots, many early-divergent angiosperms and most gymnosperms, but tricolpate (or tricolpate-derived) pollen is characteristic of eudicots (e.g. Furness and Rudall, 2004). The developmental mechanisms underlying this derived condition are increasingly well-understood; tricolpate apertures, which are associated with simultaneous microsporogenesis, have evolved at least three times independently, but apparently became fixed only once in an arrangement termed 'Fischer's rule' (where apertures are formed in pairs at six points in the developing tetrad), which is the primary eudicot condition (Furness and Rudall, 2004).

Integument number as a potential deep-node character dates from Eugen Warming's (1878) observation of a correlation between the unitegmic versus bitegmic ovule conditions and other angiosperm characters. Although the homologies of angiosperm integuments with gymnospermous structures remain obscure, bitegmy is probably plesiomorphic within angiosperms (Doyle, 2006). On the other hand, unitegmy, though probably iterative to some extent, is a defining feature of asterid eudicots (e.g. Nandi et al., 1998). The apparent correlation between unitegmy and a tenuinucellate nucellus (see also Philipson, 1974; Dahlgren, 1975) would merit further review in a modern phylogenetic context.

Perhaps the best example of a floral deep-node character is the number of organs that occur within each floral whorl, termed flower merism or merosity (Ronse De Craene and Smets, 1994), especially five (pentamery) versus three (trimery), since other organ numbers are apparently derived from one or other of these conditions. Trimery is very common in monocots and fairly common in early-divergent angiosperms, whereas eudicot flowers are predominantly pentamerous (e.g. Soltis et al., 2003). As with tricolpate pollen apertures, comparative ontogenetic studies suggest that pentamery evolved more than once in early-divergent eudicots (Wanntorp and Ronse De Craene, 2007). Nevertheless, the combination of these two characters makes eudicots perhaps the best-defined of all major seed-plant groups in terms of morphology. Remarkably, at least in eudicots, flower merism appears to be uncoupled from other aspects of floral patterning, such as symmetry and whorl number. For example, trimery is a feature that occurs frequently in both early-divergent angiosperms and monocots, but only in monocots is trimery apparently intimately linked with whorl number; trimerous-pentacyclic flowers are highly characteristic of monocots (e.g. Endress, 1995). Detailed comparative studies in monocots focused on investigating this association indicate that the trimerous-pentacyclic condition could be governed by a constraint related to sectorial differentiation and spatial patterning (Remizowa et al., 2010).

6.3 Centrifugal growth in flowers

Centrifugal growth in flowers occurs when at least some of the organs within a single flower develop in a distal-to-proximal sequence, either within a single organ zone (intrazonal development) or between organ zones (interzonal development). Such development is described as centrifugal in radial structures such as polysymmetric flowers, and as basipetal in dorsiventral structures such as compound leaves, monosymmetric flowers or organ fascicles.

Table 6.1 lists some examples of centrifugal growth in flowers, arranged in the context of a classification based on molecular phylogenetic data (Angiosperm Phylogeny Group III, 2009). Both intrazonal and interzonal types of centrifugal

Table 6.1 Examples of centrifugal initiation in flowers (see also reviews by Corner, 1946; Tucker, 1972; Sattler, 1972 a, b). Note that this list is not exhaustive.

Family/order	Genus/species	Mode	References
Early-divergent angiosperms			
Hydatellaceae (Nymphaeales)	*Trithuria*	Centrifugal reproductive unit development	Rudall et al., 2007, 2009
Eudicots			
Early-divergent eudicots			
none			
Core eudicots			
Dilleniales (Dilleniaceae)	Several genera, e.g. *Dillenia*, *Hibbertia*	Mostly intrazonal: centrifugal stamens, often in fascicles; sometimes temporal overlap with carpel initiation	Sattler, 1972b
Caryophyllales: Cactaceae, Caryophyllaceae, Portulacaceae	Several genera	Intrazonal: centrifugal stamens	Payer, 1857
Asterid eudicots			
Asterales: Apiaceae, Asteraceae, Stylidiaceae	Several genera	Inter- and intrazonal: centrifugal corolla in some species; stamens often before petals; centrifugal stamens in some species	Sattler, 1972b; Erbar and Leins, 1997 (Apiaceae)
Cornales: Hydrangeaceae, Loasaceae	*Platycrater arguta*	Inter- and intrazonal: centrifugal stamens; carpels before stamens in *Platycrater*	Payer, 1857; Ge et al., 2007
Dipsacales (Valerianaceae)	*Valeriana officinalis*	Interzonal: stamens before other organs	Sattler, 1972b

Table 6.1 (*cont.*)

Family/order	Genus/species	Mode	References
Ericales: Actinidiaceae, Lecythidaceae, Myrsinaceae, Primulaceae, Theaceae, Theophrastaceae	Several genera, e.g. *Lysimachia*	Inter- and intrazonal: centrifugal corolla; stamens often before petals (e.g. in *Lysimachia, Primula*)	Corner, 1946; Sattler, 1972b; Cusick, 1959
Saxifragales (Paeoniaceae)	*Paeonia*	Intrazonal: centrifugal stamens	Corner, 1946; Leins and Erbar, 2000
Rosid eudicots			
Brassicales: Brassicaceae, Capparidaceae	Several genera (e.g. *Arabidopsis*)	Intrazonal: centrifugal stamens	Payer, 1857; Hill and Lord, 1989; Smyth et al., 1990; Erbar and Leins, 1997
Crossomatales (Crossomataceae)		Intrazonal: centrifugal stamens	Eames, 1953 (but see Matthews and Endress, 2005, for discussion of contradictory reports)
Fabales (Leguminosae)	*Pisum sativum, Swartzia* spp.	*Pisum*: carpel initiated after first (abaxial) stamen primordia. Centrifugal stamens in some Swartzia spp. Carpel before stamens in some Papilionoideae	Sattler, 1972b; Tucker, 2003; Prenner, 2003, 2004a, b
Malvales: Bixaceae, Cistaceae, Cochlospermaceae, Malvaceae s.l.	Many polyandrous genera	Mostly intrazonal: centrifugal stamens; often in fascicles; sometimes temporal overlap with carpel initiation	Corner, 1946; Van Heel, 1966; Tucker, 1999 (*Pavonia–Malvaceae*)

Table 6.1 (*cont.*)

Family/order	Genus/species	Mode	References
Malpighiales: Clusiaceae, Flacourtiaceae, Hypericaceae, Ochnaceae, Salicaceae	Many genera e.g. *Hypericum*	Intrazonal: centrifugal stamens	Payer, 1857; Corner, 1946; Kaul, 1995; Leins, 1964 (Hypericum); Bernhard and Endress, 1999 (Flacourtiaceae); Hochwallner and Weber, 2006 (Clusia)
Myrtales: Lythraceae, Myrtaceae	*Lythrum, Lagerstroemia, Punica*	*Lythrum salicaria,* interzonal others intrazonal: centrifugal stamens	Sattler, 1972b; Mayr, 1969
Monocots			
Alismatales	Several polyandrous alismatids, e.g. *Alisma, Butomus, Hydrocleis, Limnocharis*	Many genera intrazonal: centrifugal stamens. Common tepal/stamen primordia in some genera (e.g. *Alisma*)	Kaul, 1967; Sattler and Singh, 1978
Arecaceae (palms)	Several polyandrous genera, especially calamoid palms, e.g. *Eugeissona, Palandra*	Intrazonal: centrifugal stamens	Uhl and Moore, 1977; Uhl, 1988
Commelinaceae	*Tradescantia*	Intrazonal: centrifugal stamens	Payer, 1857; Hardy and Stevenson, 2000
Poaceae	*Hordeum vulgare*	Intrazonal: stamens before lodicules (putative petal homologues)	Sattler, 1972b
Triuridaceae (Pandanales)		Centrifugal carpels in fascicles; central stamens initiated first in *Lacandonia*	Rudall, 2008

development are evident in many major eudicot and monocot groups (Table 6.1), often correlated with large flowers with relatively high organ number, especially polyandry. The most common type of centrifugal interzonal growth occurs in the stamen zone of polyandrous species, such as some early-divergent monocots (Sattler and Singh, 1978), some calamoid palms (Uhl, 1988) and the eudicots *Dillenia* and *Populus* (Endress, 1997; Bernhard and Endress, 1999), in which stamen primordia are initiated on the floral apex in a centrifugal sequence. This latter type encompasses Corner's (1946) 'centrifugal stamens'. In addition to polyandrous taxa, centrifugal stamen development also occurs in some species with only two stamen whorls, such as *Tradescantia* (Hardy and Stevenson, 2000) and *Arabidopsis* (Hill and Lord, 1989; Smyth et al., 1990; Erbar and Leins, 1997). A different type of centrifugal stamen development occurs in some other polyandrous species, in which stamens are initiated together in fascicles, which are themselves arranged in a whorl (Ronse De Craene and Smets, 1987, 1992; Prenner et al., 2008). Within each fascicle, organ development can be either acropetal or basipetal. Basipetal stamen fascicles occur in many eudicots, and are common in some families, notably Malvaceae (Van Heel, 1966).

In many eudicot groups, different patterns of centrifugal organ development can occur in the same order, family or even genus, indicating possible common underlying genetic factors. On the other hand, with the exception of Hydatellaceae (see below), centrifugal floral organ initiation is rare in early-divergent (ANA-grade) angiosperms. It is also apparently absent from the magnoliid clade, which includes some polyandrous species (e.g. Annonaceae: Leins and Erbar 1996; *Magnolia*: Xu and Rudall, 2006) and from early-divergent eudicot lineages, which also include several polyandrous species (e.g. many Ranunculaceae: Ren et al., 2010). This distribution suggests that a centripetal sequence of floral organ initiation, similar to the sequence that occurs in the developing vegetative apex, represents the ancestral condition in angiosperms.

In Hydatellaceae, organ development is centrifugal within a single reproductive unit (Rudall et al.,2007a), but reproductive units have been variously interpreted as flowers (Rudall et al., 2009) or inflorescences (e.g. Doyle and Endress, 2011, this volume). Delayed (retarded) petal growth (rather than delayed petal initiation) occurs in *Cabomba* (Tucker and Douglas, 1996; Endress, 2001; Rudall et al., 2009). The distinction between early inhibition or later retardation could be a relatively trivial one in this context, especially if (as seems likely) the arrangement of floral organs is established on the apex before organ primordia start to develop (Rudall, 2010). Aloni et al. (2006) noted that, although organs are often initiated acropetally, free auxin often becomes concentrated at organ tips, especially in anthers, in which free auxin tends to retard both acropetal and basipetal subsequent development and growth of neighbouring floral organs. They suggested that high concentrations of free IAA in young fertile floral organs (stamens and carpels) inhibits or retards organ-primordium initiation and development at the floral apex.

Many species with centrifugal stamens were formerly grouped into the subclass Dilleniidae (e.g. Cronquist, 1988; Takhtajan, 1996), which incorporated many of the polyandrous dicots that did not belong in Magnoliidae. Takhtajan's (1996) Dilleniidae included 39 orders, many of them containing only a single family. However, in recent molecular-based classifications (Angiosperm Phylogeny Group III, 2009) these former dilleniids have been dispersed among several larger eudicot orders, especially Brassicales, Caryophyllales, Crossomatales, Ericales, Malpighiales and Malvales (Table 6.1). In turn, these eudicot orders are dispersed through both the asterid and rosid clades, though Dilleniales *sensu stricto* remain in a relatively isolated position among the core eudicots.

In contrast with centrifugal intrazonal growth, centrifugal interzonal growth is probably a relatively atypical morphogenetic phenomenon in flowering plants. The most common type of centrifugal interzonal growth is early stamen initiation with respect to petals. Sattler noted this feature in some common garden plants such as *Lythrum salicaria* and *Valeriana officinalis* (Cheung and Sattler, 1967; Sattler, 1972a, b); these apparently serendipitous discoveries in plants that are readily available could indicate that centrifugal interzonal growth is more common than generally supposed. Indeed, *Lythrum salicaria* is remarkable in that three organ whorls are formed centrifugally: the outer sepal whorl (epicalyx) is initiated after the inner sepal whorl, the inner stamen whorl is initiated after the gynoecium (which consists of two congenitally united carpels), and the petals are initiated after the gynoecium and androecium. In the monotypic genus *Platycrater* (Hydrangeaceae), the entire carpel zone is initiated long before the multiple stamens, which are themselves initiated in centrifugal or irregular sequence on the hypanthium (Ge et al., 2007). Prenner (2011) noted that in the unusual polyandrous and polycarpous mimosoid legume *Acacia celastrifolia*, carpel initiation starts before the last stamens are formed, which is probably a relatively common feature in polyandrous species with an androecial ring meristem (Endress, 1994).

Centrifugal development is also frequently implicated in combination with some other atypical and iterative floral phenomena, such as obdiplostemony and development of a corona. These conditions are also highly homoplastic in angiosperms and appear to have evolved iteratively, though they can define groups, at least at the family level. The petaloid corona that occurs in some plant families (e.g. Amaryllidaceae) is late-developing with respect to adjacent floral structures (e.g. Endress and Matthews, 2006). Obdiplostemony occurs when the expected (typical) alternation of whorls is disrupted so that the stamens in two adjacent whorls occur in unexpected locations with respect to the carpels (Endress, 1994; Ronse De Craene and Smets, 1995). Thus, an obdiplostemous flower possesses two stamen whorls, of which the outer whorl lies opposite the petals rather than the sepals (Ronse De Craene, 2010). Obdiplostemony is a relatively common phenomenon among core eudicots. It can develop in different ways in different angiosperm groups, probably with different causal explanations, though this phenomenon requires review. For example, regular whorl alternation can become disturbed when sterile structures arise after the fertile whorl (centrifugal obdiplostemony; e.g. *Theobroma*: Ronse De Craene and Smets, 1995). Similarly, obhaplostemony, a condition where there is a single stamen whorl that is inserted opposite the petals, is often correlated with petal retardation (Ronse De Craene, 2010).

6.4 Conclusions

Few studies have explicitly tested the deep-node relationships between molecular and morphological data in angiosperms in cladistic analyses, though some have usefully employed a topology constrained by molecular trees to explore character evolution (e.g. Doyle, 2008). Two exceptions that combined morphological and molecular data on a broad scale are the angiosperm-wide analysis by Nandi et al. (1998), though these authors admitted that their broad study did not allow them to evaluate many features in detail, and the combined analysis by Doyle and Endress (2000), which focused on early-divergent angiosperms. Given the inherent problems in taxon sampling, comparing molecular and morphological data sets at broad taxonomic levels is probably most readily achieved by comparing topologies rather than combining data sets (e.g. Bateman et al., 2006).

One problem with inclusion of morphogenetic phenomena in phylogenetic analyses is that many of them represent character complexes that cannot readily be simplified. At least some of these character linkages are due to developmental constraints that currently are not well understood. Furthermore, many morphogenetic characters are iterative, and therefore non-homologous in the sense of true phylogenetic synapomorphies, which evolve only once and require a unique causal explanation (Patterson, 1988; Bateman and DiMichele, 1994). On the other hand, homoplasy can enhance the significance of these characters in investigating the genetic bases for morphological evolution. Indeed, detailed comparative investigations of similar phenomena between relatively distantly related species can help to indicate cases where common inherited genetic factors control similar morphological phenomena; this type of study represents one of the roots of the new science of evolutionary-developmental genetics (evo-devo).

Tracing the phylogenetic distribution of unusual and anomalous types of development can help us to understand the evolution of different growth patterns and allow us to predict the heritable genetic factors that control them. The primary topic of this paper, centrifugal growth in flowers, is a good example of a morphogenetic character that appears to represent more than one (interlinked) phenomenon. Corner's (1946) prediction that centrifugal stamens would prove to be of taxonomic significance at deep nodes in the angiosperm tree is only partially upheld in a molecular phylogenetic context. Centrifugal stamens are restricted to particular clades, mostly within the tricolpate (eudicot) clade, but are not restricted to particular groups, and probably evolved more than once within each group, often in association with polyandry.

Acknowledgements

Gerhard Prenner, Louis Ronse De Craene and Jürg Schönenberger helpfully provided some further examples, and Richard Bateman made useful comments on the manuscript.

6.5 References

Aloni, R., Aloni, E., Langhans, M. and Ullrich, C. (2006). Role of auxin in regulating *Arabidopsis* flower development. *Planta* **223**, 315–328.

Angiosperm Phylogeny Group III (2009). An update of the Angiosperm Phylogeny Group classification for the orders and families of flowering plants: APG III. *Botanical Journal of the Linnean Society*, **161**, 105–121.

Bateman, R. M. and DiMichele, W. A. (1994). Heterospory: the most iterative key innovation in the evolutionary history of the plant kingdom. *Biological Reviews*, **69**, 345–417.

Bateman, R. M., Hilton, J. and Rudall, P. J. (2006). Morphological and molecular phylogenetic context of the angiosperms: contrasting the 'top-down' and 'bottom-up' approaches to inferring the likely characteristics of the first flowers. *Journal of Experimental Botany*, **57**, 3471–3503.

Bernhard, A. and Endress, P. K. (1999). Androecial development and systematics in Flacourtiaceae. *Plant Systematics and Evolution*, **215**, 141–155.

Cheung, M. and Sattler, R. (1967). Early floral development of *Lythrum salicaria*. *Canadian Journal of Botany*, **45**, 1609–1618.

Corner, E. J. H. (1946). Centrifugal stamens. *Journal of the Arnold Arboretum*, **27**, 423–437.

Cronquist, A. (1988). *The Evolution and Classification of Flowering Plants*. New York: New York Botanical Garden.

Cusick, F. (1959). Floral morphogenesis in *Primula bulleyana* Forrest. *Journal of the Linnean Society of London, Botany*, **56**, 262–268.

Dahlgren, R. M. T. (1975). The distribution of characters within an angiosperm system. I. Some embryological characters. *Botaniska Notiser*, **128**, 181–197.

Doyle, J. A. (2006). Seed ferns and the origin of angiosperms. *Journal of the Torrey Botanical Society*, **133**, 169–209.

Doyle, J. A. (2008). Integrating molecular phylogenetic and paleobotanical evidence on origin of the flower. *International Journal of Plant Sciences*, **169**, 816–843.

Doyle, J. A. and Endress, P. K. (2000). Morphological phylogenetic analysis of basal angiosperms: comparison and combination with molecular data. *International Journal of Plant Sciences*, **161**, S121–S153.

Eames, A. J. (1953). Floral anatomy as an aid in generic limitation. *Chronica Botanica* **14**, 126–132.

Eichler, A. W. (1875). *Blüthendiagramme, I.* Leipzig: Engelmann.

Eichler, A. W. (1878). *Blüthendiagramme, II.* Leipzig: Engelmann.

Endress, P. K. (1994). *Diversity and Evolutionary Biology of Tropical Flowers.* Cambridge: Cambridge University Press.

Endress, P. K. (1995). Major evolutionary traits of monocot flowers. pp. 43–79 in Rudall, P. J., Cribb, P. J., Cutler, D. F., and Humphries, C. J. (eds.), *Monocotyledons: Systematics and Evolution.* Kew: Royal Botanic Gardens, Kew.

Endress, P. K. (1997). Relationships between floral organization, architecture, and pollination mode in *Dillenia* (Dilleniaceae). *Plant Systematics and Evolution*, **206**, 99–118.

Endress, P. K. (2001). Evolution of floral symmetry. *Current Opinion in Plant Biology*, **4**, 86–91.

Endress, P. K. and Matthews, M. L. (2006). Elaborate petals and staminodes in eudicots: diversity, function, and evolution. *Organisms, Diversity and Evolution*, **6**, 257–293.

Erbar, C. and Leins, P. (1997). Different patterns of floral development in whorled flowers, exemplified by Apiaceae and Brassicaceae. *International Journal of Plant Sciences*, **158**, S49–S64.

Erdtman, G. (1960). Pollen walls and angiosperm phylogeny. *Botaniska Notiser*, **113**, 41–45.

Furness, C. A. and Rudall, P. J. (2004). Pollen aperture evolution – a crucial factor for eudicot success? *Trends in Plant Science*, **9**, 1360–1385.

Ge, L. P., Lu, A. M. and Gong, C. R. (2007). Ontogeny of the fertile flower in *Platycrater arguta* (Hydrangeaceae). *International Journal of Plant Sciences*, **168**, 835–844.

Hardy, C. R. and Stevenson, D. W. (2000). Floral organogenesis in some species of *Tradescantia* and *Callisia* (Commelinaceae). *International Journal of Plant Sciences*, **161**, 551–562.

Hill, J. P. and Lord, E. M. (1989). Floral development in *Arabidopsis thaliana*: a comparison of the wild type and the homeotic *pistillata* mutant. *Canadian Journal of Botany*, **67**, 2922–2936.

Hochwallner, H. and Weber, A. (2006). Flower development and anatomy of *Clusia valerioi*, a Central American species of Clusiaceae offering floral resin. *Flora*, **201**, 407–418.

Kaul, R. B. (1967). Ontogeny and anatomy of the flower of *Limnocharis flava* (Butomaceae). *American Journal of Botany*, **54**, 1223–1230.

Kaul, R. B. (1995). Reproductive structure and organogenesis in a cottonwood, *Populus deltoides* (Salicaceae). *International Journal of Plant Sciences*, **156**, 172–180.

Leins, P. (1964). Das zentripetale und zentrifugale Androeceum. *Berichte der Deutschen Botanischen Gesellschaft*, **77**, 22–26.

Leins, P. and Erbar, C. (1996). Early floral developmental studies in Annonaceae. pp. 1–27 in Morawetz, W. and Winkler, H. (eds.), *Reproductive Morphology in Annonaceae.* Wein: Osterreich Akademie der Wissenschaften.

Leins, P. and Erbar, C. (2000). *Blüte und Frucht: Aspekte der Morphologie, Entwicklungsgeschichte, Phylogenie, Funktion und Ökologie.* Germany: E. Schweizerbartsche.

Matthews, M. L. and Endress, P. K. (2005). Comparative floral structure and systematics in Crossosomatales (Crossosomataceae, Stachyuraceae, Staphyleaceae, Aphloiaceae, Geissolomataceae, Ixerbaceae, Strasburgeriaceae). *Botanical Journal of the Linnean Society* **147**, 1–46.

Mayr, B. (1969). Ontogenetische Studien an Myrtales-Bluten. *Botanische Jahrbücher für Systematik*, **89**, 210–271.

Nandi, O. I., Chase, M. W. and Endress, P. K. (1998). A combined cladistic analysis of angiosperms using *rbcL* and non-molecular data sets. *Annals of the Missouri Botanical Garden*, **85**, 137–212.

Patterson, C. (1988). Homology in classical and molecular biology. *Molecular Biology and Evolution*, **5**, 603–625.

Payer, J. B. (1857). *Traité d'organogénie comparée de la fleur*. Paris: J. Cramer.

Philipson, W. R. (1974). Ovular morphology and the major classification of the dicotyledons. *Botanical Journal of the Linnean Society*, **68**, 89–108.

Prenner, G. (2003). A developmental analysis of the inflorescence and the flower of *Lotus corniculatus* (Fabaceae-Loteae). *Mitteilungen Naturwissenschaftlicher Verein Steiermark*, **133**, 99–107.

Prenner, G. (2004a). Floral development in *Daviesia cordata* (Leguminosae: Papilionoideae: Mirbelieae) and its systematic implications. *Australian Journal of Botany*, **52**, 285–291.

Prenner, G. (2004b). Floral ontogeny in *Lespedeza thunbergii* (Leguminosae: Papilionoideae: Desmodieae): variations from the unidirectional mode of organ formation. *Journal of Plant Research*, **117**, 297–302.

Prenner, G. (2011). Floral ontogeny of *Acacia celastrifolia*: an enigmatic mimosoid legume pronounced polyandry and multiple carpels per flower. pp. 260–282 in Wanntorp, L. and Ronse De Craene, L. P. (eds.), *Flowers on the Tree of Life*. Cambridge: Cambridge University Press.

Prenner, G., Box, M. S., Cunniff, J. and Rudall, P. J. (2008). The branching stamens of *Ricinus* and the homologies of the angiosperm stamen fascicle. *International Journal of Plant Sciences*, **169**, 735–744.

Ray, J. (1686–1704). *Historia Plantarum*. London.

Remizowa, M. V., Sokoloff, D. D. and Rudall, P. J. (2010). Evolutionary history of the monocot flower. *Annals of Missouri Botanical Garden*, **97**, 617–645.

Ren, Y., Chang, H. L. and Endress, P. K. (2010). Floral development in Anemoneae (Ranunculaceae). *Botanical Journal of the Linnean Society*, **162**, 77–100.

Ronse De Craene, L. P. (2010). *Floral Diagrams*. Cambridge: Cambridge University Press.

Ronse De Craene, L. P. and Smets, E. F. (1987). The distribution and systematic relevance of the androecial characters oligomery and polymery in the Magnoliophytina. *Nordic Journal of Botany*, **7**, 239–253.

Ronse De Craene, L. P. and Smets, E. F. (1992). Complex polyandry in the Magnoliatae: definition, distribution, and systematic value. *Nordic Journal of Botany*, **12**, 621–649.

Ronse De Craene, L. P. and Smets, E. F. (1994). Merosity in flowers: definition, origin, and taxonomic significance. *Plant Systematics and Evolution*, **191**, 83–104.

Ronse De Craene, L. P. and Smets, E. F. (1995). The distribution and systematic relevance of the androecial character oligomery. *Botanical Journal of the Linnean Society*, **118**, 193–247.

Rudall, P. J. (2008). Fascicles and filamentous structures: comparative ontogeny of morphological novelties in Triuridaceae. *International Journal of Plant Sciences*, **169**, 1023–1037.

Rudall, P. J. (2010). All in a spin: Centrifugal organ formation and floral patterning. *Current Opinion in Plant Biology*, **13**, 108–114.

Rudall, P. J., Remizowa, M. V., Prenner, G. et al. (2009). Non-flowers near the base of extant angiosperms? Spatiotemporal arrangement of organs in reproductive units of Hydatellaceae, and its bearing on the origin of the flower. *American Journal of Botany*, **96**, 67–82.

Rudall, P. J., Sokoloff, D. D., Remizowa, M. V. et al. (2007). Morphology of Hydatellaceae, an anomalous aquatic family recently recognized as an early-divergent angiosperm lineage. *American Journal of Botany*, **94**, 1073–1092.

Sattler, R. (1972a). *Organogenesis of Flowers. A Photographic Text-Atlas.* Toronto: University of Toronto Press.

Sattler, R. (1972b). Centrifugal primordial inception in floral development. *Advances in Plant Morphology*, **1972**, 170–178.

Sattler, R. and Singh, V. (1978). Floral organogenesis of *Echinodorus amazonicus* Rataj and floral construction of the Alismatales. *Botanical Journal of the Linnean Society*, **77**, 141–156.

Smyth, D. R., Bowman, J. L. and Meyerowitz, E. (1990). Early flower development in *Arabidopsis. The Plant Cell*, **2**, 755–767.

Sokoloff, D. D., Remizowa, M. V., Macfarlane, T. D. et al. (2008).

Seedling diversity in Hydatellaceae: implications for the evolution of angiosperm cotyledons. *Annals of Botany*, **101**, 153–164.

Soltis, D. E., Senters, A.E., Zanis, M. J., Kim, S. et al. (2003). Gunnerales are sister to other core eudicots: implications for the evolution of pentamery. *American Journal of Botany*, **90**, 461–470.

Takhtajan, A. (1996). *Diversity and Classification of Flowering Plants.* New York: Columbia University Press.

Tucker, S. C. (1972). The role of ontogenetic evidence in floral morphology. *Advances in Plant Morphology*, **1972**, 359–369.

Tucker, S. C. (1984). Evolutionary lability of symmetry in early floral development. *International Journal of Plant Sciences*, **160** (6 Supplement), S25–S39.

Tucker, S. C. (1999). Unidirectional organ initiation in leguminous flowers. *American Journal of Botany*, **71**, 1139–1148.

Tucker, S. C. (2003). Floral development in legumes. *Plant Physiology*, **131**, 911–926.

Tucker, S. C. and Douglas, A. W. (1996). Floral structure, development, and relationships of paleoherbs: *Saruma, Cabomba, Lactoris* and selected Piperales. pp. 141–175, in Taylor, D. W. and Hickey, L. J. (eds.), *Flowering Plant Origin, Evolution and Phylogeny.* New York: Chapman and Hall.

Uhl, N. W. (1988). Floral organogenesis in palms. pp. 24–44 in Leins, P., Tucker, S. C. and Endress, P. K. (eds.), *Aspects of Floral Development.* Berlin: J. Cramer.

Uhl, N. W. and Moore, H. E. (1977). Centrifugal stamen initiation in phytelephantoid palms. *American Journal of Botany*, **64**, 1152–1161.

Wanntorp, L. and Ronse De Craene, L. P. (2007). Flower development of

Meliosma (Sabiaceae): evidence for multiple origins of pentamery in the eudicots. *American Journal of Botany,* **94**, 1828–1836.

Warming, E. (1878). De l'ovule. *Annales des Sciences Naturelles; Botanique,* series 6, **5**, 177–266.

Van Heel, W. A. (1966). Morphology of the androecium in Malvales. *Blumea,* **13**, 177–394.

Xu, F. and Rudall, P. J. (2006). Comparative floral anatomy and ontogeny in Magnoliaceae. *Plant Systematics and Evolution,* **258**, 1–15.

7

Evolution of the palm androecium as revealed by character mapping on a supertree

SOPHIE NADOT, JULIE SANNIER, ANDERS BARFOD AND
WILLIAM J. BAKER

7.1 Introduction

Over the last two decades, our insight into the phylogenetic relationships among groups of living organisms has increased significantly (see, for example, the Tree of Life Project, Maddison et al., 2007). Since the first burst in phylogenetic analyses occurred in the early nineties triggered by the discovery of the PCR technique (Saiki et al., 1988), constant improvements in laboratory techniques have made it easier to reveal patterns of molecular variation across organisms (e.g. McCombie et al., 1992; Ronaghi, 2001). At the same time, computer power, access to online data and analytical tools have rapidly improved (see, for example, Guindon et al., 2003 and 2005). The most popular methods are based on most parsimonious reconstructions (MP) or Bayesian inference (BI), the latter allowing for molecular dating and therefore gaining in popularity (Huelsenbeck and Ronquist, 2001; Ronquist and Huelsenbeck, 2003). The methods have been implemented in user-friendly software such as MacClade (Maddison and Maddison, 2001), Bayestraits (available from http://www.evolution.rdg.ac.uk/BayesTraits.html, see also Pagel, 1999; Pagel et al., 2004) and the more recently developed Mesquite (Maddison and

Flowers on the Tree of Life, ed. Livia Wanntorp and Louis P. Ronse De Craene. Published by Cambridge University Press. © The Systematics Association 2011.

Maddison, 2009). As a consequence of these recent developments a large number of robust and highly resolved phylogenies are now available for various taxonomic levels, providing excellent frameworks for exploring character evolution through space and time.

The palms (Arecaceae or Palmae) are an iconic family of flowering plants comprising around 2400 species distributed worldwide. Palms constitute a highly distinctive component of tropical rain forests and often have major ecological impacts in the plant communities where they occur. At the same time they are of immense economic significance, both at the international level (e.g. oil palm, date palm, coconut, rattan) and at the village level, where they provide shelter and food. Research interest in the palm family has greatly increased in the last three decades. The results have recently been synthesized into a monograph, which describes the morphology, ecology and geographical distribution of all palm genera (Dransfield et al., 2008a). Several authors have contributed to unravelling the relationships among genera in the family (Asmussen and Chase, 2001; Hahn, 2002; Lewis and Doyle, 2002; Asmussen et al., 2006). The results have been summarized in a robust and comprehensive supertree phylogeny by Baker et al. (2009), including all genera of the family but the newly discovered *Tahina* (Dransfield et al., 2008b). This represents an excellent opportunity for studying evolutionary trends in morphological and ecological traits.

Palm flowers are usually small and trimerous, with a perianth typically consisting of three sepals and three petals. They are actinomorphic to slightly asymmetric and usually rather inconspicuous. However, in other features there is wide variation. This applies particularly to floral arrangement, sexual expression in space (hermaphroditism versus monoecy, dioecy or polygamy) and time (dichogamy), as well as number, size and synorganization of floral organs. The androecium, which we will focus on here, typically includes six stamens, but this number can be reduced to three, like *Nypa* Wurmb (flowers with a single stamen have been observed in *Dypsis lantzeana* Baill.: Rudall et al., 2003), it can vary between 6 and 12 in a more or less stable manner, like *Roystonea* O.F.Cook, or it may attain very high numbers, such as tribe Phytelepheae (subfamily Ceroxyloideae), where hundreds of stamens are packed together tightly on the floral apex (Fig 7.1). In some species with unisexual flowers, noticeable differences in number, size and shape exist between the functional androecium of the male flower and the sterile androecium of the pistillate flower (Fig 7.1). The ontogenetic development of selected polyandrous palm species has been studied by Barfod and Uhl (2001), Uhl (1976), Uhl and Moore (1977), Uhl and Moore (1980) and Uhl and Dransfield (1984). Other studies concerned with the palm androecium have focused on anther attachment and dehiscence, pollen wall ornamentation and pollen aperture type (Harley, 1999; Harley and Baker, 2001), as well as microsporogenesis (Sannier et al., 2006).

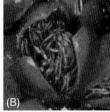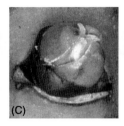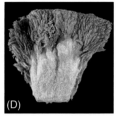

Fig 7.1 Various flower morphologies in palms. (A) Hermaphroditic flower of *Licuala peltata* (tribe Trachycarpeae, subfam. Coryphoideae) with six stamens. (B) Male flower of *Howea balmoreana* (tribe Areceae, subfam. Arecoideae) with approximately 30 to 50 stamens. (C) Female flower of *Howea balmoreana* (only the three stigmas of the pistil are visible, protruding from the flower; the three staminodes are not visible). (D) Male flower of *Aphandra natalia* (tribe Phytelepheae, subfam. Ceroxyloideae) with approximately 500 stamens.

In this chapter we optimize characters relating to the palm androecium on the recently published supertree by Baker et al. (2009). This provides insights into the role of stamen synorganization and morphology, in the diversification of palms in particular, and in monocots and flowering plants in general.

7.2 Materials and methods

7.2.1 Choice of characters and character coding

Table 7.1 shows the list of characters examined and the coding into discrete character states. All characters are related to the androecium except for 'Petal connation', which was recorded to enable a comparison with 'Stamen connation', petals being the part of the perianth that is the closest to the androecium. All coding is based on the information presented in the genus descriptions of Dransfield et al. (2008a). In most cases character variation was easily broken down into a limited number of states. To avoid polymorphism the variation found for 'Stamen connation' and 'Petal connation' was coded using an additional character state that corresponds to the presence of two different states within the same genus. Defining character states for stamen number proved to be problematic, both because of the range of variation throughout the family (from three to hundreds) and because of the range of variation within certain genera. We therefore chose to use a binary coding for the character 'Stamen number', in which state 0 is oligandry (defined here as six or less stamens) and state 1 is polyandry (more than six stamens). The number six corresponds to twice the merism of all palm flowers. Stamen numbers recorded for polyandrous genera are given in Table 7.2.

Table 7.1 List of characters examined in this study. All characters were treated as discrete and coded as such.

Characters	Character states
Characters examined for genera bearing bisexual or unisexual flowers	
Stamen number	Oligandry (6 stamens or less) = 0; Polyandry = 1
Anther dehiscence	Latrorse = 0; Extrorse = 1; Introrse = 2
Anther attachment	Dorsifixed = 0; Basifixed = 1
Stamen connation	Free = 0; Connate = 1; Free or connate* = 2
Petal connation	Free = 0; Connate = 1; Free or connate* = 2; Connate in one ring = 3
Stamen adnation (to petals)	Free = 0; Adnate = 1
Characters examined only for genera bearing unisexual flowers	
Staminode number (female flowers) versus functional stamen number (male flowers)	Identical = 0; Different = 1; Lacking = 2
Pistillode in male flowers	Present = 0; Lacking = 1; Minute = 2

*Infrageneric variation

7.2.2 Phylogenetic background and character optimization

Optimization of androecium characters was performed on a recent supertree of palms (Baker et al., 2009). It includes all genera accepted by Govaerts and Dransfield (2005) plus those recognized later by Baker et al. (2006) and Lewis and Zona (2008). We slightly modified the supertree (see Fig 3 in Baker et al., 2009) by considering only accepted genera according to Dransfield et al. (2008a). The recently discovered genus *Tahina* J.Dransf. and Rakotoarin. (Dransfield et al., 2008b) was not included in this tree. The original supertree included an outgroup of 13 genera of commelinid monocots from which we selected the following six to include in our study: *Anigozanthos* Labill. (Haemodoraceae), *Tradescantia* Ruppius ex L. (Commelinaceae), *Costus* L. (Costaceae), *Dasypogon* R.Br. (Dasypogonaceae), *Fargesia* Franchet (Poaceae) and *Vriesea* Lindl. (Bromeliaceae).

We used Maximum Parsimony (MP) for character optimization as implemented in the Mesquite software package (Maddison and Maddison, 2009). The default settings were used (character states were considered as unordered).

Table 7.2 Palm genera including at least one species producing polyandrous flowers (i.e. bearing more than six stamens). Genera in which all species are polyandrous are in bold and the number of species (from www.palmweb.org) is indicated. Genera with unisexual and strongly dimorphic flowers are marked with an asterisk. Genera including polyandrous species in which developmental studies have been conducted are marked with a plus sign. First and second columns: subfamily and tribe according to Dransfield et al. (2005). Data were obtained from Dransfield et al. (2008a) and from www.palmweb.org.

SUBFAMILY	TRIBE	Genera including polyandrous species	Stamen number	Flowers unisexual (0); bisexual (1)	Inflorescence unisexual (0); bisexual (1)	Plant unisexual (0); bisexual (1)
Calamoideae	Eugeissoneae	**Eugeissona**+ Griff. (6 sp)	**20–70**	0/1	1	1
	Lepidocaryeae	Raphia P.Beauv.	6–30	0	1	1
	Calameae	Calamus L.	6 (12 in C. ornatus)	0	0	0
		Korthalsia Blume	6–9	1	1	1
		Plectocomia Mart. and Blume	Usually 6, rarely to 12	0	0	0
Coryphoideae	Caryoteae	Arenga Labill. ex DC.	Rarely 6–9, usually more than 15	0	0/1	1
		Caryota+ L.	6–100	0	1	1
		Wallichia Roxb.	3–15	0	0	1
	Borasseae	Borassodendron Becc.	6–15	0	0	0
		Latania Comm. ex Juss. (3 sp)	**15–30 or more**	0	0	0
		Lodoicea+ Comm. ex DC. (1 sp)	**17–22**	0	0	0
	Cryosophileae	Chelyocarpus Dammer	5–9	1	1	1
		Coccothrinax Sarg.	9(6–13)	1	1	1
		Hemithrinax Hook.f.	6–8	1	1	1
		Itaya H.E.Moore (1 sp)	**18–24**	1	1	1
		Thrinax L.f. ex Sw.	5–15	1	1	1
		Zombia L.H.Bailey (1 sp)	**9–12**	1	1	1
	Phoeniceae	Phoenix L.	Usually 6, rarely 3 or 9	0	0	0

SUBFAMILY	TRIBE	Genera including polyandrous species	Stamen number	Flowers unisexual (0); bisexual (1)	Inflorescence unisexual (0); bisexual (1)	Plant unisexual (0); bisexual (1)
Ceroxyloideae	Phytelepheae	*Ammandra*[+] O.F.Cook* (1 sp)	300–1200	0	0	0
		Aphandra[+] Barfod* (1 sp)	400–650	0	0	0
		Phytelephas[+] Ruiz and Pav.* (6 sp)	36–900	0	0	0
	Ceroxyleae	*Ceroxylon*[+] Bonpl. ex DC.	6–15(–17)	0	0	0
Arecoideae	Areceae	*Acanthophoenix* H.Wendl.	6–12	0	1	1
		Actinokentia Dammer (2 sp)	19–50	0	1	1
		Actinorhytis H.Wendl. and Drude (1 sp)	24–33	0	1	1
		Adonidia Becc. (1 sp)	45–50	0	1	1
		Archontophoenix H.Wendl. and Drude (6 sp)	12–14	0	1	1
		Areca L.	3,6,9 or up to 30 or more	0	1	1
		Balaka Becc. (11 sp)	24–50	0	1	1
		Brassiophoenix Burret (2 sp)	100–230	0	1	1
		Calyptrocalyx Blume (26 sp)	6–140	0	1	1
		Carpentaria Becc. (1 sp)	ca 33	0	1	1
		Chambeyronia Vieill. (2 sp)	19–55	0	1	1
		Cyphokentia Brongn.	6–12	0	1	1
		Cyrtostachys Blume (11 sp)	9–15	0	1	1
		Deckenia H.Wendl. ex Seem.	6–9	0	1	1
		Dransfieldia W.J.Baker and Zona (1 sp)	Numerous, up to 19	0	1	1
		Drymophloeus Zipp. (8 sp)	24–320 or more	0	1	1
		Hedyscepe H.Wendl. and Drude (1 sp)	9–10(12)	0	1	1
		Heterospathe Scheff.	6–36	0	1	1

Table 7.2 (cont.)

SUBFAMILY	TRIBE	Genera including polyandrous species	Stamen number	Flowers unisexual (0); bisexual (1)	Inflorescence unisexual (0); bisexual (1)	Plant unisexual (0); bisexual (1)
		Howea Becc. (2 sp)	**30–70 or more**	0	1	1
		Hydriastele H.Wendl. and Drude	6–24	0	1	1
		Kentiopsis Brongn. (4 sp)	**11–38**	0	1	1
		Laccospadix H.Wendl. and Drude (1 sp)	**9–12**	0	1	1
		Lemurophoenix J.Dransf. (1 sp)	**52–59**	0	1	1
		Linospadix H.Wendl.	6–12	0	1	1
		Loxococcus H.Wendl. and Drude (1 sp)	**12**	0	1	1
		Nephrosperma Balf.f. (1 sp)	**40–50**	0	1	1
		Normanbya F.Muell. ex Becc. (1 sp)	**24–40**	0	1	1
		Oncosperma Blume	6–9	0	1	1
		Phoenicophorium H.Wendl. (1 sp)	**15–18**	0	1	1
		Pinanga Blume	Rarely 6, usually 12–68	0	1	1
		Ponapea Becc. (3 sp)	**Ca. 100**	0	1	1
		Ptychococcus Becc. (2 sp)	**Up to 100**	0	1	1
		Ptychosperma[+] Labill. (29 sp)	**9–100 or more**	0	1	1
		Rhopaloblaste Scheff.	6–9	0	1	1
		Solfia Rech. (1 sp)	**35–37**	0	1	1
		Tectiphiala H.E.Moore	6–7	0	1	1
		Veitchia H.Wendl. (8 sp)	**Numerous to 100 or more**	0	1	1
		Wodyetia A.K.Irvine (1 sp)	**60–71**	0	1	1
	Cocoseae	*Allagoptera* Nees	6–100	0	1	1
		Astrocaryum G.Mey.	(3–)6(–12)	0	1	0
		Attalea Kunth	3–75	0	0/1	1

SUBFAMILY	TRIBE	Genera including polyandrous species	Stamen number	Flowers unisexual (0); bisexual (1)	Inflorescence unisexual (0); bisexual (1)	Plant unisexual (0); bisexual (1)
		Bactris Jacq. ex Scop.	(3–)6(–12)	0	1	1
		Beccariophoenix Jum. and H.Perrier (2 sp)	**15–21**	0	1	1
		Desmoncus Mart.	6–9	0	1	1
		Jubaea Kunth (1 sp)	**18**	0	1	1
		Jubaeopsis Becc. (1 sp)	**(7–)8–16**	0	1	1
		Parajubaea Burret (3 sp)	**13–15**	0	1	1
		Voanioala J.Dransf. (1 sp)	**12(–13)**	0	1	1
	Euterpeae	Oenocarpus Mart.	6(7–8)9–20	0	1	1
	Geonomateae	Asterogyne H.Wendl. ex Hook.f.	6–c. 24	0	1	1
		Geonoma Willd.	(3) 6(rarely more)	0	1	1
		Welfia[+] H.Wendl. (1 sp)	**36(27–42)**	0	1	1
	Iriarteeae	**Iriartea** Ruiz and Pav. (1 sp)	**9–20**	0	1	1
		Socratea[+] H.Karst. (5 sp)	**17–145**	0	1	1
		Wettinia[+] Poepp. ex Endl.	6–20	0	0	1
	Manicarieae	**Manicaria** Gaertn. (1 sp)	**30–35**	0	1	1
	Oranieae	Orania Zipp.	3,4,6 or 9–32	0	1	1
	Reinhardtieae	**Reinhardtia** Liebm. (6 sp)	**8–140**	0	1	1
	Roystoneeae	Roystonea O.F.Cook	6–12	0	1	1
	Sclerospermeae	**Sclerosperma** G.Mann and H.Wendl. (3 sp)	**60–100**	0	1	1

7.3 Results

Figure 7.2 shows the MP optimization of the character 'Stamen number' on the palm supertree. It strongly supports the widely accepted idea that six stamens represent the ancestral state in palms. Ten genera include species in which the androecium is reduced to three stamens (names followed by stars in Fig 7.2): *Areca* L., *Astrocaryum* G.Mey., *Attalea* Kunth, *Bactris* Jacq. ex Scop., *Dypsis* Noronha ex Mart., *Geonoma* Willd., *Nypa* Wurmb., *Orania* Zipp., *Synechanthus* H.Wendl. and *Wallichia* Roxb. In the genera *Areca* and *Orania*, the number of stamens varies among species between three and several dozen. Polyandrous flowers (with an androecium composed of more than six stamens) have been recorded within 82 genera out of 183 (Table 7.2). Forty-five of these genera produce exclusively polyandrous flowers (Table 7.2; black rectangles in Fig 7.2), the remaining genera being polymorphic (grey boxes in Fig 7.2). According to the optimization, polyandry has evolved numerous times independently during the diversification of palms, within all subfamilies except Nypoideae. Due to the high number of polymorphic genera, it is difficult to identify precisely the number of transitions towards polyandry. However, it can be noted that the highest number of transitions, together with the widest range of variation in terms of stamen number, occurs in Arecoideae, the largest subfamily, containing about half of the species and genera recognized in the family.

Basifixed anthers represent the ancestral condition for the family as a whole (optimization not shown) and for the subfamilies Calamoideae, Nypoideae and Ceroxyloideae. Dorsifixed stamens evolved five times within Calamoideae and once in Ceroxyloideae; it represents the ancestral condition for the subfamilies Coryphoideae and Arecoideae except the early diverging tribe Iriarteeae. In both subfamilies there are reversals towards the basifixed condition in higher-order branches. The vast majority of palm genera possess latrorse anthers that open by longitudinal slits lateral to the filament, which most probably represents the ancestral condition for the family. According to our optimization (not shown), introrse anthers have evolved more than 20 times independently. It should be noticed that this mode of dehiscence is synapomorphic for the Geonomateae + Manicarieae clade.

Sixty-six or approx. one third of the genera include species with connate stamens. This condition has evolved at least 29 times independently from ancestral free stamens, throughout all subfamilies (Fig 7.3). More than one third of the genera with connate stamens belong to the Coryphoideae. The optimization of petal connation shows that sympetaly is also derived (Fig 7.3). This character state is a synapomorphy for the subfamilies Calamoideae and Coryphoideae, as well as for a few genera within the Arecoideae. Several reversals towards free petals have occurred within Calamoideae and Coryphoideae. The single member of Nypoideae, *Nypa fruticans* Wurmb, has the ancestral choripetalous corolla.

Stamen and petal connation only partially co-occur throughout the palms, thus several genera are either sympetalous or synandrous. In many genera with unisexual flowers, the degree of petal fusion differs between staminate and pistillate flowers. This is the case in Ceroxyloideae, tribe Borasseae (in subfamily Coryphoideae) and tribe Calameae (in subfamily Calamoideae). The optimization furthermore indicates that stamens adnate to petals are ancestral in subfamily Calamoideae and perhaps represent the original condition in palms overall. It is considered a derived feature in most families of flowering plants. Adnation of stamens to petals is a synapomorphy for the tribes Trachycarpeae and Borasseae in subfamily Coryphoideae, and for a clade within the Arecoideae which includes *Calyptronoma* Griseb. + *Calytrogyne* H.Wendl. + *Asterogyne* H.Wendl. ex Hook.f. + *Geonoma* Willd.

In genera with unisexual flowers we paid special attention to the difference between the number of staminodes in female flowers and the number of stamens in male flowers respectively. Our results (Fig 7.4) show identical numbers in the Calamoideae, and part of the Coryphoideae and Ceroxyloideae. Within the Coryphoideae, the difference in the number of androecium parts is pronounced in the tribes Borasseae + Caryoteae + Corypheae that form a clade. Within the Ceroxyloideae the dioecious tribe Phytelepheae is characterized by a strong dimorphism between male and female flowers and a more than tenfold difference between functional stamens and staminodes in terms of number. In the Arecoideae our optimization reveals an overall trend towards fewer staminodes relative to the number of stamens. The presence of three staminodes in the female flowers is characteristic of the clade composed of the Pacific subtribes Ptychospermatinae + Archontophoenicinae + Basseliniinae + Carpoxylinae + Clinospermatinae + Linospadicinae + *Loxococcus*. The male flowers in these groups may have various numbers of functional stamens, depending on the species. A staminodial ring is found in subtribe Attaleinae, in which the number of stamens varies between 12 and 24.

We also optimized presence versus absence of sterile organs in genera with unisexual flowers (Fig 7.4) to check whether any trend in reduction of organs could be detected. Most of the dioecious or monoecious genera have staminate flowers with a pistillode and pistillate flowers with staminodes. Complete abortion of the gynoecium occurs in 17 genera throughout the family and most likely evolved from unisexual flowers with a pistillode. In several cases the character transformation passed through a transitional state, which involves a minute pistillode, a state commonly found throughout the family. This is the case for the [*Salacca* Reinw. + *Eleiodoxa* (Becc.) Burret] clade in Calamoideae, for subtribe Rhapidinae in Coryphoideae, for the [*Phytelephas* Ruiz and Pav. + *Aphandra* Barfod] clade in Ceroxyloideae and for *Howea* in Arecoideae. Optimization of 'Lack of staminodes' showed that this character state evolved ten times independently.

(A)

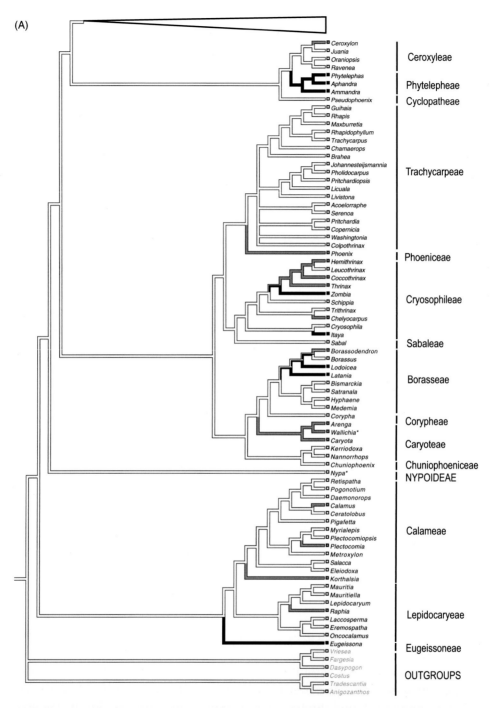

Fig 7.2 MP optimization of 'Stamen number' on the palm supertree. (a) all subfamilies except Arecoideae. (b) Arecoideae (continuation of (a)). Upper left in (b): categories (character states) defined here for the character. Branches are coloured according to the inferred ancestral states. Boxes at the tip of branches correspond to the actual observations. White = oligandry, Black = polyandry, Grey = infrageneric polymorphism (oligandry + polyandry). Genera including species with three stamens only are indicated by a star. Right hand side: tribe names according to Dransfield et al. (2005).

(B)

Stamen number
☐ Oligandry (six or less stamens)
■ Polyandry (more than six stamens)
▨ Both

Fig 7.2 (cont.)

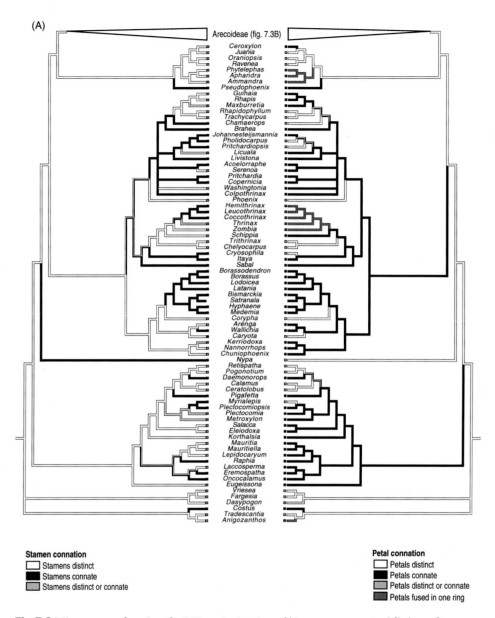

(A)

Arecoideae (fig. 7.3B)

Ceroxylon
Juania
Oraniopsis
Ravenea
Phytelephas
Aphandra
Ammandra
Pseudophoenix
Guihaia
Rhapis
Maxburretia
Rhapidophyllum
Trachycarpus
Chamaerops
Brahea
Johannesteijsmannia
Pholidocarpus
Pritchardiopsis
Licuala
Livistona
Acoelorraphe
Serenoa
Pritchardia
Copernicia
Washingtonia
Colpothrinax
Phoenix
Hemithrinax
Leucothrinax
Coccothrinax
Thrinax
Zombia
Schippia
Trithrinax
Chelyocarpus
Cryosophila
Itaya
Sabal
Borassodendron
Borassus
Lodoicea
Latania
Bismarckia
Satranala
Hyphaene
Medemia
Corypha
Arenga
Wallichia
Caryota
Kerriodoxa
Nannorrhops
Chuniophoenix
Nypa
Retispatha
Pogonotium
Daemonorops
Calamus
Ceratolobus
Pigafetta
Myrialepis
Plectocomiopsis
Plectocomia
Metroxylon
Salacca
Eleiodoxa
Korthalsia
Mauritia
Mauritiella
Lepidocaryum
Raphia
Laccosperma
Eremospatha
Oncocalamus
Eugeissona
Vriesea
Fargesia
Dasypogon
Costus
Tradescantia
Anigozanthos

Stamen connation
☐ Stamens distinct
■ Stamens connate
▨ Stamens distinct or connate

Petal connation
☐ Petals distinct
■ Petals connate
▨ Petals distinct or connate
▨ Petals fused in one ring

Fig 7.3 Mirror trees showing the MP optimization of 'Stamen connation' (left tree) versus 'Petal connation' (right tree) on the palm supertree. (A) all subfamilies except Arecoideae. (B) Arecoideae (continuation of (a)). Upper left and upper right: character states defined here for each character. Left: White = distinct stamens, Black = connate stamens, Grey = distinct or connate stamens. Right: White = distinct petals, Black = connate petals, Grey = distinct or connate petals, Dark grey = petals fused in a ring. Branches are coloured according to the inferred ancestral states. Boxes at the tip of branches correspond to the actual observations.

(B)

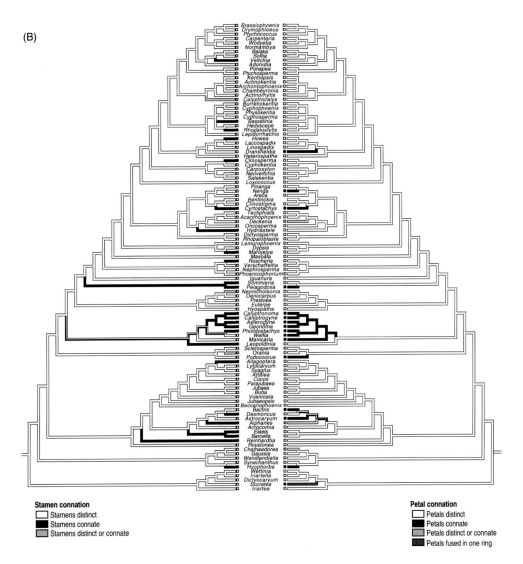

Stamen connation
- ☐ Stamens distinct
- ■ Stamens connate
- ▩ Stamens distinct or connate

Petal connation
- ☐ Petals distinct
- ■ Petals connate
- ▩ Petals distinct or connate
- ▦ Petals fused in one ring

Fig 7.3 (cont.)

(A)

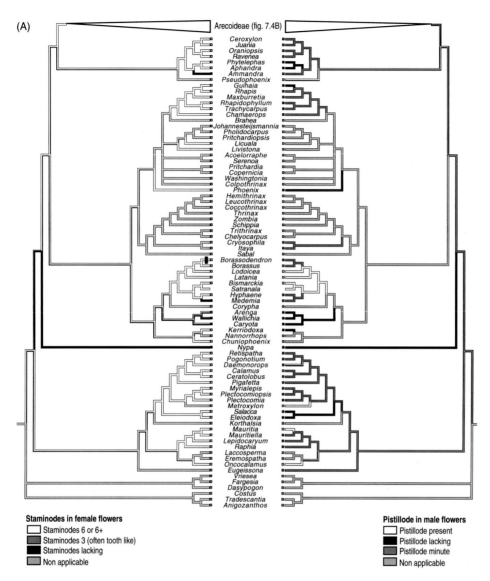

Staminodes in female flowers

☐ Staminodes 6 or 6+
▨ Staminodes 3 (often tooth like)
■ Staminodes lacking
▨ Non applicable

Pistillode in male flowers

☐ Pistillode present
■ Pistillode lacking
▨ Pistillode minute
▨ Non applicable

Fig 7.4 Mirror trees showing the MP optimization of 'Number of staminodes versus functional stamens' (left tree) versus 'Pistillode in male flowers' (right tree) on the palm supertree. (A) All subfamilies except Arecoideae. (B) Arecoideae (continuation of (a)). Upper left and upper right: character states defined here for each character. Left: White = six or more staminodes, Dark grey = three staminodes, Black = staminodes lacking, Light grey = not applicable (bisexual flowers only). Right: White = pistillode present, Black = pistillode lacking, Dark grey = pistillode minute, Light grey = not applicable (bisexual flowers only). Branches are coloured according to the inferred ancestral states. Boxes at the tip of branches correspond to the actual observations. Black bars indicate a transition towards a staminodial ring.

(B)

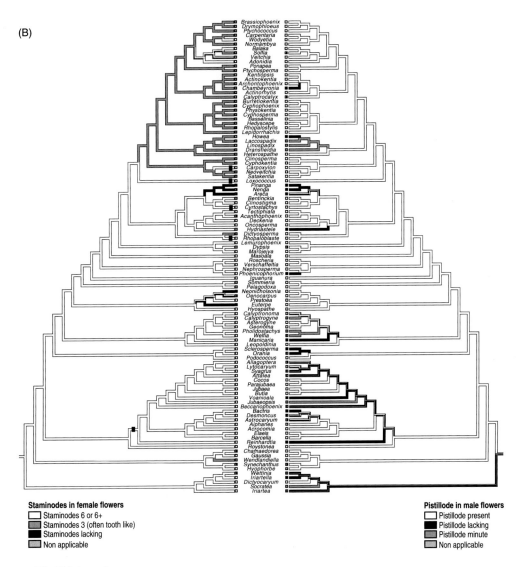

Staminodes in female flowers
☐ Staminodes 6 or 6+
☐ Staminodes 3 (often tooth like)
☐ Staminodes lacking
☐ Non applicable

Pistillode in male flowers
☐ Pistillode present
☐ Pistillode lacking
☐ Pistillode minute
☐ Non applicable

Fig 7.4 (cont.)

7.4 Discussion

7.4.1 Stamen number

An androecium composed of numerous stamens has long been considered an ancestral feature in angiosperms. Both in monocots and eudicots evolution has proceeded towards a reduction of the stamen number. However, secondary increases have evolved several times in both clades. According to traditional perception, ancestral polyandry is associated with a spiral organization of stamens, whereas derived polyandry is associated with cyclic organization (Endress and Doyle, 2007), although this view was recently challenged by Endress and Doyle (2009).

Although palm flowers follow the basic monocot arrangement of organs, they stand out in several aspects. This applies to the androecium, which is a focus of this chapter, but also to other morphological characters which display a variation unequalled or almost so within the monocots. Polyandry (defined here as more than twice merism, i.e. more than six stamens) occurs in 19 monocot families representing six different orders (Fig 7.5). In most families except palms, only one or a few genera display polyandry (data extracted from Dahlgren et al., 1985), and in most polyandrous genera the number of stamens does not exceed 12. Palms clearly stand out compared to all other monocot families, since polyandrous flowers are present in almost half of the 183 genera. Furthermore the range of variation in stamen number is exceptionally wide in palms, ranging from only one

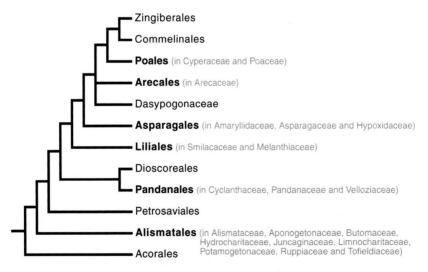

Fig 7.5 Tree of the monocot orders (topology as from Stevens, 2001 onwards) showing the orders (bold) and families (grey) in which flowers with androecia including more than six stamens are found.

(Dransfield et al., 1995) to more than one thousand (Dransfield et al., 2008a). The ancestral stamen number inferred for the palm family is six, which is also the ancestral number found for monocots as a whole (Nadot, unpubl. data). In four out of five subfamilies polyandry evolved several times from this trimerous androecium (with six stamens). The highest number of transitions towards polyandry, but also towards reduction in stamen number (from six to three) is found within subfamily Arecoideae, which include approx. half of all species of palms. It should be noted that flowers of all arecoid palms are unisexual and that almost all species are monoecious.

The spatial separation of sexual expression is highly variable in palms. Hermaphroditism, monoecy and dioecy are widespread in the family. Cases of polygamy (which corresponds to various combinations of sexual expression, such as andromonoecy or androdioecy) are also found. All three types of sexual expression are represented in the subfamilies Calamoideae and Coryphoideae. The subfamily Arecoideae is almost entirely monoecious, except for two dioecious genera (*Chamaedorea* Willd. + *Wendlandiella* Dammer), whereas Ceroxyloideae are mostly dioecious, with only one hermaphroditic to polygamous genus (*Pseudophoenix* H.Wendl. ex Sarg.). *Nypa* is monoecious, but highly unusual morphologically with the proximal position of the male flowers and the distal position of the female flowers (acrogyny). Inflorescence acrogyny is found in *Arenga* (Coryphoideae) in which sometimes distal inflorescences bear female flowers and proximal inflorescences bear male flowers. Polyandry coincides in most genera with unisexual flowers (Table 7.2), but whether this pattern results from adaptation in relation to pollination mechanisms or results from some constraints imposed by shared ancestry remains to be explored.

7.4.2 Anther features

Orientation of anther dehiscence is a relatively stable character in palms. There are only five transitions from latrorse to extrorse dehiscence, all in unrelated genera (*Nypa, Wallichia, Hemithrinax* Hook.f., *Allagoptera* Nees and *Orania*), and some 20 transitions from latrorse to introrse, also almost all in unrelated genera, with the exception of the [Manicarieae + Geonomateae] clade, for which introrse anthers are a synapomorphy. An unusual case of apical opening by pores is found within the genus *Areca*. Studies of pollination mechanisms in palms have revealed the existence of interactions with a number of pollinating insects, especially curculionid, nitidulid and staphylinid beetles, halictid bees and various groups of flies. The ecological and evolutionary significance of dehiscence remains unclear, since only a few studies have focused on the mechanisms of pollen transfer.

We are well aware that variation in the character 'anther attachment' is almost continuous. For convenience however, we use here the classical main categories recognized in the botanical literature, namely dorsifixed and basifixed anthers. Our character optimization suggests that basally attached anthers represent the ancestral condition for the family and that the dorsal type of attachment has evolved several times during the diversification of palms. It represents the ancestral condition for subfamilies Coryphoideae and Arecoideae, in which reversals to the basifixed type have occurred several times. For the same reasons as cited above, the ecological and evolutionary significance of this variation is unknown, but would be well worth exploring, considering the diversity of palm pollinators. The importance of anther dehiscence, opening and attachment in pollination processes was underlined by D'Arcy (1996).

7.4.3 Organ synorganization

The connation of petals and stamens is another variable feature within the palm family. Our results suggest that both free petals and free stamens are ancestral states in the family. Partial fusions in the corolla and androecium have evolved several times. Fusions in the androecium evolved predominantly in taxa with connate petals (Calamoideae, Coryphoideae and the clade composed of Manicarieae + Geonomateae within Arecoideae). Although the two phenomena are correlated, it should be noted that not all genera with connate anthers have connate petals and vice versa. In lineages with unisexual flowers, petal connation in staminate and pistillate flowers is similar, whereas fusions are less closely linked in pistillate and staminate androecia. The number of staminodes can be quite different from the number of functional stamens, especially in polyandrous genera, which suggests further sex divergence of the genetic control mechanisms underlying the formation of the androecium.

Adnation of stamens to petals has evolved several times and predominantly in lineages with connate petals and typically connate stamens, such as Coryphoideae, Calamoideae and Geonomateae. Only in the two sister genera, *Oraniopsis* (Becc.) J.Dransf., A.K.Irvine and N.W.Uhl, and *Ravenea* H.Wendl. ex C.D.Bouché (Ceroxyloideae) does adnation occur between free petals and free stamens. Increase in stamen number is apparently unconstrained by stamen synorganization, since polyandry is almost equally represented in groups with free or fused stamens respectively.

7.4.4 Sterile organs

Character optimization shows that in Calamoideae, Ceroxyloideae and Coryphoideae (except *Medemia* Wurttenb. ex H.Wendl.), the number of staminodes in female flowers is equal to the number of functional stamens in the male flowers, if this is six. This pattern is contrasted in several arecoid genera that have six functional stamens and a different number of staminodes. In one particular clade that includes genera with six stamens and genera with more than six stamens, the number of staminodes is reduced to three. Overall, the number of staminodes is often lower than the number of functional stamens in taxa with more than six stamens, and staminodes are completely lacking in a few hexandrous genera, such as in *Nenga* H.Wendl. and Drude (Arecinae) (some species of *Nenga* do produce minute staminodes, however) and the closely related genera in tribe Euterpeae, *Neonicholsonia* Dammer, *Oenocarpus* Mart. and *Euterpe* Mart. Interestingly, in our optimization no straightforward connection appears between the absence of a pistillode in staminate flowers and the absence of staminodes in female flowers. This may be a result of diverging selective pressures acting upon male and female flowers. The fact that species deviating from the ancestral hexandrous condition tend to produce less staminodes suggests a loss in the adaptive value of staminodes, since they do not produce pollen, and therefore are under a different selective pressure compared to the functional stamens.

7.4.5 How and why has polyandry evolved in palms?

Increase in stamen number appears to have occurred in different ways and perhaps in response to different factors in different groups of palms.

(Uhl and Moore, 1980).

The increase in stamen number in palms may reflect adaptation to different pollen transferring agents. In palms the framework of interaction with the pollinating agents as set by the plant is typically rather loose. This means that closely related species and even populations of the same species are visited by different taxonomic groups of potential pollinators, as revealed in genera such as *Euterpe* (Bovi and Cardoso, 1986; Reis et al., 1993; Kuchmeister et al., 1997;), *Aiphanes* Willd. (Listabarth, 1992; Borchsenius, 1993) and *Licuala* Wurmb (Barfod et al., 2003). Since most studies of palms are only dealing with the pollination mechanism at one specific site at one specific time, it is difficult to generalize about the co-evolutionary relationships. Both beetles and bees, which are the predominant pollinators of palms (Sannier et al., 2009 for a review), are attracted to flowers or inflorescences that produce copious amounts of pollen, and polyandry is one way that this can be potentially achieved. Another way is by close insertion of the flowers along the rachillae, whereby parts of or the entire inflorescence may constitute the functional unit in the interaction with the potential pollinators. Both strategies co-exist in palms, which can display both highly polyandrous flowers and densely packed flowers. Therefore, although polyandry may be of key importance for our understanding of the diversification of the palm flower, it is not straightforward to understand the underlying evolutionary processes that led to its multiple appearances in the evolution of palms.

Several polyandrous palm flowers belonging to all four subfamilies in which polyandry occurs have been the subject of developmental studies (see Table 7.2)

(Uhl, 1976; Uhl and Moore, 1977, 1980; Uhl and Dransfield, 1984; Barfod and Uhl, 2001). In all groups but tribe Phytelepheae, which is exceptional in having a centrifugal stamen development, the androecium of polyandrous taxa exhibits underlying trimery and the stamen initiation follows the order of inception of the perianth parts, with antesepalous stamens being formed before antepetalous ones, in distinct sectors (Uhl and Dransfield, 1984). Besides these basic common features, there is nevertheless considerable variation in the arrangement of stamens and in the way the floral apex expands to accommodate numerous stamens. In *Ptychosperma mooreanum* Essig (Uhl, 1976), *Lodoicea maldivica* (J.F.Gmel.) Pers. ex H.Wendl. and *Caryota mitis* Lour. (Uhl and Moore, 1980), the high and rather variable number of stamens is a consequence of the floral apex varying in width and height, as well as of differences in the number of primordia that can form in the outer, and quite wide, antepetalous whorl. In these species, never more than one stamen occurs in the antesepalous position (Uhl and Moore, 1980), whereas several stamens can form in antesepalous positions in *Socratea exorrhiza*, *Wettinia castanea* and *Welfia georgii*. In *Eugessiona utilis*, which has the highest number of stamens in the genus, an additional row of antepetalous stamens develop centrifugally outside of and alternating with the first-formed row (Uhl and Dransfield, 1984), perhaps due to the presence of residual meristem. Increase in stamen number through duplication of the primordia is probably made possible by expansion of the floral apex, which may be released from spatial constraints and hormonal control by the apical meristem. Trimery is lost in Phytelepheae, which display the highest stamen numbers in monocots. In this tribe, centrifugal expansion is believed to represent an alternative way to increase the size of the floral apex thereby accommodating numerous stamens (Uhl and Moore, 1977). Apical expansion and polyandry are thought to have arisen following changes in the morphology of inflorescence bracts and perianth segments (Uhl and Moore, 1980), a point that could be further explored within a phylogenetic framework.

7.5 Conclusion

The present chapter highlights the outstanding diversity of the palm androecium compared to other monocots, opening the way to various future studies. As mentioned above, the observed pattern of variation across the phylogeny raises questions about the ecological and evolutionary significance of the variation in stamen number. It also raises the question of the molecular processes that underlie such variation. The molecular basis of floral development has been rather thoroughly investigated across angiosperms including palms over the last 20 years (Adam et al., 2007a, b). Although relatively few studies have focused on the molecular basis of floral organ variation, the *SUPERMAN* gene has been shown to cause an increase in stamen number in *A. thaliana* and *Petunia hybrida* when mutated (Bowman et al., 1992; Nakagawa et al. 2004), and mutated forms of the *FON* gene (Floral Organ Number) have led to an increase in stamen number in conjunction with an increase in floral meristem size in rice (Suzaki, 1991; Nagasawa et al. 1996; Suzaki, 2006). These studies provide excellent candidate genes for further exploration of the molecular mechanisms underlying the extraordinary variation in floral apex expansion and stamen number in palms.

Acknowledgements

We are gratefully indebted to Peter Endress and John Dransfield for carefully reviewing this manuscript and suggesting valuable improvements. The Aarhus University Research Foundation is acknowledged for funding SN as guest researcher at Aarhus University, Denmark.

7.6 References

Adam, H., Jouannic, S., Morcillo, F. et al. (2007a). Determination of flower structure in *Eleais guineensis*: do palms use the same homeotic genes as other species? *Annals of Botany*, **100**, 1–12.

Adam, H., Jouannic, S., Orieux, Y. et al. (2007b). Functional characterization of MADS box genes involved in the determination of oil palm flower structure. *Journal of Experimental Botany*, **58**, 1245–1259.

Asmussen, C. B. and Chase, M. W. (2001). Coding and noncoding plastid DNA in palm family systematics. *American Journal of Botany*, **88**, 1103–1117.

Asmussen, C. B., Dransfield, J., Deickmann, V. et al. (2006). A new subfamily classification of the palm family (Arecaceae): evidence from plastid DNA phylogeny. *Botanical Journal of the Linnean Society*, **151**, 15–38.

Baker, W. J., Savolainen, V., Asmussen-Lange, C. B. et al. (2009). Complete generic-level phylogenetic analyses of Palms (Arecaceae) with comparisons of supertree and supermatrix approaches. *Systematic Biology*, **58**, 240–256.

Baker, W. J., Zona, S., Heatubun, C. D. et al. (2006). *Dransfieldia* (Arecaceae) – a new palm genus from western New Guinea. *Systematic Botany*, **31**, 61–69.

Barfod, A. S. and Uhl, N. W. (2001). Floral development in *Aphandra* (Arecaceae). *American Journal of Botany*, **88**, 185–195.

Barfod, A., Burholt, T. and Borchsenius, F. (2003). Contrasting pollination modes in three species of *Licuala* (Arecaceae: Coryphoideae). *Telopea*, **10**, 207–223.

Borchsenius, F. (1993). Flowering biology and insect visitors of the three Ecuadorean *Aiphanes* species. *Principes*, **37**, 139–150.

Bovi, M. L. A. and Cardoso, M. (1986). Biologia floral do açaizeiro (*Euterpe oleracea* Mart.) in *Congresso da Sociedade Botânica de Sao Paulo*, **6**, Campinas 1986.

Bowman, J. L., Sakai, H., Jack, T. et al. (1992). SUPERMAN, a regulator of floral homeotic genes in *Arabidopsis*. *Development*, **114**, 599–615.

D'Arcy, W.G. (1996).Anthers and stamens and what they do. pp. 1–24 in D 'Arcy, W. G. and Keating, R. C. (eds.), *The Anther – Form, Function and Phylogeny*, Cambridge: Cambridge University Press.

Dahlgren, R. M. T., Clifford, H. T. and Yeo, P. F. (1985). *The Families of the Monocotyledons. Structure, Evolution, and Taxonomy*. Berlin: Springer-Verlag.

Dransfield, J., Beentje, H., Tebbs, M. et al. (1995). *The Palms of Madagascar*. Kew: Royal Botanic Gardens, Kew.

Dransfield, J., Rakotoarinivo M., Baker, W. J. et al. (2008b). A new Coryphoid palm genus from Madagascar. *Botanical Journal of the Linnean Society*, **156**, 79–91.

Dransfield, J., Uhl, N. W., Asmussen, C. B., Baker, W. J., Harley M. M. and Lewis C. E. (2005). A new phylogenetic classification of the palm family, Arecaceae. *Kew Bulletin*, **60**, 559–569.

Dransfield, J., Uhl, N. W., Asmussen, C. B. et al. (2008a). *Genera Palmarum – The Evolution and Classification of Palms*. Royal Botanic Gardens Kew: Kew Publishing.

Endress, P. K. and Doyle, J. A. (2007). Floral phyllotaxis in basal angiosperms: development and evolution. *Current Opinion in Plant Biology*, **10**, 52–57.

Endress, P. K. and Doyle, J. A. (2009). Reconstructing the ancestral angiosperm flower and its initial specializations. *American Journal of Botany*, **96**, 22–66.

Govaerts, R. and Dransfield, J. (2005). *World Checklist of Palms*. Kew: Royal Botanic Gardens Press.

Guindon, S. and Gascuel, O. (2003). A simple, fast and accurate algorithm to estimate large phylogenies by maximum likelihood. *Systematic Biology*, **52**, 696–704.

Guindon, S., Lethiec, F., Duroux, P. and Gascuel, O. (2005). PHYML Online – a web server for fast maximum likelihood-based phylogenetic inference. *Nucleic Acids Research*, **33**(web server issue), W557-W559.

Hahn, W. J. (2002). A phylogenetic analysis of the Arecoid Line of palms based on plastid DNA sequence data. *Molecular Phylogenetics and Evolution*, **189**, 189–204.

Harley, M. M. (1999). Palm pollen: overview and examples of taxonomic value at species level. pp. 95–120 in Henderson, A. and Borchsenius, F. (eds.), *Evolution, Variation and Classification of Palms*, Vol. 83. Memoirs of the New York Botanical Garden, New York: New York Botanical Garden.

Harley, M. M. and Baker, W. J. (2001). Pollen aperture morphology in Arecaceae: application within phylogenetic analyses, and a summary of the fossil record of palm-like pollen. *Grana*, **40**, 45–77.

Huelsenbeck, J. P and Ronquist, F. (2001). Mr Bayes: Bayesian inference of phylogeny. *Bioinformatics*, **17**, 754–755.

Kuchmeister, H., Silberbauer-Gottsberger, I. and Gottsberger, G. (1997). Flowering, pollination, nectar standing crop, and nectaries of *Euterpe precatoria*, an Amazonian rain forest palm. *Plant Systematics and Evolution*, **206**, 71–97.

Lewis, C. E. and Doyle, J. J. (2002). A phylogenetic analysis of tribe Areceae (Arecaceae) using two low-copy nuclear genes. *Plant Systematics and Evolution*, **236**, 1–17.

Lewis, C. E. and Zona, S. (2008). *Leucothrinax morrisii*, a new name for a familiar Caribbean palm. *Palms*, **52**, 84–88.

Listabarth, C. (1992). A survey of pollination strategies in the Bactridinae. *Bulletin de l'Institut Français d'Etudes Andines*, **21**, 699–714.

Maddison, D. R. and Maddison, W. P. (2001). *MacClade 4: Analysis of phylogeny and character evolution.* Version 4.02. Sinauer Associates, Sunderland, MA.

Maddison, D. R., Schulz, K.-S. and Maddison W. P. (2007). The Tree of Life Web Project. *Zootaxa*, **1668**, 19–40.

Maddison, W. P. and Maddison, D. R. (2009). Mesquite: a modular system for evolutionary analysis. Version 2.7 [http://mesquite.project.org].

McCombie, W. R., Heiner, C., Kelley, J. M., Fitzgerald, M. G. and Cocayne J. D. (1992). Rapid and reliable fluorescent cycle sequencing of double-stranded templates. *DNA Sequencing and Mapping*, **2**, 289–296.

Nagasawa, N., Miyoshi, M., Kitano, H. et al. (1996). Mutations associated with floral organ number in rice. *Planta*, **198**, 627–633.

Nakagawa, H., Ferrario, S., Gerco, C. et al. (2004). The petunia ortholog of *Arabidopsis SUPERMAN* plays a distinct role in floral organ morphogenesis. *The Plant Cell*, **16**, 920–932.

Pagel, M. (1999). The maximum likelihood approach to reconstructing ancestral character states of discrete characters on phylogenies. *Systematic Biology*, **48**, 612–622.

Pagel, M., Meade, A. and Barker, D. (2004). Bayesian estimation of ancestral character states on phylogenies. *Systematic Biology*, **53**, 673–684.

Reis, M., Guimaraes, E. and Oliveira, G. (1993). Estudos preliminares da biologia reprodutiva do palmiteiro (*Euterpe edulis*) em mata residual do Estado de Sao Paulo. *Anais 70 Congreso Florestal Brasileiro*, 358–360.

Ronaghi, M. (2001). Pyrosequencing sheds light on DNA sequencing. *Genome Research*, **11**, 3–11.

Ronquist, F. and Huelsenbeck, J. P. (2003). MrBayes 3: Bayesian phylogenetic inference under mixed models. *Bioinformatics*, **19**, 1572–1574.

Rudall, P. J., Abranson, K., Dransfield, J. and Baker, W. (2003). Floral anatomy in *Dypsis* (Arecaceae-Areceae): a case of complex synorganization and stamen reduction. *Botanical Journal of the Linnean Society*, **143**, 115–133.

Saiki, R. K., Gelfand, D. H., Stoffel, S. et al. (1988). Primer-directed enzymatic amplification of DNA with a thermostable DNA polymerase. *Science*, **239**, 486–491.

Sannier, J., Baker, W. J., Anstett, M. C. and Nadot, S. (2009). A comparative analysis of pollinator type and pollen ornamentation in the Araceae and the Arecaceae, two unrelated families of the monocots. *BMC Research Notes*, **2**:145 (online publication).

Sannier, J., Nadot, S., Forchioni, A., Harley, M. M. and Albert, B. (2006). Variation in the microsporogenesis of monosulcate palm pollen. *Botanical Journal of the Linnean Society*, **151**, 93–102.

Stevens, P. F. (2001 onwards). Angiosperm Phylogeny Website. Version 9, June 2008. http://www.mobot.org/MOBOT/research/APweb/.

Suzaki, T., Sato, M., Ashikari, M. et al. (1991). The gene *FLORAL ORGAN NUMBER1* regulates floral meristem size in rice and encodes a leucine-rich repeat receptor kinase orthologous to *Arabidopsis CLAVATA1. Development*, **131**, 5649–5657.

Suzaki, T., Toriba, T., Fujimoto, M. et al. (2006). Conservation and diversification of meristem maintenance mechanism in *Oryza sativa*: Function of the *FLORAL ORGAN NUMBER2* Gene. *Plant Cell Physiology*, **47**, 1591–1602.

Uhl, N. W. (1976). Developmental studies in *Ptychosperma* (Palmae). II The staminate and pistillate flowers. *American Journal of Botany*, **63**, 97–109.

Uhl, N. W. and Moore Jr, H. E. (1977). Centrifugal stamen initiation in phytelephantoid palms. *American Journal of Botany*, **64**, 1152–1161.

Uhl, N. W. and Moore Jr, H. E. (1980). Androecial development in six polyandrous genera representing five major groups of palms. *Annals of Botany*, **45**, 57–75.

Uhl, N.W. and Dransfield, J. (1984). Development of the inflorescence, androecium and gynoecium with reference to palms. pp. 397–449 in White, R. A. and Dickinson, W. C. (eds.), *Contemporary Problems in Plant Anatomy*. New York: Academic Press.

8

Comparative floral structure and development of Nitrariaceae (Sapindales) and systematic implications

JULIEN B. BACHELIER, PETER K. ENDRESS AND
LOUIS P. RONSE DE CRAENE

8.1 Introduction

For the last 20 years, the development and improvement of molecular methods, based mostly on the comparison of DNA sequences, have been increasingly successful in reconstructing the phylogenetic tree of plants at all hierarchical levels. Consequently, they have contributed greatly to the recent improvement of angiosperm systematics. In addition, they have shown that earlier classifications, based mostly on plant vegetative and reproductive structures, had sometimes been misled by homoplastic characters, and a number of orders and families have had to be newly circumscribed or even newly established (e.g. APG, 1998, 2003, 2009; Stevens, 2001 onwards). These new results provide a novel basis for comparative structural studies to characterize the newly recognized clades and to evaluate clades that have only limited molecular support. However, because such comparative studies are time-consuming and the systematic classification has different hierarchical levels, they can only be done in a stepwise fashion (for eudicots, e.g. Matthews and Endress, 2002, 2004, 2005a, b, 2006, 2008; von Balthazar et al., 2004; Schönenberger and Grenhagen, 2005; Endress and Matthews, 2006; Ronse De

Flowers on the Tree of Life, ed. Livia Wanntorp and Louis P. Ronse De Craene. Published by Cambridge University Press. © The Systematics Association 2011.

Craene and Haston, 2006; von Balthazar et al., 2006; Bachelier and Endress, 2008, 2009; Janka et al., 2008; Schönenberger, 2009; von Balthazar and Schönenberger, 2009; Schönenberger et al., 2010).

As part of such a comparative approach, we studied the floral structure of Nitrariaceae, a small family which has been recently reclassified in Sapindales (APG, 2009). Nitrariaceae comprise four genera and around 15 species (Stevens, 2001 onwards; APG, 2009). They are native to arid and semi-arid regions of the Old World and are small to medium-sized shrubs (*Nitraria*; Engler, 1896a, b; Bobrov, 1965; Noble and Whalley, 1978), perennial herbs (*Peganum* and *Malacocarpus*; Engler, 1896a, 1931; El Hadidi, 1975) or small annual herbs of only a few centimetres height (*Tetradiclis*; Engler, 1896b, 1931; Hamzaoğlu et al., 2005). In earlier classifications, the position and the mutual affinities of these genera varied tremendously, depending on the weight an author gave either to their vegetative or their reproductive features (Takhtajan, 1969, 1980, 1983, 2009; El Hadidi, 1975; Dahlgren, 1980; Cronquist, 1981, 1988; see Sheahan and Chase, 1996 for a detailed review of classifications). Because none of the traditional classifications was entirely satisfactory, however, most authors followed Engler's influential work (1896a, b, 1931) and *Nitraria*, *Tetradiclis* and *Peganum* (including *Malacocarpus*) remained for a long time in their own sub-families in Zygophyllaceae (Nitrarioideae, Tetradiclidoideae and Peganoideae; for more details of the history of classification, see Sheahan and Chase, 1996).

Molecular phylogenetic studies have shown that the earlier Zygophyllaceae *sensu* Engler (1896a, b, 1931) comprised a cluster of unrelated genera. They first showed that Nitrariaceae were more closely related to Sapindales of malvids than to the remainder of Zygophyllaceae, which are now in Zygophyllales of fabids (Fig 8.1A; Fernando et al, 1995; Gadek et al., 1996; Sheahan and Chase, 1996; Bakker et al., 1998; Savolainen et al., 2000; Muellner et al., 2007; Wang et al., 2009; Worberg et al., 2009). These studies also showed that *Nitraria*, *Peganum*, *Malacocarpus* and *Tetradiclis* consistently form a clade, in which *Nitraria* is always sister to the other three, and *Tetradiclis* is sister to *Peganum* + *Malacocarpus* (Fig 8.1B; Sheahan and Chase, 1996; Savolainen et al., 2000; Muellner et al., 2007). However, the position of the Nitrariaceae with regard to the other families of Sapindales is not well resolved yet, and depends on the sampling size of the outgroup and the ingroup, as well as on the DNA regions used. For instance, if *Nitraria* and *Peganum* are taken separately, they appear as sister to the remainder of Sapindales in Fernando and Quinn (1995) and Worberg et al. (2009), but *Nitraria* appears as sister to the clade of Anacardiaceae and Burseraceae in Wang et al. (2009). If *Nitraria* and *Peganum* (including *Malacocarpus*) are taken together, they appear as sister to the remainder of Sapindales (Gadek et al., 1996; Bakker et al., 1998). In contrast, when *Tetradiclis* is added, Nitrariaceae are nested within Sapindales, but they are either sister to

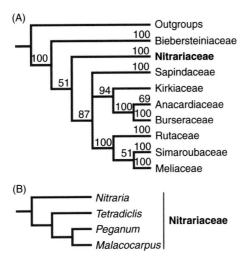

Fig 8.1 (A) Phylogeny of Sapindales, based on *rbc*L sequences (Bayesian posterior probabilities indicated above the branches; simplified from Muellner et al., 2007). (B) Phylogeny of Nitrariaceae, based on several studies using molecular and non-molecular data (Sheahan and Chase, 1996; Savolainen et al., 2000; Muellner et al., 2007).

the clade of Anacardiaceae + Burseraceae (Savolainen et al., 2000), or appear as the second basal most sapindalean family (Muellner et al., 2007).

It is now clear that the similarities in vegetative gross morphology shared by the earlier Zygophyllaceae have only limited phylogenetic significance and are just adaptive convergences due to their similar extreme habitats. Earlier comparative studies also indicated that these four genera, especially *Nitraria* and also *Peganum* (with *Malacocarpus*), were more distinct from the remainder of Zygophyllaceae than previously suggested (for vegetative anatomy, see Sheahan and Chase, 1993; for palynology, see Erdtman, 1952; Agababyan, 1964; Perveen and Qaiser, 2006; for embryology, see Mauritzon, 1934; Souèges, 1953; Kapil and Ahluwalia, 1963; Kamelina, 1985, 1994; Li and Tu, 1994; for floral structure, see Baillon, 1873; Nair and Nathawat, 1958; Ronse De Craene and Smets, 1991; Ronse De Craene et al., 1996; for chemistry, see Hussein et al., 2009). In contrast to most of the current Zygophyllaceae, which have pentamerous, isomerous and obdiplostemonous flowers, the flowers in *Tetradiclis* are tetramerous, isomerous and haplostemonous, while in *Nitraria* and *Peganum* (+ *Malacocarpus*), the flowers have a pentamerous perianth and androecium, conspicuous antepetalous stamen pairs and a tricarpellate gynoecium (Payer, 1857; Baillon, 1873; Engler, 1896a, b, 1931; Ronse De Craene and Smets, 1991; Ronse De Craene et al., 1996). Nair and Nathawat (1958) mentioned that *Peganum* and *Nitraria* have a similar vascular anatomy and a distinct androecium, with five groups of three antesepalous stamens, but some comparative developmental studies of the unusual androecium of *Nitraria* and *Peganum* suggested that the pairs were not homologous (Payer, 1857; Ronse De Craene and Smets, 1991; Ronse De Craene et al., 1996). In *Nitraria*, each pair belongs to a triplet of antesepalous stamens and is thus derived from a haplostemonous pattern (Payer, 1857; Ronse De Craene and Smets, 1991) whereas, in *Peganum*, each pair has been interpreted to be formed by duplication of the antepetalous stamen primordium and, therefore, derived from an obdiplostemonous pattern (Payer, 1857; Eckert, 1966; Ronse De Craene and Smets, 1996; Ronse De Craene et al., 1996).

Except for the opinion that *Nitraria* and *Peganum* + *Malacocarpus* may not belong to the Zygophyllaceae, it has neither been suggested that these genera are closely related, nor that they would be part of the Sapindales (except for Cronquist, 1981, who, in any case, placed the entire Zygophyllaceae in Sapindales). The new light shed on the relationships between these genera and their affinities by molecular studies provides new ground to perform a comparative study of their flower structure and development. This enables us to re-evaluate the unusual androecium of *Nitraria* and *Peganum* and their affinities with *Tetradiclis*, as well as with other sapindalean families.

8.2 Material and methods

The following material has been studied:

> *Nitraria retusa*: P. Endress, s.n., 07 May 1971, Algeria; L. Ronse De Craene 43LS, Senegal; 302LT, Tunisia.
>
> *Peganum harmala*: P. Endress, s.n., 01 May 1971, Algeria.
>
> *Tetradiclis tenella:* G. Woronow and P. Popow, s.n., 18 May 1971, Azerbaijan [Z-ZT]; J. Bornmüller 640, Iraq [Z-ZT]; P. Sintenis 1537, Turkmenistan [Z-ZT].

The material was studied using light microscopy (LM) and scanning electron microscopy (SEM). Fresh flowers of *Nitraria* and *Peganum* were collected and stored in 70% ethanol, while material of *Tetradiclis* was taken from herbarium collections. The dry flower buds and anthetic flowers of *Tetradiclis* were soaked in dioctyl sodium sulfosuccinate solution for three days and stored in 70% ethanol.

For LM investigations, part of the material of *Peganum* and *Nitraria* was prepared according to the standard paraplast embedding protocol, whereas the material of *Tetradiclis* plus additional material of *Peganum* and *Nitraria* were embedded in plastic using Kulzer's Technovit 7100 (2-hydroxyethyl methacrylate) following a protocol adapted from Igersheim (1993) and Igersheim and Cichocki (1996). Serial microtome sections were made at 5, 7, 10 or 15 µm. Sections in paraplast were stained with safranin and Astrablue and mounted in Eukitt, whereas sections in plastic were stained with ruthenium red and toluidine blue and mounted in Histomount (protocol adapted from Weber and Igersheim, 1994).

For SEM investigations, specimens were stained with 2% osmium tetroxide, dehydrated in ethanol and acetone, critical-point dried and sputter-coated with gold, and studied at 20 kV with a Hitachi S-4000 scanning electron microscope. The liquid-fixed material and the permanent slides of serial microtome sections are deposited at the Institute of Systematic Botany of the University of Zurich (Z).

8.2.1 General comments on illustrations

Figures 8.2 to 8.8 represent structural analyses of the genera studied. In all drawings, the morphological surfaces are drawn with thick continuous lines; pollen tube transmitting tracts are shaded dark grey (only in anthetic gynoecia); nectaries are shaded light grey. In schematic longitudinal sections of gynoecia, (A) shows a median section of two carpels projected onto the drawing plane because of the odd number of carpels, while (A′) shows a longitudinal section through the middle of one carpel and the area between the two other carpels. Outlines of parts outside the median plane are drawn with thick broken lines; postgenitally fused surfaces are hatched; embryo sacs are drawn with thin continuous lines. In transverse microtome section series, sections are ordered from top, downwards; postgenitally fused surfaces and pollen sacs are drawn with thick broken lines; vascular bundles are drawn with thin continuous lines.

8.2.2 Glossary

Angiospermy type 3 Carpels closed by postgenital fusion at the entire periphery, but with an open canal in the inner angle of the ventral slit (Endress and Igersheim, 2000).

Angiospermy type 4 Carpels completely closed by postgenital fusion (Endress and Igersheim, 2000).

Antitropous Ovule, in which the direction of curvature is opposite to the direction of carpel closure (Fig 8.3B, see also Endress, 1994), more or less corresponding to epitropous (*sensu* Agardh, 1858).

PTTT Pollen tube transmitting tract.

Syntropous Ovule, in which the direction of curvature is the same as the direction of carpel closure (Fig 8.3A, see also Endress, 1994), more or less corresponding to apotropous (*sensu* Agardh, 1858).

8.3 Results

8.3.1 *Peganum harmala*

Morphology

The flowers are relatively large (2 to 4 cm across), morphologically bisexual and entomophilous. The floral organ whorls are pentamerous, except for the trimerous gynoecium (Figs 8.2–8.5). The androecium is seemingly two-whorled and obdiplostemonous; the antepetalous stamen whorl gives the impression of being the outer one and the antesepalous the inner one (Figs 8.2B–D, 8.5). In addition, the androecium is unusual because the outer whorl comprises five stamen pairs instead of five single stamens (Figs 8.2B–D, 8.5).

The free sepals are long and narrow and have one or two (rarely more) small basal lateral appendages (Figs 8.2, 8.4A). In bud, their aestivation varies during development (Figs 8.2, 8.4A; see Section 8.3.1, Development). The petals are elliptical and mucronate. They have a pronounced median dorsal rib (Figs 8.2A–D, 8.4B), and rarely two or three tips. Their aestivation is somewhat open at the base, and further up, it is contort. Postgenital coherence is formed between their overlapping parts by interdentation of their papillate epidermis and cuticular ornamentation, and by interlocking glandular hairs on the margins of the petal tips (Fig 8.4A). At anthesis, however, the perianth is wide open and the inner reproductive organs are entirely exposed.

All stamens are free and similar in shape (Fig 8.2B–D). The filament bases are broad and thick with conspicuously flattened margins (Fig 8.4C, D). They overlap slightly with each other and form a collar around the ovary. Further up, the filaments become narrower and more rounded, and the transition from filament to anther is constricted (Fig 8.4C, D). The anthers are sagittate and (slightly dorsally) basifixed (Fig 8.4C). The dorsal side of the thecae is slightly larger and much longer than the ventral one, and the anthers are thus slightly introrse and apiculate (Figs 8.2B, 8.4C, E). Each theca has a longitudinal dehiscence line, which encompasses its upper and lower shoulders (Fig 8.4C). Between the stamens and gynoecium base there is a lobed nectary disc with five conspicuous antepetalous depressions (Figs 8.2E, 8.3O).

The gynoecium is syncarpous and polysymmetric and has a superior ovary (Figs 8.2B–D, 8.3, 8.4F). However, because of the difference in merism of gynoecium and the outer whorls of organs, the flower is in some way monosymmetric (Fig 8.2). The gynoecium has a short and stout gynophore, a globose ovary with conspicuous dorsal bulges surrounding and hiding the base of a long and slender style (Figs 8.3A, 8.4F, G), which is narrower and more rounded at the base than further up (Fig 8.3A–G) and becomes triangular at mid-length (Figs 8.3B–D, 8.4F, H). The upper part of the gynoecium is apocarpous for a long distance, but with

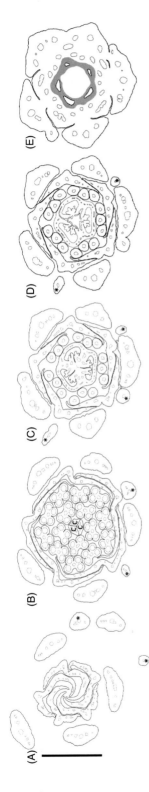

Fig 8.2 *Peganum harmala*. Floral bud, transverse section series. Sepal basal lateral appendages marked with asterisks [*]. (A) Distal zone, five sepals and five petals with contort aestivation. (B) Fifteen stamens surrounding three carpel tips [c], free but postgenitally united. (C) Stamen filament bases in five antesepalous triplets, transition from symplicate to synascidiate zone of the ovary at the level of the placentae. (D) Synascidiate zone of the ovary. (E) Floral base, with five antesepalous nectariferous depressions. Scale bar: (A)–(E) = 500 μm.

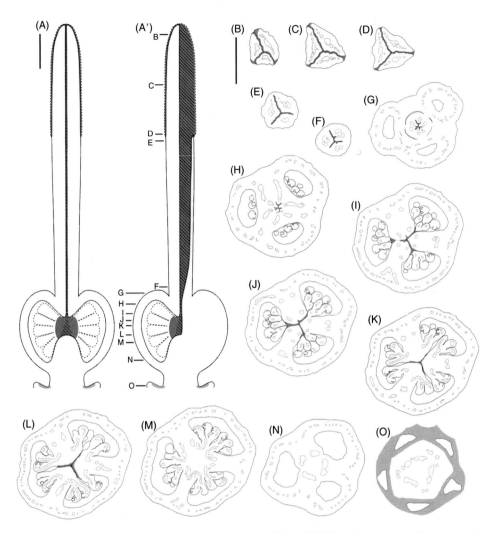

Fig 8.3 *Peganum harmala*. Anthetic gynoecium. (A) and (A′) (see Section 8.2.1) schematic median longitudinal Section. (B)–(O) Transverse section series. (B)–(D) Apocarpous zone, carpel tips postgenitally united, stigmas decurrent along carpel margins. (E)–(L) Symplicate zone. (E)–(G) Style. (G–H) Apical septum. (K)–(M) Transition between symplicate and synascidiate zone, with axile lateral placentae protruding into the locules and bearing many ovules. (M), (N) Synascidiate zone. (O) Below the ovary, five antepetalous depressions surrounded by nectariferous tissue. Scale bars: = 1 mm.

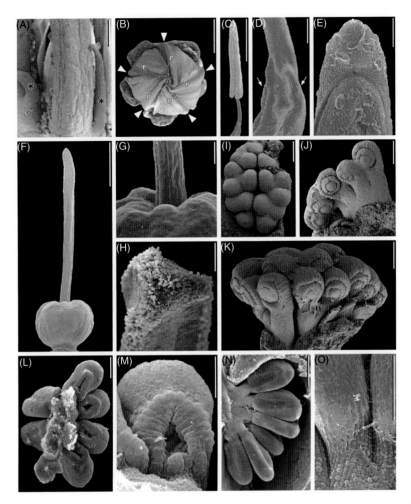

Fig 8.4 *Peganum harmala*. Floral organs. (A) Floral bud, lateral view. Sepal bases marked with [S] and basal lateral appendages with an asterisk [*], (B) Floral bud, older than (A), sepals removed, from above. Petal contort aestivation, with multicellular hairs on tips. Petals [P], dorsal median ridge indicated by white arrowheads. (C)–(E) Stamen. White arrows point to flattened margins of the filament. (C) Stamen, ventral view. (D) Close-up of filament base. (E) Close-up of apical beak. (F)–(H) Gynoecium. (F) Lateral view. (G) Close-up of style base and ovary dorsal bulges. (H) Close-up of three-angled stigmatic head from above, with three papillate stigmatic crests decurrent on edges. (I)–(O) Ovules. (I) Early stage of development, two collateral placentae, each with two vertical rows of ovule primordia. (J) Later in development, young ovules on branched placenta, with circular inner integument and hood-shaped outer integument. (K) Two collateral placentae bearing slightly older ovules. (L) Anthetic ovules borne on a single placenta. (M) Close-up, micropyle, with loosely closed exostome. (N) Anthetic placenta still in position, with smooth PTTT and dorsal sides of ovules plus funicles. (O) Close-up, papillate dorsal side of funicle covered with secretion. Scale bars: (C), (F) = 3 mm; (B) = 1 mm; (G), (H) = 500 μm; (A), (D), (E) = 300 μm; (L), (N) = 200 μm; (J), (K), (M), (O) = 100 μm, (I) = 50 μm.

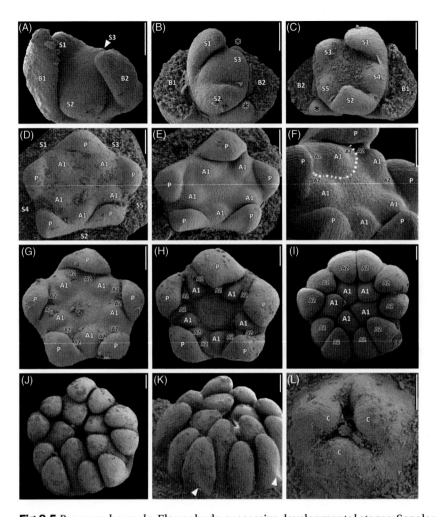

Fig 8.5 *Peganum harmala*. Flower buds, successive developmental stages. Sepals marked with S, petals with P, antesepalous stamens with A1, antepetalous stamens with A2, carpels with C. Orientation of buds with sepal S2 in median lower position. Prophyll and sepal initiation is counterclockwise in (A), (B) and (D), and clockwise in (C). (A) Two lateral prophylls [B1 and B2], and first and third sepals [S1 and S3] in anterior position, and second sepal [S2] posterior. (B) Prophylls B1 and B2 removed, basal appendages on prophylls marked with an asterisk [*]. (C) Initiation of lateral sepals [S4 and S5]. (D)–(H) Sepals removed. (D) Pentagonal inner part of young flower with five petal at the angles [P], and five antesepalous stamen primordia [A1]. (E), (F) Slightly older stage. Appearance of smaller antepetalous pairs of stamen primordia between the antesepalous ones [A2]. (E) From above. (F) Close-up, dotted line around a triplet of stamen primordia. (G) All stamens initiated, showing the differential development of A2 pairs relative to the neighbouring A1. (H) Syncarpous gynoecium primordium rising in the centre, compressed by the development of surrounding triplets of stamens. (I)–(K) Sepals and petals removed, showing differential growth of A2 pairs affected by the development of closest A1, and by the petals (white arrowheads). (L) Sepals, petals and stamens removed, syncarpous gynoecium with three carpels [C]. Scale bars: (A)–(L) = 50 μm.

the carpels postgenitally united; the carpels are flat in this zone (Figs 8.3A–D, 8.4H). Along the carpel margins stigmatic tissue is decurrent for a long distance as three wavy lines along the style (Figs 8.3A–D, 8.4F, H). Each of these three lines represents a partial external compitum (Figs 8.3A–D, 8.4H). The stigmatic lines exhibit unicellular, slightly secretory papillae (Figs 8.4H, 8.9A). The gynoecium is of angiospermy type 4 (Fig 8.3A–L). Below the stigmatic lines the carpels are congenitally united. The synascidiate zone extends up to the mid-length of the ovary (Fig 8.3A, M, N) and the symplicate zone up to mid-length of the style. The bulged zone of the ovary contains an apical septum (Fig 8.3A, E–L). Below the stigmas, the PTTTs differentiate along the ventral side of the carpels, and they are one cell layer thick and secretory (Figs 8.3A–L, 8.9A, B). They form an internal compitum, which is continuous with the external one and extends down to the transition zone between the symplicate and synascidiate zone of the ovary (Fig 8.3A–L). There, the carpel ventral sides are only loosely postgenitally united and a placenta differentiates on each of their margins (Fig 8.3A, K–M). There are two axile lateral placentae which develop into each locule, and each placenta bears *c.* 10 ovules (Figs 8.3J–M, 8.4I–N). The ovules are densely packed, but do not entirely fill the locule at anthesis (Fig 8.4N). Each ovule is anatropous, syntropous, crassinucellar and bitegmic (Figs 8.3J–M, 8.4J–M). The inner integument is 2–3 cell layers thick and is shorter than the outer one, which is 3–4 cell layers thick. The outer integument is hood-shaped, whereas the inner integument entirely surrounds the nucellus. Both integuments are thickened above the nucellus and form a micropyle, which faces the base of the funicle and comprises a circular endostome hidden by an exostome, which is not completely closed on the adaxial side (Fig 8.4L, M). At anthesis, the dorsal side of the funicle is papillate and covered with secretion (Fig 8.4N, O).

Development

Each flower develops in the axil of a subtending bract (pherophyll) and has two lateral prophylls which appear successively, opposite each other (Fig 8.5A–C). They are quickly followed by a first sepal primordium, which develops close to the larger (first) prophyll, and four additional sepal primordia which appear in a regular (clockwise or anticlockwise) spiral sequence (Fig 8.5A–C). The free sepals become valvate to slightly imbricate further up, depending on whether or not lateral basal appendages are present. When all sepals are initiated the floral apex becomes almost pentagonal, while five petal primordia appear more or less simultaneously, alternating with the sepals (see Figs 3–4 in Ronse De Craene et al., 1996). Initiation of the androecium quickly follows that of the corolla. The first whorl of five stamens alternates with the petals (Fig 8.5D–E, see Fig 6 in Ronse De Craene et al., 1996). Shortly thereafter, ten smaller stamen primordia appear more or less simultaneously in pairs between the antesepalous stamens (Fig 8.5E, F). However, separation soon appears to be deeper between the stamens of each antepetalous pair than between the pair and the adjacent antesepalous stamens (Fig 8.5G). In addition, the difference in size and position between the stamen primordia of the two apparent whorls, as well as between the two stamens in each antepetalous position, strongly suggest that the androecium is primarily haplostemonous and that the antesepalous triplets develop centrifugally, with a first median and inner (or upper) stamen and two lateral and slightly outer (or lower) stamens (Figs 8.2C, D, 8.5F, G). In the floral centre, the syncarpous gynoecium first appears as a rounded bulge (Fig 8.5E), but while the surrounding pentamerous androecium enlarges, sometimes it becomes slightly pentagonal (Fig 8.5H; see also Fig 10 in Ronse De Craene et al., 1996). However, only three carpels clearly develop to anthesis (Fig 8.5L). The gynoecium was trimerous in all the material studied here. Petal development is not delayed, and they become longer than the sepals in bud, and thus protect the inner floral organs up to anthesis (Fig 8.2). The contort petal aestivation affects the growth of the outer stamens, which may reach different sizes at mid-development (Fig 8.5I–K). The outer antepetalous stamens are at first shorter than the antesepalous stamens, but become longer before anthesis.

At anthesis, the sepals and petals are fully expanded. The petals are no longer postgenitally connected. Nectaries are situated between the androecium and gynoecium, where they form five depressions in the petal radii (Fig 8.2E). After anthesis, the sepals are persistent. The petals and stamens fall off. The long style dries up distally, but also persists for some time and no abscission zone can be seen at anthesis.

Anatomy and histology

Each sepal has three main vascular traces at the base, but further up, there can be up to seven branches formed by successive lateral branching (Fig 8.2A–E). Sometimes, a main vascular trace and additional smaller ones are also present in the teeth at the base of the largest sepals. Petals and stamens have a single trace (Fig 8.2A–E). In petals there may be up to 14 vascular bundles (Fig 8.2A–D). In carpels, the dorsal vasculature forms a more or less continuous band along the entire length of the gynoecium (Figs 8.2B–D, 8.3B–N). The lateral vascular bundles are conspicuous only in the ovary septa. They differentiate from the dorsal vasculature at the base of the style (Fig 8.3F, G) and extend downward in the postgenitally united septa (Fig 8.3I, J). At the transition between the symplicate and synascidiate zone, each lateral bundle connects to a placenta (Fig 8.3J–M). Each placental bundle appears to be 'inverted' (xylem peripheral and phloem central) and splits into branches, which supply the ovules (Fig 8.3K–M). Only below the placentae, in the synascidiate zone, do the lateral bundles of adjacent carpels merge and form synlateral bundles, arranged around the floral centre, and extend downwards into the floral base (Fig 8.3N, O).

On the ventral side of the sepal bases, there is a dense carpet of hairs with a uniseriate multicellular stalk and a round, multiseriate and multicellular head.

Smaller, but similar, hairs are also found on the petal tips. No tanniferous cells or special mucilage cells (for term, see Matthews and Endress, 2006) have been observed in buds or anthetic gynoecia. During development, the mesophyll becomes lacunar in all organs.

8.3.2 *Nitraria retusa*

Morphology

The flowers are relatively small (*c.* 1 cm across), morphologically bisexual and entomophilous. The floral organ whorls are pentamerous, except for the trimerous (sometimes dimerous) gynoecium (Figs 8.6–8.8). The androecium is seemingly two-whorled and obdiplostemonous; the antepetalous stamen whorl gives the impression of being the outer one and the antesepalous the inner one. However, the androecium is unusual because the outer whorl appears to comprise five stamen pairs instead of only five single stamens (Figs 8.6A–D, 8.8D, E).

A difference in size of the sepals indicates that they are initiated in a spiral sequence. They are free in early stages of development but later their expanded bases become united. Their free tips are acute and become contiguous. They are valvate or may even be slightly overlapping (Fig 8.8A). The petals are free for their entire length (Fig 8.6). They have a narrow base, but become larger further up, with their margins folded inwards around a pair of antepetalous stamens. Their acute tips are strongly bent inwards around the thecae of adjacent antesepalous stamens (Fig 8.6B). The petals are thus hood-shaped. Their aestivation is shortly open at the base and induplicate-valvate further up (Figs 8.6, 8.8B, C). The contiguous margins are covered with interlocking hairs and are postgenitally coherent (Figs 8.6B–D, 8.8C). In young buds, when the floral base and sepal bases enlarge, the aestivation of the sepal free tips becomes open and the corolla takes over the protection of the inner organs up to anthesis (Figs 8.6A–D, 8.8A–C). In older buds, however, the corolla is not fully closed in the floral centre and the carpel tips are exposed before the flower opens (Figs 8.6A, B, 8.8D). It is unclear whether these flowers are bisexual and strongly protogynous, or functionally female. At anthesis, the calyx is wide open, but not the corolla because the petals remain erect, and the inner reproductive organs remain partly hidden.

The stamens are free (Fig 8.8E). In our material the ten antepetalous stamens are shorter and smaller than the five antesepalous ones (Figs 8.6A–C, 8.8E). The filaments are broad and irregularly thickened at the base, but become narrower and rounder further up. The transition from filament to anther is constricted (Figs 8.6A–D, 8.8E–G). The anthers are X-shaped and (slightly dorsally) basifixed, and versatile at anthesis. The dorsal side of the thecae is slightly larger and longer than the ventral one, and the anthers are almost latrorse (Figs 8.6B, C, 8.8F, G). Each theca has a longitudinal dehiscence line, which encompasses its upper and lower

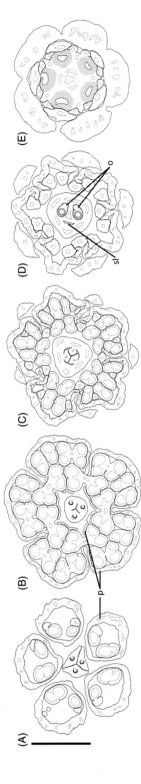

Fig 8.6 *Nitraria retusa*. Floral bud, transverse section series. (A) Distal zone, five petals [p] with valvate aestivation and tips bent inwards surrounding postgenitally united carpel tips [c]. (B) Fifteen stamens surrounding the style. (C) Ovary at the level of the placentae. (D) Ovary, with two fertile locules, each with one ovule [o], and one sterile, empty locule [sl]. (E) Floral base, with five antesepalous depressions (nectaries). Scale bars (A)–(E) = 1 mm.

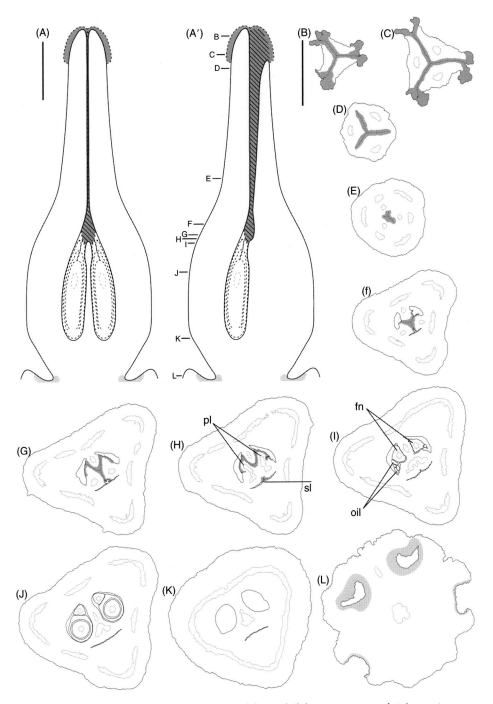

Fig 8.7 *Nitraria retusa*. Anthetic gynoecium. (A) and (A') (see section 8.2.1) Schematic median longitudinal section. (B)–(L) Transverse section series. (B), (C) Apocarpous zone, carpel tips postgenitally united, stigmas decurrent along carpel margins. (D)–(G) Symplicate zone, style. (H) Transition between symplicate and synascidiate zone, with two fertile locules, each having one apical axile lateral placenta [pl], and one sterile locule [sl]. (I)–(K) (Synascidiate zone.) (I) Showing funicles [fn] and contiguous outer integument lobes of the ovules [oil]. (L) Below the ovary, five (four visible) antepetalous depressions (nectaries). Scale bars: = 2 mm.

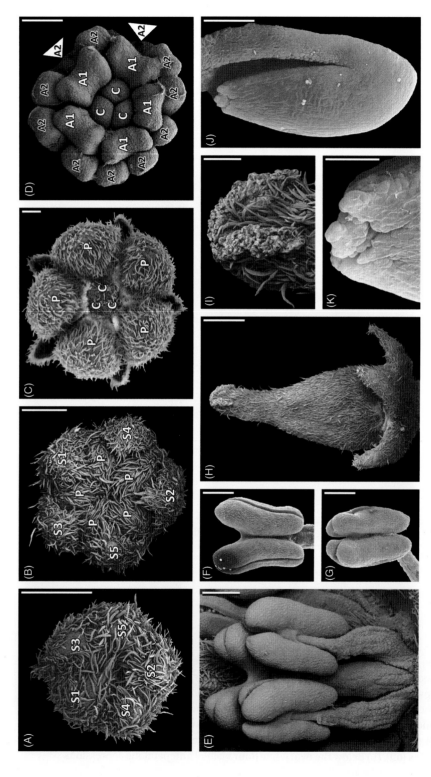

Fig 8.8 *Nitraria retusa.* Flower buds and floral organs. (A) Apical view of young closed bud showing the valvate sepal lobes (S1–S5). (B) Apical view of older bud showing separating sepals and valvate petals [P], carpels [c]. (C) Bud with expanding petals, carpels visible. (D) Sepals and petals removed. Fifteen stamens surrounding three carpel tips, positions of missing antepetalous stamens (one of them not developed, one removed) indicated with arrowheads. (E) Dorsal view of a stamen triplet. (F), (G) Ventral view of anther. (F) Antepetalous stamen. (G) Antesepalous stamen. (H) Lateral view of anthetic gynoecium, petals and stamens removed. (I) Lateral view of stigmatic head, with two stigmatic crests. (J) Ovule, lateral view. (K) Detail of micropyle. Scale bars: (H) = 2 mm; (D), (E), (F), (G) = 1 mm; (A), (B), (C), (I), (J) = 500 μm; (K) = 200 μm.

shoulders (Fig 8.8F, G). Between androecium and gynoecium in the petal radii there is a small lobed nectary disc with five depressions (Figs 8.6E, 8.7L).

The gynoecium is syncarpous, and commonly trimerous. Sometimes it is dimerous and then disymmetric, or one of the three carpels is sterile and the gynoecium appears monosymmetric (Figs 8.6A–D, 8.7, 8.8C, D). The superior ovary is cylindrical and the style is conical and slightly swollen distally. The uppermost part of the gynoecium is apocarpous for a short distance, but with the carpels postgenitally united; the carpels are flat in this zone. Along the carpel margins, stigmatic tissue is decurrent for a short distance as three wavy lines along the style forming a short stigmatic head; stigmatic tissue has unicellular, slightly secretory papillae (Figs 8.7A–C, 8.8H, I, 8.9C). These three lines form three partial external compita (Figs 8.7, 8.8H, I, 8.9C). The gynoecium is of angiospermy type 3 or 4 (Fig 8.7).

The syncarpous zone encompasses the ovary and a large part of the style. The ovary is synascidiate for most of its length (Fig 8.7A, H–L). In the symplicate zone in the style the inner morphological surfaces of the carpels are partly contiguous, but not postgenitally united, thus forming a narrow stylar canal (Figs 8.7D–F, 8.9D). Only in the lowermost part (in the ovary) are they postgenitally united (Fig 8.7G, H). Below the stigmas, the PTTTs differentiate along the ventral side of the carpels. They consist of the epidermis, which is papillate, and fills the stylar canal with secretion (Fig 8.9D); the papillae point downward (Fig 8.7A–H).

The internal compitum is continuous with the external one and extends down to the transition between the symplicate and the synascidiate zone of the ovary (Fig 8.7A–H). At the transition between the synascidiate and the symplicate zones, only one of the two margins of each carpel differentiates into an axile lateral placenta and bears a single ovule. Sometimes a locule is empty at anthesis (Fig 8.7A, G–J). The ovule is inserted apically. It is long and cylindrical and fills the locule at anthesis (Figs 8.7A, H–J, 8.8J). It is anatropous, syntropous, crassinucellar and bitegmic (Figs 8.7A, H–J, 8.8J). The inner integument is 2–3 cell layers thick and is shorter than the outer one, which is 3–4 cell layers thick. As the placenta is lateral, the ovule has a somewhat oblique position in the locule (Fig 8.7A, H–J). Although the ovule is anatropous, the outer integument is relatively well developed on the concave side of the ovule (Fig 8.8J). Above the nucellus the inner integument is thicker than the outer one. The endostome is slit-shaped, whereas the exostome is closed by irregular lobes (Figs 8.7A, 8.8J, K). Both integuments together form a long and S-shaped micropyle (Fig 8.7A).

Anatomy and histology

The sepals have three main vascular traces, and there are up to seven branches formed by successive lateral branching (Fig 8.6A–E). The petals and stamens have a single vascular trace (Fig 8.6A–E). In petals, up to ten vascular bundles may be present (Fig 8.6A–D). In carpels the median dorsal vasculature forms a more or less continuous band throughout the length of the gynoecium (Figs 8.6A–D, 8.7B–K). The lateral bundles form 'inverted' (xylem peripheral and phloem central) synlateral bundles in the septa of the ovary and each connects to a placenta (Fig 8.7E–H). In the synascidiate zone they become united in the floral centre, from where they extend downwards into the floral base (Figs 8.6E, 8.7I–L).

The perianth and gynoecium are covered by an indumentum of lignified unicellular hairs orientated towards the tips of the organs. Tanniferous cells are present in the hypodermis and mesophyll of almost all floral organs, including the ovule outer integument, and they are only lacking in stamens (Fig 8.9C, D, F, G). Special mucilage cells (for term, see Matthews and Endress, 2006) are present in all floral organs, in the stamen filaments and anther connectives, and in the gynoecium in the style and outer integument of the ovules (Fig 8.9F, G). They are lacking in the epidermis, but occur in the mesophyll (Fig 8.9F). They are differentiated successively from the outer to the inner organs during floral development, and from tip to base in petals and stamens.

8.3.3 *Tetradiclis tenella*

Morphology

Herbarium material does not allow for a description as extensive as for the other two species. The flowers are very small (less than 5 mm across), morphologically bisexual, tetramerous (also in the gynoecium), with a double perianth and haplostemonous (Fig 8.9H). The sepals appear free for most of their length; their aestivation is imbricate (decussate) (Eichler, 1878). The petals are free. In bud their aestivation is open at the base and imbricate further up. They become longer than the sepals during development and protect the inner organs up to anthesis. At anthesis, the perianth is wide open and the inner reproductive organs are entirely exposed.

The stamens are antesepalous and free. They have a broad filament base, which narrows further up. The transition from filament to anther is slightly constricted. The anthers are X-shaped, dorsally basifixed and likely versatile. The dorsal side of the thecae is slightly larger and longer than the ventral one. The anthers are slightly introrse. Each theca has a longitudinal dehiscence line, which encompasses its upper and lower shoulder. It is unclear whether a nectary is present between stamens and gynoecium.

The gynoecium is syncarpous and polysymmetric and has a superior ovary. The carpels are antepetalous (Fig 8.9H). The ovary is short and squared (Fig 8.9E, H, I). It gives the impression of being apocarpous because the carpel dorsal sides are strongly bulged and have deep longitudinal furrows running between them (Fig 8.9H, I). However, there is a single gynobasic, squared style (Fig 8.9E). Only the upper part of the style is apocarpous for a short distance, but the carpels are

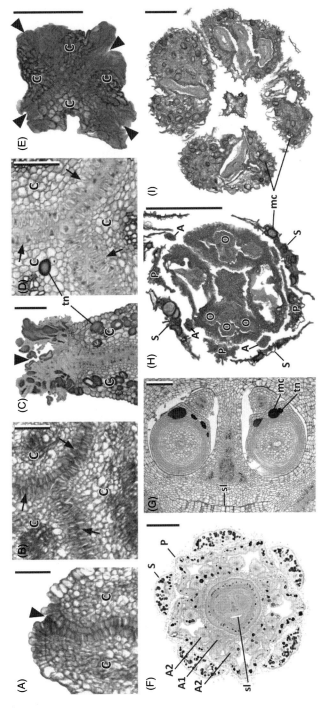

Fig 8.9 Nitrariaceae. Flower buds, and anthetic and postanthetic flowers, transverse sections. Signatures: Sepal [S], petal [P], antepalous stamen [A1], antesepalous stamen [A2], carpel [C], ovule [O], sterile locule [sl], tanniferous cells [tn] are blue, special mucilage cells [mc] are pink to red. Arrowheads point to stigma and partial compitum, arrows point to carpel morphological surfaces covered with one cell-layered PTTT. (A)–(E) Styles and stigmas. (A), (B) *Peganum harmala*. (C), (D) *Nitraria retusa*. (E) *Tetradiclis tenella*. (F), (G) *Nitraria retusa*. (F) Preanthetic floral bud. (G) Anthetic gynoecium, with two locules with a single ovule and a sterile locule lacking an ovule. (H), (I) *Tetradiclis tenella*. (H) Anthetic flower. (I) Postanthetic flower. Scale bars: (F), (G), (H), (I) = 250 μm; (A), (B), (C), (D) = 100 μm; E = 50 μm. For colour illustration see plate section.

postgenitally connected in this zone and form a stigmatic head. Here the carpels are flat and their margins (flanks) are decurrent as four wavy stigmatic lines with unicellular papillae, which form four partial compita (Fig 8.9E). The gynoecium is of angiospermy type 4. In the syncarpous region, the synascidiate zone appears only short. The bulged part of the ovary forms an apical septum which extends almost down to the floor of the ovary. Below the stigmas, the PTTTs differentiate along the ventral side of the carpels; they are one cell layer thick and may be secretory (Fig 8.9E). The morphological surfaces of the carpel ventral sides are contiguous for the entire length of the style and form an internal compitum (Fig 8.9E, H, I), which is continuous with the external one and extends down to the base of the symplicate zone in the ovary. Because the synascidiate zone is so short, the two lateral placentae of each carpel each appear to be located at the base of each locule. The placentae are erect and each bears three (sometimes fewer) ovules. The ovules fill the locule at anthesis. The ovules are anatropous, syntropous, crassinucellar and bitegmic. Both integuments are 2–3 cell layers thick. The outer integument is longer than the inner and is hood-shaped, whereas the inner integument completely surrounds the nucellus. Both integuments are thickened above the nucellus and together form the micropyle, which faces the base of the funicle.

Histology

Special mucilage cells (for term, see Matthews and Endress, 2006) are present in sepals, petals and gynoecium (Fig 8.9H, I). Because of the poor preservation of the herbarium material used, it is unclear whether they are in the epidermal or subepidermal or even deeper cell layers of the perianth organs. During floral development these cells appear successively, from the sepals to the ovules. In the gynoecium they are present only after anthesis and last in the outer integument of the developing seeds (compare Fig 8.9H and I).

8.4 Discussion

8.4.1 Flower morphology, merism and symmetry

Flower size is diverse: large in *Peganum* (2 to 4 cm), small in *Nitraria* (*c.* 1 cm) and tiny in *Tetradiclis* (<5 mm). Flowers are (at least morphologically) bisexual and probably entomophilous in all three genera, although nectaries have been unambiguously reported only in *Nitraria* and *Peganum* (this study; Nair and Nathawat, 1958; Kapil and Ahluwalia, 1963; Ronse De Craene and Smets, 1991; Ronse De Craene et al., 1996; see discussion on Nectary). Flowers with a reduced number or incomplete development of organs of one of the two sexes have occasionally been observed both in *Nitraria* and in *Peganum*, and with reduction in ovule number per locule in *Tetradiclis* (this study; Baillon, 1873; Engler, 1931; Noble and Whalley,

1978). Partial functional unisexuality has only been clearly reported in *Nitraria* (Noble and Whalley, 1978). For *Peganum* and *Tetradiclis* it is unknown whether there are functionally unisexual flowers.

Floral *Bauplan* and merism are more similar between *Nitraria* and *Peganum* than between either of them and *Tetradiclis*. In *Tetradiclis*, the flowers are always completely isomerous (and therefore completely polysymmetric), independently of whether they are tetramerous or (more rarely) trimerous, and they have a single whorl of (antesepalous) stamens and antepetalous carpels (this study; Engler, 1931). In contrast, in *Nitraria* and in *Peganum*, the flowers have a pentamerous perianth and androecium, with fifteen stamens and a trimerous gynoecium (this study; Payer, 1857; Baillon, 1873; Eichler, 1878; Engler, 1931; Nair and Nathawat, 1958; Kapil and Ahluwalia, 1963; Ronse De Craene and Smets, 1991; Ronse De Craene et al., 1996; Hussein et al., 2009). Most notably, in both genera the 15 stamens are not arranged in two whorls as described by some earlier authors, but they form five triplets with a primary antesepalous stamen flanked by two secondary stamens. This can be interpreted as a specialized haplostemonous pattern. In addition, because of the difference in merism between the gynoecium and the outer organs, the flowers in *Nitraria* and *Peganum* exhibit an (inconspicuous) monosymmetry (in *Peganum* along the median plane; Eichler, 1878). The variation in organ number observed in both genera is also similar. In *Nitraria* the number of antepetalous stamens may vary and the androecium may even be reduced to a single whorl of five antesepalous stamens (Baillon, 1873; Engler, 1931). More commonly, one of the three carpels is sterile and lacks an ovule in the locule, or it may even be so reduced that the gynoecium appears dimerous (this study; Baillon, 1873); Baillon (1873) also mentions cases with up to six carpels. In *Peganum*, the perianth and androecium may be tetramerous (Engler, 1931). Independently of the other organ whorls, the gynoecium may be reduced to two carpels or, more rarely, have four carpels, or even five (Baillon, 1873; Engler, 1931; Shukla, 1955; Li and Tu, 1994). It has been suggested that the trimerous gynoecium in *Peganum* may be derived from a pentamerous one by reduction of two carpels because of the pentagonal shape of the syncarpous gynoecium ring primordium in early stages of development (Ronse De Craene et al., 1996). We also observed that the shape of the trimerous gynoecium may not be as triangular as would be expected, but we did not observe the initiation or the development of more than three carpels. Thus, because of the phylogenetic position of *Tetradiclis*, the idea cannot be ruled out that the flowers of *Peganum* are derived from isomerous, fully pentamerous ancestral forms. It is more likely, however, that the pentagonal shape of the gynoecium is superimposed by the pentamerous androecium as in analogous cases of other families discussed by Endress (2006, 2008).

Small-sized entomophilous flowers with a nectary are common in Sapindales and other rosids (see discussion on Nectary). Morphologically bisexual flowers with one of the two sexes more or less reduced or aborted, and thus functionally unisexual (as partly in *Nitraria*), have also been reported in almost all other sapindalean families, except for Biebersteiniaceae about which little is known (i.e. Anacardiaceae, Burseraceae, Kirkiaceae, Meliaceae, Rutaceae and Simaroubaceae; see Bachelier and Endress, 2008, 2009). In Sapindales, flowers are commonly isomerous and polysymmetric as in *Tetradiclis*. Flowers in which the gynoecium is not isomerous with the other floral whorls exhibit a kind of inconspicuous (oblique) monosymmetry (such as in Nitrariaceae, except for *Tetradiclis*). Isomerous and polysymmetric flowers are pentamerous in Biebersteiniaceae (Reiche, 1889), tetramerous in Kirkiaceae (Stannard, 1981; Bachelier and Endress, 2008), and can variously have three, four or five organs per whorl in Anacardiaceae, Burseraceae, Meliaceae, Rutaceae or Simaroubaceae (Bachelier and Endress, 2009). Such inconspicuously monosymmetric flowers with an oblique symmetry also occur in Anacardiaceae and Burseraceae (Bachelier and Endress, 2009). However, pronounced monosymmetry with unequal differentiation of certain floral sectors is less common and not present in Nitrariaceae. It is known from Sapindaceae, some Rutaceae and exceptionally in Meliaceae (Ronse De Craene et al., 2000; Weckerle and Rutishauser, 2003, 2005; Endress and Matthews, 2006; Ronse De Craene and Haston, 2006; Endress, accepted).

Perianth

The perianth has a calyx and corolla in all three genera. In *Nitraria* and *Peganum*, the sepals are initiated in a clockwise or anticlockwise spiral sequence. In contrast, the five petals are initiated more or less simultaneously. In both genera, the second sepal is in median adaxial position as is common in core eudicots (this study; Payer, 1857; Ronse De Craene, 2010; for *Nitraria*, also Baillon, 1873; for *Peganum*, also Eichler, 1878). The abaxial petal between sepals 1 and 3 (= petal 1) is slightly larger than the others at first, in both genera (this study; Ronse De Craene et al., 1996). In *Tetradiclis* the first pair of sepals is in the median plane (Eichler, 1878).

The sepals are free in early stages of development in both genera, but later they appear (congenitally) united basally in *Nitraria*. Sepal aestivation changes during development: at first the margins become contiguous and may overlap later, which results in valvate or imbricate (quincuncial) aestivation. This may account for the different descriptions of sepal aestivation in the literature: valvate in *Nitraria* and *Peganum* (Baillon, 1873; Nair and Nathawat, 1958; Kapil and Ahluwalia, 1963), or imbricate in *Nitraria* (Baillon, 1873; Nair and Nathawat, 1958; Liu and Zhou, 2008) and in *Peganum* (Payer, 1857; Baillon, 1873; Ronse De Craene et al., 1996). In *Tetradiclis* sepal aestivation is imbricate (decussate) (Eichler, 1878).

The petals are free in all genera throughout development. Their aestivation is somewhat open at the base, but induplicate-valvate further up in *Nitraria* (and the petal tips are strongly bent inwards), whereas it is imbricate in *Tetradiclis* and *Peganum* (contort in *Peganum*, cochlear in *Tetradiclis*) (this study; Payer, 1857; Baillon, 1873; Eichler, 1878; Nair and Nathawat, 1958; Kapil and Ahluwalia, 1963; Ronse De Craene and Smets, 1991; Ronse De Craene et al., 1996). In all three genera, the aestivation of the sepals becomes open during development; the petals become longer than the sepals, and thus, protect the inner floral organs up to anthesis (this study; Ronse De Craene and Smets, 1991). Postgenital coherence between petals was found in *Nitraria* and *Peganum* (this study), but is unknown in *Tetradiclis*. After anthesis, the calyx is persistent in all genera, while the corolla and androecium are persistent only in *Tetradiclis*, but caducous in *Nitraria* and *Peganum* (this study; Baillon, 1873; Noble and Whalley, 1978; Ronse De Craene and Smets, 1991; Hamzaoğlu et al., 2005; Liu and Zhou, 2008).

Flowers with a pentamerous calyx, in which the sepals are initiated in a quincuncial spiral sequence with the second sepal in abaxial (posterior) position, such as in *Nitraria* and *Peganum*, are common in Sapindales and other rosids (Ronse De Craene, 2010). That the corolla protects the inner floral organs up to anthesis, with a postgenital coherence between the petals (formed by the papillate epidermis or interlocking hairs), also occurs in other Sapindales (Bachelier and Endress, 2009; for details, see Ronse De Craene and Haston, 2006, p. 468), and in other rosids, especially in malvids (Endress and Matthews, 2006; Matthews and Endress, 2006; Endress, 2010).

Androecium

The androecium is the most puzzling and therefore most frequently discussed part of the flowers in Nitrariaceae. It has one stamen whorl in *Tetradiclis*, but in *Nitraria* and *Peganum*, it seems to have two whorls. In addition, the androecium in *Nitraria* and *Peganum* does not follow the rule of organ alternation (as in diplostemonous flowers), in that the antesepalous whorl of stamens appears to be the inner one, and the antepetalous the outer one. This condition has generally been referred to as obdiplostemony (Payer, 1857; Beille, 1902; Saunders, 1937; Eckert, 1966; Ronse De Craene and Smets, 1991; Ronse De Craene and Smets, 1995; Ronse De Craene et al., 1996). However, what is most unusual in Nitrariaceae is that the apparent antepetalous whorl commonly comprises ten stamens arranged in antepetalous pairs (Payer, 1857; Eckert, 1966; Ronse De Craene and Smets, 1991; Ronse De Craene et al., 1996). The structure of the androecium and the origin of the antepetalous stamens in *Nitraria* and *Peganum* have been differently interpreted. Earlier comparative studies interpreted the androecium of *Nitraria* as derived from a haplostemonous pattern and thus each antesepalous stamen plus the two closest antepetalous stamens should be seen as a triplet of antesepalous stamens (Baillon, 1873; Ronse De Craene and Smets, 1991). As an exception, haplostemonous flowers with only five antesepalous stamens were reported in *Nitraria* (Baillon, 1873). In contrast, in *Peganum*, the androecium was interpreted as derived from a diplostemonous androecium pattern, with the antepetalous stamen pairs having doubled antepetalous positions (Payer, 1857; Beille, 1902; Ronse De Craene et al., 1996). Only Nair and Nathawat (1958) suggested that the antepetalous stamens in *Peganum* belong to antesepalous triplets as in *Nitraria*, but only based on mature stages. In both *Nitraria* and *Peganum* the first stamen primordia to appear are antesepalous (this study; Payer, 1857; Beille, 1902; Ronse De Craene and Smets, 1991; Ronse De Craene et al., 1996). Our developmental studies show that the antepetalous stamens of *Peganum* arise close to the antesepalous ones. The further development of the androecium in *Peganum* is indeed very similar to that in *Nitraria*, contrary to earlier claims of a difference. Therefore the origin of the antepetalous stamens is very likely the same in *Nitraria* and *Peganum*, and probably derived from a haplostemonous pattern as found in *Tetradiclis*. Ronse De Craene et al. (1996) erroneously interpreted the antepetalous stamens as double on the evidence that the antepetalous stamens have a separate vascular connection from the antesepalous stamen and developmental evidence that the antepetalous primordia converge towards each other and appear to be similar in size, contrary to stamens belonging to a triplet. However, pressure by the petals may have caused differences in the development of individual stamens (Fig 8.5K).

Stamen structure is similar in all three genera. The filament is broad at the base and rounder further up. The filament tip is constricted and the anther, which is basifixed and sagittate or X-shaped, is shorter than the filament. The filament base thickens at anthesis in both *Nitraria* and *Peganum*, and its margins become more or less flattened (this study; Payer, 1857; Nair and Nathawat, 1958). Only in *Peganum* are the anthers apiculate. A notable difference in our material is that at anthesis the antepetalous stamens are longer than the antesepalous ones in *Peganum*, but the other way around in *Nitraria*.

In Sapindales, flowers are commonly diplostemonous, whereas haplostemonous flowers are much less common. Nevertheless, they have a scattered occurrence in almost all families, except for the unigeneric Biebersteiniaceae (Bachelier and Endress, 2008, 2009). In addition, when the flowers are haplostemonous, only the antesepalous stamens are present as in *Tetradiclis*. This condition may be the result of a reduction of the antepetalous stamens (see Bachelier and Endress, 2009). In contrast, the development of polyandrous flowers with more stamens than the expected number based on floral merism, is much rarer in Sapindales, and has been reported in a very few genera such as *Gluta*, *Sclerocarya* and *Sorindeia*, in Anacardiaceae (Ding Hou, 1978; von Teichman and Robbertse, 1986; Breteler, 2003), *Canarium* in Burseraceae (Lam, 1932), *Chisocheton*, *Clemensia*, *Turraea* and *Vavaea* in Meliaceae (Harms, 1940), *Mannia* in Simaroubaceae (Engler, 1931),

Deinbollia in Sapindaceae (Eichler, 1878) and *Citrus* and *Aegle* in Rutaceae (Leins, 1967; Lord and Eckard, 1985). In such cases, there are commonly more than two whorls of stamens. Triplets of antesepalous stamens as in *Nitraria* and *Peganum* are, to our knowledge, not known in any other sapindalean family. Among other rosids they are known from Geraniaceae in Geraniales (*Hypseocharis, Monsonia, Sarcocaulon*, Payer, 1857; Saunders, 1937; Narayana and Arora, 1963; Rama Devi, 1991; Ronse De Craene and Smets, 1996), Elaeocarpaceae in Oxalidales (Matthews and Endress, 2002) and Ixonanthaceae in Malpighiales (Narayana and Rao, 1966; Ronse De Craene and Smets, 1996). In addition, antepetalous stamens are in general smaller than antesepalous ones across the eudicots (e.g. Bachelier and Endress, 2009). Thus, longer antepetalous stamens, such as in *Peganum*, are unusual and have been reported elsewhere in Sapindales only in *Protium* in Burseraceae (Daly, 1992; Bachelier and Endress, 2009).

Stamens such as those in Nitrariaceae, with a sagittate and basifixed anther, and longitudinal dehiscence slits extending for the entire length of the thecae, are common in other rosids in general (Endress and Stumpf, 1991). The exact phylogenetic position of Nitrariaceae in Sapindales is not yet completely clear and it is therefore uncertain whether the unusual androecium structure of *Nitraria* and *Peganum* is a synapomorphy for the family, with a secondary reduction in *Tetradiclis*, or whether it is a convergence derived from a haplostemonous pattern in both *Nitraria* and *Peganum*. However, it is a good feature to support the monophyly of Nitrariaceae.

Nectary

In *Nitraria* and *Peganum*, floral nectaries are present between the androecium and the gynoecium (this study; Nair and Nathawat, 1958). They have been described as a more or less developed intrastaminal lobed disc in *Peganum* (Baillon, 1873; Kapil and Ahluwalia, 1963; Ronse De Craene et al., 1996) and as antepetalous pits in *Nitraria* (Ronse De Craene and Smets, 1991). However, we found that the nectaries are similar in both genera in consisting of five depressions in the petal radii. Nectaries have not been reported as yet in *Tetradiclis*.

An intrastaminal floral nectary is common in Sapindales and most other rosids (Ronse De Craene and Haston, 2006). However, the presence of conspicuous nectariferous depressions in the petal radii, as in *Peganum* and *Nitraria*, is unusual. This feature could thus be another synapomorphy for Nitrariaceae.

Gynoecium

The gynoecium is syncarpous in all Nitrariaceae (this study; Baillon, 1873; Nair and Nathawat, 1958; Kapil and Ahluwalia, 1963; Ronse De Craene and Smets, 1991; Ronse De Craene et al., 1996). The synascidiate zone is present in variable proportions (longest in *Peganum*, shortest in *Tetradiclis*). The symplicate zone extends to the base of the stigmas. The stigmatic zone is apocarpous, but with the carpels contiguous or postgenitally connected into a stigmatic head. The carpels are flat in this zone and their margins form three decurrent stigmatic lines (if the gynoecium is tricarpellate). In *Tetradiclis* and *Peganum* the dorsal parts of the carpels are bulged in the ovary, extremely so in *Tetradiclis*, which results in a morphologically very short ovary, an apical septum and therefore also a very short synascidiate zone. In *Peganum* and *Tetradiclis* the base of the style is hidden by the dorsal bulges of the ovary. However, in all three genera, the style is (morphologically) apical and it is rounder at the base than further up, and distally, the style and stigmatic head are conspicuously angled, with as many angles as there are carpels (this study; Baillon, 1873; Kapil and Ahluwalia, 1963). The stigmatic lines along the edges of the stigmatic head are covered with unicellular papillae in all three genera. They form an external compitum which is continuous with the internal compitum formed by the contiguous ventral sides of the carpels. In *Nitraria*, the gynoecium is of angiospermy type 3 or 4, and the style is hollow in transverse section, with as many slits as there are carpels, and it is filled with secretion. In contrast, in *Peganum*, the gynoecium is of angiospermy type 4 and the ventral sides of the carpels are tightly appressed to each other and postgenitally connected. In *Tetradiclis*, the inner part of the style appears as in *Peganum* at anthesis, but later becomes hollow as in *Nitraria* and is filled with secretion or remains of the former PTTT (Fig 8.9I).

In Sapindales, the gynoecium is commonly syncarpous (Bachelier and Endress, 2009), especially in the putative basal groups (Biebersteiniaceae, Nitrariaceae). Also, the presence of a stigmatic head formed by the free, but postgenitally united, carpel tips is relatively more widespread in Sapindales (Endress et al., 1983; Endress and Matthews, 2006; Bachelier and Endress, 2009) and malvids in general than in other eudicots (Endress, 2010). However, in contrast to Nitrariaceae, in other Sapindales the receptive surface is not restricted to the carpel margins. Stigmatic lines decurrent along the edges of an angled stigmatic head may thus be a synapomorphy for Nitrariaceae. In addition, continuity between the external compitum formed by the stigmatic head and an internal compitum, such as in Nitrariaceae, is probably not common in Sapindales. It has been poorly studied, but it occurs in core Burseraceae (*Canarium*, Bachelier and Endress, 2009), and may also be present in some Rutaceae (Ramp, 1988). Unicellular stigmatic papillae, such as in Nitrariaceae, are common in Sapindales. Only in the clade of Kirkiaceae and Anacardiaceae + Burseraceae are the papillae multicellular and uniseriate (Bachelier and Endress, 2009). The presence of a gynophore, such as in *Peganum*, is also common in Sapindales and other malvids (Endress and Matthews, 2006; Endress, 2010). In all three genera the placentae are lateral, even in *Nitraria*, which has only one ovule per carpel.

Ovules

The ovules are crassinucellar and bitegmic (this study; Mauritzon, 1934; Kapil and Ahluwalia, 1963; Kamelina, 1994; Li and Tu, 1994). Both integuments contribute to the formation of the micropyle. In *Malacocarpus* the micropyle is zigzag-shaped (Kamelina, 1985). In *Nitraria* and *Peganum* the outer integument is slightly thicker than the inner one (this study; Mauritzon, 1934; Kapil and Ahluwalia, 1963; Li and Tu, 1994). In *Tetradiclis*, the integuments are equally thick (this study; Kamelina, 1994). The ovules are anatropous (Payer, 1857; Baillon, 1873; Nair and Nathawat, 1958; Kapil and Ahluwalia, 1963; Kamelina, 1994). However, in *Nitraria* and *Tetradiclis* the outer integument is relatively long on the concave side of the ovules, in contrast to those in *Peganum* and *Malacocarpus* (this study; Kamelina, 1985, 1994). It would be interesting to know whether this long part of the outer integument develops early or only appears late in development, as an adjustment to the available space in *Nitraria* and *Tetradiclis*. The direction of the curvature is syntropous in all genera. Also in all genera, the funicle is relatively long, especially in *Nitraria* and in *Tetradiclis* (Fig 8.10; this study; Kamelina, 1985, 1994).

Ovules are most commonly anatropous, bitegmic and crassinucellar in rosids. In Sapindales and other malvids there is a tendency for the inner integument to be thicker than the outer (Endress and Matthews, 2006; Endress, 2010), a feature not present in Nitrariaceae. A long funicle, such as in Nitrariaceae, is especially frequent in Anacardiaceae. Interestingly, both families have syntropous ovules and

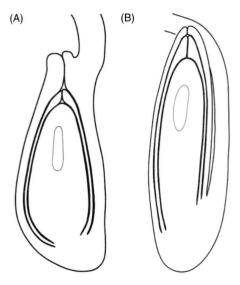

Fig 8.10 Ovules in Nitrariaceae. Schematic median longitudinal sections. (A) *Peganum harmala*. (B) *Nitraria retusa*.

a long funicle may thus be functionally linked to the direction of ovule curvature (Bachelier and Endress, 2009).

8.4.2 Anatomy

In *Nitraria* and *Peganum*, the sepals have three vascular traces and the petals one vascular trace, as is common in eudicots. The dorsal vasculature of the carpels appears as broad reticulate bands of anastomosing bundles. The synlateral bundles are inverted and have the xylem in peripheral position (for *Nitraria*, see also Shukla, 1955; Nair and Nathawat, 1958; for *Peganum*, see also Beille, 1902).

A diffuse or weakly differentiated carpel dorsal vasculature, such as in *Nitraria* and *Peganum*, is a feature that appears to be relatively pronounced in Sapindales (Ronse De Craene and Haston, 2006; Bachelier and Endress, 2008, 2009). Also, inverted synlateral vascular bundles have been reported in Sapindales, such as in some Burseraceae (Stevens, 2001 onwards; Bachelier and Endress, 2009).

8.4.3 Histology

Tanniferous cells are present in *Nitraria* and are lacking in *Tetradiclis* and *Peganum*. Special mucilage cells (for term, see Matthews and Endress, 2006) are present in the hypodermis and mesophyll of all floral organs in *Nitraria* and *Tetradiclis* (this study; Kamelina, 1994). In both genera their differentiation proceeds centripetally, in parallel to the differentiation of the organs. They are already present in all organs at anthesis in *Nitraria*, whereas in *Tetradiclis*, they become conspicuous only after anthesis in the wall of the developing fruit, and even later in the seed coat (this study; Kamelina, 1994). In addition, in flowers of *Nitraria*, differentiation also clearly proceeds from the tip of the organs downwards in petals and anthers, and special mucilage cells may thus also be present in the ovary after anthesis. Therefore, the idea cannot be ruled out that mucilage cells also develop in the reproductive organs of *Peganum* or *Malacocarpus* after anthesis, since they are present in their leaves, as in *Nitraria* and *Tetradiclis* (Metcalfe and Chalk, 1950; Sheahan and Cutler, 1993). Special mucilage cells occur in many other orders of eudicots (but sometimes only in leaves) and their presence is more or less consistent at the familial level (survey in Matthews and Endress, 2006).

8.4.4 Floral structure and systematics

Features supporting the exclusion of *Nitraria, Tetradiclis,* and *Peganum* from Zygophyllaceae

- Leaves with spiral phyllotaxis (versus opposite in Zygophyllaceae; El Hadidi, 1975; Liu and Zhou, 2008)
- Flowers with pentamerous perianth and androecium but trimerous gynoecium, or flower isomerous and tetramerous but haplostemonous (versus isomerous,

pentamerous and diplostemonous in Zygophyllaceae; this study; Engler, 1931; Nair and Nathawat, 1958)

■ Petal development not delayed and corolla protecting the inner organs in bud (versus delayed in most Zygophyllaceae; Ronse De Craene, pers. obs. in *Fagonia*, *Zygophyllum* and *Tribulus* – not delayed in *Balanites*; this study)

■ Ovules crassinucellar and without endothelium (versus sometimes weakly crassinucellar [for definition see Endress, 2010], and almost, always with endothelium, even if crassinucellar, in Zygophyllaceae; this study; Mauritzon, 1934; Nair and Gupta, 1961; Kapil and Ahluwalia, 1963; Masand, 1963; Kamelina, 1985; Li and Tu, 1994)

■ Ovules syntropous (versus antitropous in Zygophyllaceae; this study; Nair and Jain, 1956; Narayana and Prakasa Rao, 1962, 1963; Li and Tu, 1994)

■ Micropyle formed by both integuments (versus only inner integument in Zygophyllaceae, but formed by both integuments also in *Balanites* and *Seetzenia*; this study; Kamelina, 1985)

■ Funicle long and outer integument unusually well developed on the concave side of the ovule in *Nitraria* and *Tetradiclis* (not well developed at anthesis in *Peganum* and *Malacocarpus*) (versus short and not well developed in Zygophyllaceae; this study; Kamelina, 1985, 1994).

Features shared by *Nitraria*, *Tetradiclis*, and *Peganum* supporting their inclusion in Sapindales

■ Carpel tips free and flat, but postgenitally connected forming a stigmatic head with external compitum and stigmas decurrent along the carpel margins (*Kallstroemia* in Zygophyllaceae, which has a superficially similar gynoecium tip, whose structure is unknown in detail, also needs to be analysed and compared to Nitrariaceae)

■ Ovary bulged dorsally so that the style appears 'gynobasic' in *Peganum* and *Tetradiclis* (not in *Nitraria*)

■ Nectaries intrastaminal (perhaps lacking in *Tetradiclis*)

■ Special mucilage cells present in floral organs.

Potential synapomorphies for a broad circumscription of Nitrariaceae (including *Nitraria*, *Tetradiclis* and *Peganum* + *Malacocarpus*) (features rather uncommon in other sapindalean families)

■ Flowers haplostemonous, or derived from a haplostemonous pattern (antesepalous triplets)

- Stigmatic head angled, with as many angles as there are carpels
- PTTT one cell layer thick
- Compitum extending from stigmas to placentae, and external and internal compitum continuous
- Ovules syntropous with a long funicle
- Nectaries forming antepetalous depressions (perhaps lacking in *Tetradiclis*).

Potential apomorphies for *Nitraria* if considered as a monogeneric family (excluding *Tetradiclis* and *Peganum* + *Malacocarpus*)

- Style hollow filled with secretion and unicellular papillate PTTT
- Ovary without dorsal bulges and with apical placenta
- Ovule one per carpel.

Potential apomorphies for the clade of *Tetradiclis* + *Peganum* (excluding *Nitraria*)

- Angiospermy of type 4
- Carpels with conspicuous dorsal bulges and placentae basal or at mid-length
- Ovules more than one per carpel (and more than one seed per fruit)
- Funicle much longer than the remainder of the ovule.

8.5 Conclusions

This study shows the interdependence of 'modern' phylogenetic reconstructions and 'traditional' comparative approaches for the study of angiosperm systematics and evolution. Using the example of Nitrariaceae, we have demonstrated how the new light shed on the relationships by molecular-based methods has changed our understanding of the evolution of the flower structure and unusual androecium of Nitrariaceae as circumscribed now in APG (2009). In addition, this study shows that the exclusion of *Nitraria*, *Tetradiclis* and *Peganum* from Zygophyllaceae, and their inclusion in Sapindales, is well supported by floral structure. The morphological variation observed between the three genera of Nitrariaceae is reminiscent of that found in the clade of Kirkiaceae and Anacardiaceae + Burseraceae (Bachelier and Endress, 2008, 2009). In addition, the estimated age of divergence between these genera is similar to that estimated between these three sapindalean families (Muellner et al., 2007), and the divergence between their *rbc*L sequences is as extensive as that between other families of Sapindales (Sheahan and Chase, 1996). Therefore, this study also shows that the separation of the Nitrariaceae, as circumscribed in APG (2009), into three families (e.g. Takhtajan, 2009), with Nitrariaceae being unigeneric, Tetradiclidaceae being monotypic and Peganaceae comprising two genera, of which *Malacocarpus* would be monotypic, could be supported. However, this study also shows that these three genera share features of the androecium and gynoecium which are unusual in other Sapindales and thus could as well be included in a single family. Additional molecular phylogenetic studies and comparative morphological studies of other families of Sapindales are thus urgently needed and we can only hope for more interdisciplinary collaboration in future to understand the systematics and evolution of Sapindales and angiosperms.

Acknowledgements

We thank Livia Wanntorp for her help in co-organizing the symposium 'Flowers on the Tree of Life' in Leiden. The Systematics Association and the Sibbald Trust at the Royal Botanic Garden Edinburgh are acknowledged for their financial support to JBB. Peter Linder is acknowledged for allowing JBB to use the laboratory facilities of the Institute of Systematic Botany and Botanic Garden of the University of Zurich and for covering the fees for some of the SEM micrographs, as well as Urs Ziegler and the staff of the Centre for Microscopy and Image Analysis of the University of Zurich for their help with the SEM. Ruth Jacob is thanked for the preparation of some of the microtome section series and Joana Meyer for her careful help in preparing valuable material for SEM. We also thank an anonymous reviewer for helpful comments.

8.6 References

Agababyan, V. S. (1964). Morphological analysis of pollen and systematic classification of Zygophyllaceae. *Izvestiya Akademii Nauk Armyanskoi, SSR, Biologicheskikh Nauk*, **17**, 39–45.

Agardh, J. G. (1858). *Theoria Systematis Plantarum*. Lund: Gleerup.

APG (Angiosperm Phylogeny Group). (1998). An ordinal classification for the families of flowering plants. *Annals of the Missouri Botanical Garden*, **85**, 531–553.

APG (Angiosperm Phylogeny Group). (2003). An update of the Angiosperm Phylogeny Group classification for the orders and families of flowering plants: APG II. *Botanical Journal of the Linnean Society*, **141**, 399–436.

APG (Angiosperm Phylogeny Group). (2009). An update of the Angiosperm Phylogeny Group classification for the orders and families of flowering plants: APG III. *Botanical Journal of the Linnean Society*, **161**, 105–121.

Bachelier, J. B. and Endress, P. K. (2008). Floral structure of *Kirkia* (Kirkiaceae) and its systematic position in Sapindales. *Annals of Botany*, **102**, 539–550.

Bachelier, J. B. and Endress, P. K. (2009). Comparative floral morphology of Anacardiaceae and Burseraceae, with a special emphasis on the gynoecium. *Botanical Journal of the Linnean Society*, **159**, 499–571.

Baillon, H. (1873). *Histoire des Plantes, IV*. Paris: Hachette.

Bakker, F. R., Vassiliades, D .D., Morton, C. and Savolainen, V. (1998). Phylogenetic relationships of *Biebersteinia* Stephan (Geraniaceae) inferred from *rbc*L and *atp*B sequence comparisons. *Botanical Journal of the Linnean Society*, **127**, 149–158.

Beille, L. (1902). *Recherches sur le développement floral des disciflores*. Bordeaux: Durand.

Bobrov, E. G. (1965). On the origin of the flora of the old world deserts, as illustrated by the genus *Nitraria* L. *Botaniceskij Zhurnal* (Moscow and Leningrad), **50**, 1053–1067. [in Russian, with English summary].

Breteler, F. J. (2003). The African genus *Sorindeia* (Anacardiaceae): A synoptic revision. *Adansonia*, **25**, 93–113.

Cronquist, A. (1981). *An Integrated System of Classification of Flowering Plants*. New York: Columbia University Press.

Cronquist, A. (1988). *The Evolution and Classification of Flowering Plants*, 2nd edn. Bronx: New York Botanical Garden.

Dahlgren, R. M. T. (1980). A revised system of classification of the angiosperms. *Botanical Journal of the Linnean Society*, **80**, 91–124.

Daly, D. C. (1992). New taxa and combinations in *Protium* Burm. f. Studies in Neotropical Burseraceae VI. *Brittonia*, **44**, 280–299.

Ding Hou (1978). Anacardiaceae. pp.395–548 in van Steenis, C. G. G. J. (ed.), *Flora Malesiana*, Ser. I, 8 (3). Alphen: Sijthoff and Noordhoff.

Eckert, G. (1966). Entwicklungsgeschichtliche und blütenanatomische Untersuchungen zum Problem der Obdiplostemonie. *Botanische Jahrbücher für Systematik*, **85**, 523–604.

Eichler, A. W. (1878). *Blüthendiagramme, II*. Leipzig: Engelmann.

El Hadidi, M. N. (1975). Zygophyllaceae in Africa. *Boissiera*, **24**, 317–323.

Endress, P. K. (1994). *Diversity and Evolutionary Biology of Tropical Flowers*. Cambridge: Cambridge University Press.

Endress, P. K. (2006). Angiosperm floral evolution: Morphological developmental framework. *Advances in Botanical Research*, **44**, 1–61.

Endress, P. K. (2008). The whole and the parts: Relationships between floral architecture and floral organ shape, and their repercussions on the interpretation of fragmentary floral fossils. *Annals of the Missouri Botanical Garden*, **95**, 101–120.

Endress, P. K. (2010). Flower structure and trends of evolution in eudicots and their major subclades. *Annals of the Missouri Botanical Garden*, **97**, 541–583.

Endress, P. K. (accepted). The immense diversity of floral monosymmetry and asymmetry across angiosperms.

Endress, P. K. and Igersheim, A. (2000). Gynoecium structure and evolution in basal angiosperms. *International Journal of Plant Sciences*, **161** (Suppl.), S211–S223.

Endress, P. K. and Matthews, M. L. (2006). First steps towards a floral structural characterization of the major rosid subclades. *Plant Systematics and Evolution*, **260**, 223–251.

Endress, P. K. and Stumpf, S. (1991). The diversity of stamen structures in 'lower' Rosidae. *Botanical Journal of the Linnean Society*, **107**, 217–293.

Endress, P. K., Jenny, M. and Fallen, M. E. (1983). Convergent elaboration of apocarpous gynoecia in higher advanced angiosperms (Sapindales, Malvales, Gentianales). *Nordic Journal of Botany*, **3**, 293–300.

Engler, A. (1896a). *Über die geographische Verbreitung der Zygophyllaceen im Verhältniss zu ihrer systematischen Gliederung*. Berlin: Abhandlungen der Preussischen Akademie der Wissenschaften, Physikalisch-mathematische Klasse.

Engler, A. (1896b). Zygophyllaceae. pp. 74–93, 353–357 in Engler, A. and Prantl, K. (eds.), *Die natürlichen Pflanzenfamilien*, 1st edn, III, 4. Leipzig: Engelmann.

Engler, A. (1931). Zygophyllaceae, Rutaceae, Simaroubaceae, Burseraceae. pp. 144–184, 187–456 in Engler, A. and Prantl, K. (eds.), *Die natürlichen Pflanzenfamilien*, 2nd edn, 19a. Leipzig: Engelmann.

Erdtman, G. (1952). *Pollen Morphology and Plant Taxonomy*. Stockholm: Almqvist and Wiksell.

Fernando, E. S., Gadek, P. A. and Quinn, C. J. (1995). Simaroubaceae, an artificial construct: evidence from *rbc*L sequence variation. *American Journal of Botany*, **82**, 92–103.

Gadek, P. A., Fernando, E. S., Quinn, C. J. et al. (1996). Sapindales: molecular delimitation and infraordinal groups. *American Journal of Botany*, **83**, 802–811.

Hamzaoğlu, E., Duran, A. and Akhani, H. (2005). New genus record for the flora of Turkey: *Tetradiclis* Stev. ex M. Bieb. (Zygophyllaceae). *Turkish Journal of Botany*, **29**, 403–407.

Harms, H. (1940). Meliaceae. pp. 1–72 in Engler, A. and Prantl, K. (eds.), *Die*

natürlichen Pflanzenfamilien, 2nd edn, 19b, *1*. Leipzig: Engelmann.

Hussein, S. R., Kawashty, S. A., Tantawy, M. E. and Saleh, N. A. M. (2009). Chemosystematic studies of *Nitraria retusa* and selected taxa of Zygophyllaceae in Egypt. *Plant Systematics and Evolution*, **277**, 251–264.

Igersheim, A. (1993). The character states of the Caribbean monotypic *Strumpfia* (Rubiaceae). *Nordic Journal of Botany*, **13**, 545–559.

Igersheim, A. and Cichocki, O. (1996). A simple method for microtome sectioning of prehistoric charcoal specimens, embedded in 2-hydroxyethyl methacrylate (HEMA). *Review of Palaeobotany and Palynology*, **92**, 389–393.

Janka, H., von Balthazar, M., Alverson, W. S., Baum, D., Semir, J. and Bayer, C. (2008). Structure, development and evolution of the androecium in Adansonieae (core Bombacoideae, Malvaceae s.l.). *Plant Systematics and Evolution*, **275**, 69–91.

Kamelina, O. P. (1985). Zygophyllaceae, Nitrariaceae, Balanitaceae. pp. 145–157 in Yakovlev, M. S. (ed.), *Comparative Embryology of Flowering Plants: Brunelliaceae – Tremandraceae*. Leningrad: Nauka.

Kamelina, O. P. (1994). Embryology and systematic position of *Tetradiclis* (Tetradiclidaceae). *Botaniceskij Zhurnal* (St Petersburg), **79** (5), 11–27.

Kapil, R. N. and Ahluwalia, K. (1963). Embryology of *Peganum harmala* Linn. *Phytomorphology*, **13**, 127–140.

Lam, H. J. (1932). Morphologie der dreizähligen Burseraceae-Canarieae II. Weitere Tendenzen in Blütenstand, Blüte, Frucht und Vegetationsorganen; Anatomisches; Schlussbetrachtungen und Zusammenfassung. *Annales du*

Jardin Botanique de Buitenzorg, **42**, 23–56.

Leins, P. (1967). Die frühe Blütenentwicklung von *Aegle marmelos* (Rutaceae). *Berichte der Deutschen Botanischen Gesellschaft*, **80**, 320–325.

Li, S. W. and Tu, L. Z. (1994). The embryology of *Nitraria* and its systematic significance. *Bulletin of Botanical Research*, **14**, 255–262.

Liu, Y.-X. and Zhou, L.-H. (2008). Nitrariaceae. pp. 41–42 in Wu, Z.-Y., Raven, P. H. and Hong, D.-Y. (eds.), *Flora of China*, 11. Beijing: Science Press, St. Louis: Missouri Botanical Garden Press.

Lord, E. M. and Eckard, K. J. (1985). Shoot development in *Citrus sinensis* L. (Washington navel orange) I. Floral and inflorescence ontogeny. *Botanical Gazette*, **146**, 320–326.

Matthews, M. L. and Endress, P. K. (2002). Comparative floral structure and systematics in Oxalidales (Oxalidaceae, Connaraceae, Cephalotaceae, Brunelliaceae, Cunoniaceae, Elaeocarpaceae, Tremandraceae). *Botanical Journal of the Linnean Society*, **140**, 321–384.

Matthews, M. L. and Endress, P. K. (2004). Comparative floral structure and systematics in Cucurbitales (Corynocarpaceae, Coriariaceae, Datiscaceae, Tetramelaceae, Begoniaceae, Cucurbitaceae, Anisophylleaceae). *Botanical Journal of the Linnean Society*, **145**, 129–185.

Matthews, M. L. and Endress, P. K. (2005a). Comparative floral structure and systematics in Celastrales (Celastraceae, Parnassiaceae, Lepidobotryaceae). *Botanical Journal of the Linnean Society*, **149**, 129–194.

Matthews, M. L. and Endress, P. K. (2005b). Comparative floral structure and systematics in Crossosomatales (Crossosomatacae, Stachyuraceae, Staphyleaceae, Aphloiaceae, Geissolomataceae, Ixerbaceae, Strasburgeriaceae). *Botanical Journal of the Linnean Society*, **147**, 1–46.

Matthews, M. L. and Endress, P. K. (2006). Floral structure and systematics in four orders of rosids, including a broad survey of floral mucilage cells. *Plant Systematics and Evolution*, **260**, 199–221.

Matthews, M. L. and Endress, P. K. (2008). Comparative floral structure and systematics in Chrysobalanaceae *s.l.* (Chrysobalanaceae, Dichapetalaceae, Euphroniaceae, Trigoniaceae; Malpighiales). *Botanical Journal of the Linnean Society*, **157**, 249–309.

Mauritzon, J. (1934). Etwas über die Embryologie der Zygophyllaceen sowie einige Fragmente über die der Humiriaceen. *Botaniska Notiser*, **87**, 409–422.

Metcalfe, C. R. and Chalk, L. (1950). *Anatomy of the dicotyledons*. Oxford: Clarendon Press.

Muellner, A. N., Vassiliades, D. D. and Renner, S. S. (2007). Placing Biebersteiniaceae, a herbaceous clade of Sapindales, in a temporal and geographic context. *Plant Systematics and Evolution*, **266**, 233–252.

Nair, N. C. and Gupta, I. (1961). A contribution to the floral morphology and embryology of *Fagonia cretica* Linn. *Journal of the Indian Botanical Society*, **40**, 635–640.

Nair, N. C. and Jain, R. K. (1956). Floral morphology and embryology of *Balanites roxburghii* Planch. *Lloydia*, **19**, 269–279.

Nair, N. C. and Nathawat, K. S. (1958). Vascular anatomy of some flowers of Zygophyllaceae. *Journal of the Indian Botanical Society*, **10**, 175–180.

Fig 2.3 Teratological female cones of extant conifers showing reversion from reproductive to vegetative growth (a–e), in one case (c) followed by a switch to male expression. (a, b) *Cunninghamia lanceolata*, (c) *Cryptomeria japonica*, (d) *Sequoia sempervirens*, (e) *Abies Koreana*. Images by Richard Bateman (a–c, e), Julien Bachelier (d).

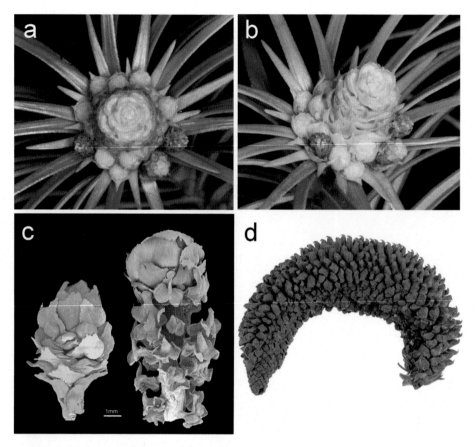

Fig 2.4 Bisporangiate cone-aggregates and cones of extant conifers. (a, b) Single terminal female cone of *Cunninghamia lanceolata* closely subtended by a pseudowhorl of lateral male cones. (c) Artificially coloured scanning electron micrographs of two teratological conifer cones that show a transition from basal male to apical female expression in *Tsuga dumosa* (right) and from basal female to indeterminate to male to apical female in *Tsuga caroliniana* (left). Colours: green = vegetative, blue = male, pink = female, purple = equivocal gender. (d) Cone of *Araucaria bidwillii* that is closer in size (10 cm long) and gross morphology to a typical male cone of the species but bears a longitudinal 'feminized' zone along the convex margin. Images by Richard Bateman (a, b), Paula Rudall (c), Raymond van der Ham (d).

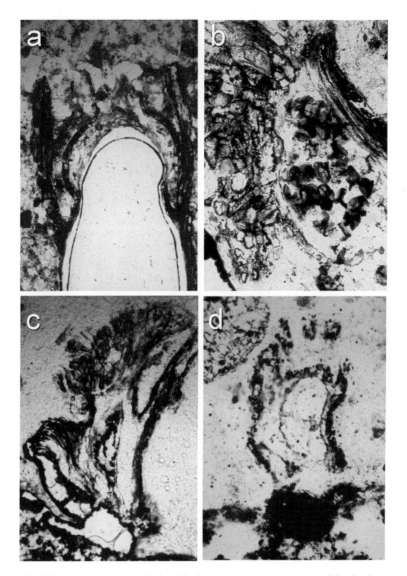

Fig 2.5 Light micrographs highlighting a putative teratos of the hydrasperman pteridosperm cupule *Pullaritheca longii* from the Mississippian of southeast Scotland that together suggest clinal control of gender. (d) Portion of one of 38 recorded dehisced specimens showing one of several abortive ovules attached to the placenta below. (b, c) Two teratological specimens first reported by Long (1977a) at the margin of a single atypical cupule; (c) largely resembles the abortive ovules, but has undergone atypical proliferation of the apical nucellar tissue responsible for capturing pre-pollen grains, (b) shows similar nucellar proliferation, but contains many microspores rather than the expected single megaspore. (a) Shows the wildtype pollen-receiving apparatus of the ovule, compressed by expansion of the ovum following successful pollination (modified after Bateman and DiMichele, 2002, Fig 7.4). Images by Richard Bateman.

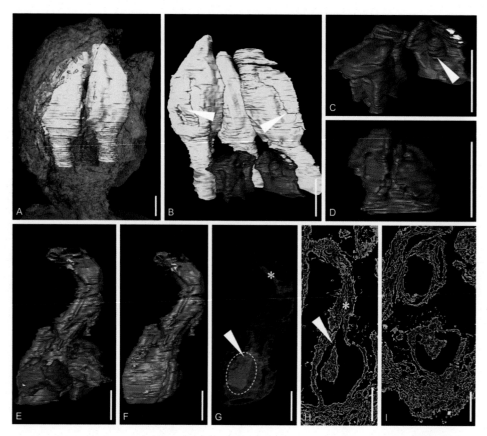

Fig 3.3 Gynoecium and androecium morphology of *Cohongarootonia hispida* (PP53716). SRXTM micrographic reconstructions and SRXTM sections. (A) Three-dimensional SRXTM reconstructions showing size and position of androecium in flower. (B) Three-dimensional SRXTM reconstructions showing three stamens (yellow) and three staminodes (pink and purple), lateral view. Arrowheads indicate apically hinged flaps. (C) Three-dimensional SRXTM reconstructions showing staminodes with paired spade-like appendages, ventral view. Arrowhead indicates short central staminodial tip. (D) Three-dimensional SRXTM reconstructions showing staminodes with paired spade-like appendages, dorsal view. (E) Three-dimensional SRXTM reconstruction showing lateral view of carpel (blue) and three staminodes. (F) Three-dimensional SRXTM reconstructions showing lateral view of carpel. (G) Three-dimensional semi-transparent SRXTM reconstructions showing size and position of single ovule and stylar canal (asterisk). Arrowhead indicates point of ovule insertion. (H) Tangential SRXTM section of carpel, showing insertion of ovule (arrowhead) and stylar canal (asterisk). (I) Transverse SRXTM section of carpel showing median and ventral-apical insertion of ovule. Scale bars: (A–H) = 100 μm.

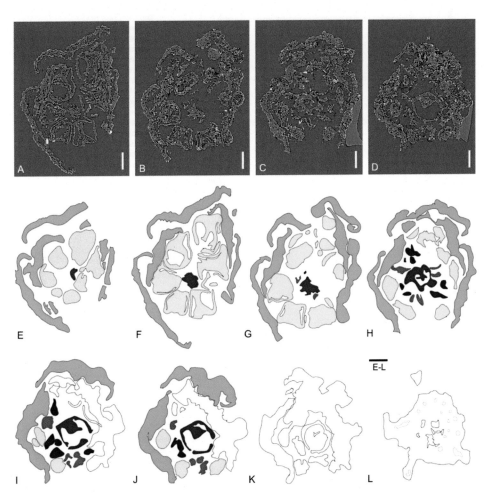

Fig 3.4 *Cohongarootonia hispida* (PP53716). Digital transverse SRXTM sections (A–D) and line drawing series of digital transverse sections (E–L), beginning at the apex and progressing to the base of the flower, indicating arrangement of organs (green = tepals, yellow = stamens, pink and purple = staminodes, blue = carpel). Scale bars = 100 μm.

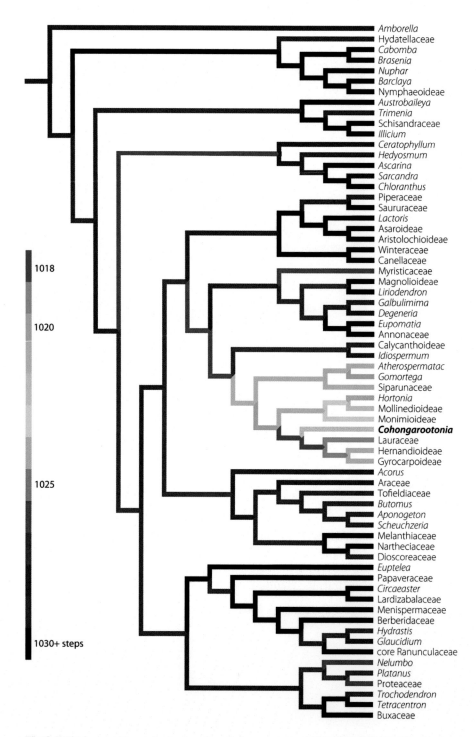

Fig 3.7 Phylogenetic analysis showing *Cohongarootonia hispida* (PP53716) added to the backbone tree of Doyle and Endress (2010). Branch colouring refers to the number of steps under parsimony (morphological character changes) that are required if *Cohongarootonia hispida* is attached to the corresponding branch.

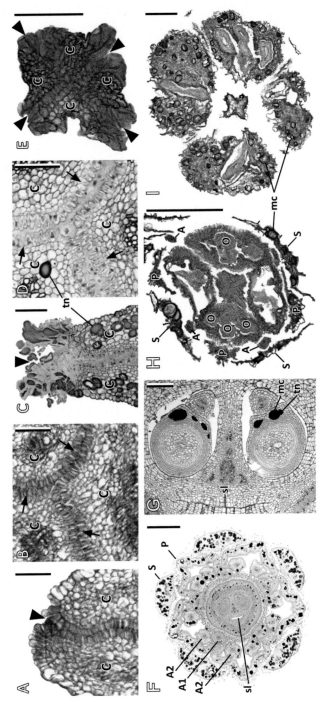

Fig 8.9 Nitrariaceae. Flower buds, and anthetic and postanthetic flowers, transverse sections. Signatures: Sepal [S], petal [P], antepetalous stamen [A1], antesepalous stamen [A2], carpel [C], ovule [O], sterile locule [sl], tanniferous cells [tn] are blue, special mucilage cells [mc] are pink to red. Arrowheads point to stigma and partial compitum, arrows point to carpel morphological surfaces covered with one cell-layered PTTT. (A)–(E) Styles and stigmas. (A), (B) *Peganum harmala*. (C), (D) *Nitraria retusa*. (E) *Tetradiclis tenella*. (F), (G) *Nitraria retusa*. (F) Preanthetic floral bud. (G) Anthetic gynoecium, with two locules with a single ovule and a sterile locule lacking an ovule. (H), (I) *Tetradiclis tenella*. (H) Anthetic flower. (I) Postanthetic flower. Scale bars: (F), (G), (H), (I) = 250 μm; (A), (B), (C), (D) = 100 μm; E = 50 μm.

Fig 9.1 Flower diversity in *Conostegia*. (A), (B). *C. centronioides*. (C) *C. macrantha*; (D) *C. oerstediana*. (A) Longitudinal section of young flower bud showing sepals fused into a calyptra (S) and developing petals (P). Scale bar: (A) = 200 mu, (B)–(D) = 5 mm.

Fig 11.1 *Acacia celastrifolia*. (A) Flowering branch. (B) Inflorescence with phyllode at its base. (C) Detail of phyllode with nectary and drop of nectar. (D) Flower with recurved petals, highly polyandrous androecium and four styles which are protruding above the level of the anthers. Scale bar = 1 cm in all.

Fig 12.2 Mature flowers of *Napoleonaea*. (A)–(B) *N. imperialis*; (C)–(F) *N. vogelii*. (A)–(D) Distal and lateral view of flower at anthesis. (E) detail of (C): central part with staminodes, stamens and ovary. (F) Floral bud prior to anthesis. Note the petals opening along the midribs. Af, fertile stamen; As, sterile stamen; Ci, inner corona; Co, outer corona; Ct, corona threads; St, stylar lobe. Scale bars = 10 mm.

Narayana, H. S. and Arora, P. K. (1963). Floral anatomy of *Monsonia senegalensis* Guill. and Pen. *Current Science*, **32**, 184–185.

Narayana, H. S. and Prakasa Rao, C. G. (1962). Floral anatomy of *Seetzenia orientalis* Decne. *Current Science*, **31**, 209–211.

Narayana H. S. and Prakasa Rao, C. G. (1963). Floral morphology and embryology of *Seetzenia orientalis* Decne. *Phytomorphology*, **13**, 197–205.

Narayana, L. L. and Rao, D. (1966). Floral morphology of Linaceae. *Journal of Japanese Botany*, **41**, 1–10.

Noble, J. C. and Whalley, R. D. B. (1978). The biology and autecology of *Nitraria* L. in Australia. I. Distribution, morphology and potential utilization. *Australian Journal of Ecology*, **3**, 141–163.

Payer, J.-B. (1857). *Traité d'organogénie comparée de la fleur*. Paris: Masson.

Perveen, A. and Qaiser, M. (2006). Pollen flora of Pakistan – XLIX, Zygophyllaceae. *Pakistan Journal of Botany*, **38**, 225–232.

Rama Devi, D. (1991). Floral anatomy of *Hypseocharis* (Oxalidaceae) with a discussion on its systematic position. *Plant Systematics and Evolution*, **177**, 161–164.

Ramp, E. (1988). Struktur, Funktion und systematische Bedeutung des Gynoeciums bei den Rutaceae und Simaroubaceae. Doctoral dissertation, University of Zurich. Zurich: ADAG.

Reiche, K. (1889). Geraniaceae. pp. 1–14 in Engler, A. and Prantl, K. (eds.), *Die Natürlichen Pflanzenfamilien*, III, 4. (1st edn) Leipzig: Engelmann.

Ronse De Craene, L. P. (2010). *Floral Diagrams. An Aid to Understanding Flower Morphology and Evolution*. Cambridge: Cambridge University Press.

Ronse De Craene, L. P. and Haston, E. (2006). The systematic relationships of glucosinolate-producing plants and related families: a cladistic investigation based on morphological and molecular characters. *Botanical Journal of the Linnean Society*, **151**, 453–494.

Ronse De Craene, L. P. and Smets, E. (1991). Morphological studies in Zygophyllaceae. I. The floral development and vascular anatomy of *Nitraria retusa*. *American Journal of Botany*, **78**, 1438–1448.

Ronse De Craene, L. P. and Smets, E. (1995). The distribution and systematic relevance of the androecial character oligomery. *Botanical Journal of the Linnean Society*, **118**, 193–247.

Ronse De Craene, L. P. and Smets, E. (1996). The morphological variation and systematic value of stamen pairs in the Magnoliatae. *Feddes Reperorium*, **107**, 1–17.

Ronse De Craene, L. P., De Laet, J. and Smets, E. (1996). Morphological studies in Zygophyllaceae. II. The floral development and vascular anatomy of *Peganum harmala*. *American Journal of Botany*, **83**, 201–215.

Ronse De Craene, L. P., Smets, E. F. and Clinckemaillie, D. (2000). Floral ontogeny and anatomy in *Koelreuteria* with special emphasis on monosymmetry and septal cavities. *Plant Systematics and Evolution*, **223**, 91–107.

Saunders, E. R. (1937). *Floral Morphology, A New Outlook, With Special Reference to the Interpretation of the Gynoecium*. I. Cambridge: Heffer and Sons.

Savolainen, V., Fay, M. F., Albach, D. C. et al. (2000). Phylogeny of the eudicots: a nearly complete familial analysis based on *rbc*L gene sequences. *Kew Bulletin*, **55**, 257–309.

Schönenberger, J. (2009). Comparative floral structure and systematics of Fouquieriaceae and Polemoniaceae (Ericales). *International Journal of Plant Sciences*, **170**, 1132–1167.

Schönenberger, J. and Grenhagen, A. (2005). Early floral development and androecium organization in Fouquieriaceae (Ericales). *Plant Systematics and Evolution*, **254**, 233–249.

Schönenberger, J., von Balthazar, M. and Sytsma, K. J. (2010). Diversity and evolution of floral structure among early diverging lineages in the Ericales. *Philosophical Transactions of the Royal Society, B, Biological Sciences*, **365**, 437–448.

Sheahan, M. C. and Chase, M. W. (1996). A phylogenetic analysis of Zygophyllaceae R.Br. based on morphological, anatomical and *rbc*L DNA sequence data. *Botanical Journal of the Linnean Society*, **122**, 279–300.

Sheahan, M. C. and Cutler, D. F. (1993). Contribution of vegetative anatomy to the systematics of Zygophyllaceae R.Br. *Botanical Journal of the Linnean Society*, **113**, 227–262.

Shukla, R. D. (1955). On the morphology of the two abnormal gynoecia of *Peganum harmala*. *Journal of the Indian Botanical Society*, **34**, 382–387.

Souèges, R. (1953). Embryogénie des Péganacées. Développement de l'embryon chez le *Peganum harmala* L. *Comptes Rendus de l'Académie des Sciences*, **236**, 2185–2188.

Stannard, B. L. (1981). A revision of *Kirkia* (Simaroubaceae). *Kew Bulletin*, **35**, 829–839.

Stevens, P. F. (2001 onwards). Angiosperm Phylogeny Website. Version 9, June 2008 [and more or less continuously updated since]. http://www.mobot. org/MOBOT/research/APweb/ [accessed December 2009].

Takhtajan, A. L. (1969). *Flowering Plants: Origin and Dispersal*. Edinburgh: Oliver and Boyd.

Takhtajan, A. L. (1980). Outline of the classification of flowering plants (Magnoliophyta). *Botanical Review*, **46**, 225–359.

Takhtajan, A.L. (1983). The systematic arrangement of dicotyledonous families. pp. 180–201 in Metcalfe, C. R. and Chalk, L. (eds.), *Anatomy of the Dicotyledons*, 2nd edn, Vol. 2. Oxford: Clarendon Press.

Takhtajan, A. L. (2009). *Flowering Plants*. Berlin: Springer.

von Balthazar, M. and Schönenberger, J. (2009). Floral structure and organization in Platanaceae. *International Journal of Plant Sciences*, **170**, 210–225.

von Balthazar, M., Alverson, W.S., Schönenberger, J. and Baum, D. A. (2004). Comparative floral development and androecium structure in Malvoideae (Malvaceae *s.l.*). *International Journal of Plant Sciences*, **165**, 445–473.

von Balthazar, M., Schönenberger, J., Alverson, W. S., Janka, H., Bayer, C. and Baum, D. A. (2006). Structure and evolution of the androecium in the Malvatheca clade (Malvaceae *s.l.*) and implications for Malvaceae and Malvales. *Plant Systematics and Evolution*, **260**, 171–197.

von Teichman, I. and Robbertse, P. J. (1986). Development and structure of the drupe in *Sclerocarya birrea* (Richard) Hochst. subsp. *caffra* Kokwaro (Anacardiaceae), with special reference to the pericarp and the operculum. *Botanical Journal of the Linnean Society*, **92**, 303–322.

Wang, H., Moore, M. J., Soltis, P. S. et al. (2009). Rosid radiation and the rapid rise of angiosperm-dominated forests. *Proceedings of the National Academy of Sciences USA*, **106**, 3853–3858.

Weber, M. and Igersheim, A. (1994). 'Pollen buds' in *Ophiorrhiza* (Rubiaceae) and their role in pollenkitt release. *Botanica Acta*, **107**, 257–262.

Weckerle, C. S. and Rutishauser, R. (2003). Comparative morphology and systematic position of *Averrhoidium* within Sapindaceae. *International Journal of Plant Sciences*, **164**, 775–792.

Weckerle, C. S. and Rutishauser, R. (2005). Gynoecium, fruit and seed structure of Paullinieae (Sapindaceae). *Botanical Journal of the Linnean Society*, **147**, 159–189.

Worberg, A., Alford, M. A., Quandt, D. and Borsch, T. (2009). Huerteales sister to Brassicales + Malvales, and newly circumscribed to include *Dipentodon, Gerrardina, Huertea, Perrottetia*, and *Tapiscia. Taxon*, **58**, 468–478.

9

Multiplications of floral organs in flowers: a case study in *Conostegia* (Melastomataceae, Myrtales)

Livia Wanntorp, Carmen Puglisi, Darin Penneys and Louis P. Ronse De Craene

9.1 Introduction

The genus *Conostegia* (Miconieae, Melastomataceae) includes shrubs and trees distributed in the Caribbean, Mexico, Central America, N Andes and Brazil (Schnell, 1996). At present, *Conostegia* contains about 40 species (Schnell, 1996; Mabberley, 1997), although over 100 names have been applied to the genus in the past. However, the elevated number of species has been explained as a probable misinterpretation of the intraspecific variation that occurs in some species (Schnell, 1996).

The name *Conostegia*, which is derived from the Greek words κονοσ = cone and στεγοσ = roof, was chosen by D. Don (1823) for grouping species characterized by flowers having their sepals fused into a cone-shaped calyptra (Fig 9.1 A–D). Despite the fact that a calyptrate calyx is present in other genera of the Melastomataceae, such as *Bellucia, Blakea, Centronia, Henriettea, Llewellynia, Miconia* and *Pternandra* (Schnell, 1996; Penneys et al., 2010), the peculiar calyx of *Conostegia* has long been regarded as a useful character for segregating *Conostegia* from other non-calyptrate species of Melastomataceae (Don, 1823). Species of *Conostegia* are immediately

Flowers on the Tree of Life, ed. Livia Wanntorp and Louis P. Ronse De Craene. Published by Cambridge University Press. © The Systematics Association 2011.

recognizable by the character combination of terminal inflorescences, flowers often multistaminate, calyx clearly circumscissily dehiscent at anthesis, anthers isomorphic and unappendaged, ovary inferior and berry fruits (Almeda, 2008).

Conostegia has long been considered a natural group (Judd and Skean, 1991; Schnell, 1996), though intrageneric phylogenetic relationships remain unresolved (Goldenberg et al., 2008; Michelangeli et al., 2004, 2008). In his unpublished revision of the genus, Schnell (1996) divided *Conostegia* into subgenera *Lobatistigma*, *Conostegia* (sections *Notostegia, Axilliflora, Conostegia, Dasystegia, Tomentostegia, Parvistigma*) and *Ossaeiformis*, mainly based on the morphology of the style and of the stigma.

The flowers in *Conostegia* vary considerably in their number of organs, making the genus an interesting plant model to study the evolutionary changes of merism. Parallel to species with a more regular number of parts (e.g. five petals, ten

Fig 9.1 Flower diversity in *Conostegia*. (A), (B). *C. centronioides*. (C) *C. macrantha*; (D) *C. oerstediana*. (A) Longitudinal section of young flower bud showing sepals fused into a calyptra [S] and developing petals [P]. Scale bar: (A) = 200 mu, (B)–(D) = 5 mm. For colour illustration see plate section.

stamens, five carpels), there are others in the genus with a disproportionate number of petals, but with a regular correlation between stamens and carpels. Finally, some other species of *Conostegia* do not show any meristic correlation between different organ whorls (Schnell, 1996).

In the present chapter, we have investigated flower development and morphology of 11 taxa of *Conostegia* and compared these with the development of three non-calyptrate genera of Melastomataceae. We also discuss the phenomenon of variations in merism throughout the angiosperms.

9.2 Materials and methods

Eleven taxa of *Conostegia* (tribe Myconieae) and three other non-calyptrate species of Melastomataceae, *Heterocentron elegans* Kuntze (tribe Myconieae), *Dissotis rotundifolia* (Sm.) Triana and *Clidemia octona* (Bonpl.) L.O.Williams (tribe Melastomeae) were investigated. Voucher details including provenance of material, collector name and number of each specimen are reported in Table 9.1. All specimens of *Conostegia* were verified against the taxonomy reported in Schnell (1996).

Flower material was fixed in FAA and stored in 70% ethanol. Flowers were prepared and dried in an ethanol (70%, 95%, 100%) and acetone 100% series before being critical-point dried in CO_2 using an Emitech K850. Finally, the material was coated with platinum and observed with a LEO Supra 55VP Scanning Electron Microscope at the Royal Botanic Garden, Edinburgh.

9.3 Results

Although not all developmental stages were available for all species, the results of the present investigation indicate a comparable flower morphology and ontogeny for all species of *Conostegia* that we have examined. We therefore describe flower development and morphology by referring to illustrations from all taxa and pointing to differences that may occur. We also present data on the flower morphology and development of *Dissotis rotundifolia*, *Heterocentron elegans* and *Clidemia octona*.

9.3.1 *Conostegia*

Conostegia are glabrescent to densely pubescent trees or shrubs with a variety of different trichome types. Leaves are usually petiolate, with only a few species having subsessile leaves. Flowers are grouped in inflorescences described as pleiothyrsoid, i.e. with cymose partial inflorescences (Fig 9.2A), which can

Table 9.1 Taxa of *Conostegia* investigated with distribution and collection details.

Species	Distribution	Coll. Date	Collector	Coll. Number
C. aff. *montana* D.Don	Monteverde, Alajuela, Costa Rica	02.05.2002	D. S. Penneys	1520
C. cf. *centronioides* var. lancifolia Markgr.	Prov. Napo, Ecuador	14.11.2005	D. S. Penneys	1857
C. icosandra Urb.	Oaxaca, Mexico	28.12.2001	D. S. Penneys	1445
C. macrantha O. Berg ex Triana	Cartago, Costa Rica	09.05.2002	D. S. Penneys	1545
	Chiriquí, Panama	08.02.2005	D. S. Penneys	1738
C. monteleagreana Cogn.	Chiriquí, Panama	02.02.2005	D. S. Penneys	1719a
C. oerstediana O. Berg ex Triana	Monteverde, Alajuela, Costa Rica	28.04.2002	D. S. Penneys	1486
C. pittieri Cogn. ex Durand	Monteverde, Alajuela, Costa Rica	28.04.2002	D. S. Penneys	1488
C. rhodopetala Donn.Sm.	Monteverde, Alajuela, Costa Rica	29.04.2002	D. S. Penneys	1505
C. setosa Triana	Cocle, Panama	13.02.2005	D. S. Penneys	1760
C. subcrustulata Triana	Bocas del Toro, Panama	04.02.2005	D. S. Penneys	1727
C. xalapensis D.Don	Oaxaca, Mexico	28.12.2001	D. S. Penneys	1444
	Monteverde, Alajuela, Costa Rica	28.04.2002	D. S. Penneys	1485
	Cocle, Panama	13.02.2005	D. S. Penneys	1758
	Chiriquí, Panama	09.02.2005	D. S. Penneys	1742
Clidemia octona (Bonpl.) L.O. Williams	Cockscomb Reserve, Belize	15.01.2007	H. R. Morris and L. P. Ronse De Craene	980La
Dissotis rotundifolia (Sm.) Triana	Royal Botanic Garden Edinburgh	14.06.2007	C. Puglisi and L. P. Ronse De Craene	1045 Led
Heterocentron elegans Kuntze	Royal Botanic Garden Edinburgh	14.06.2007	C. Puglisi and L. P. Ronse De Craene	1046 Led

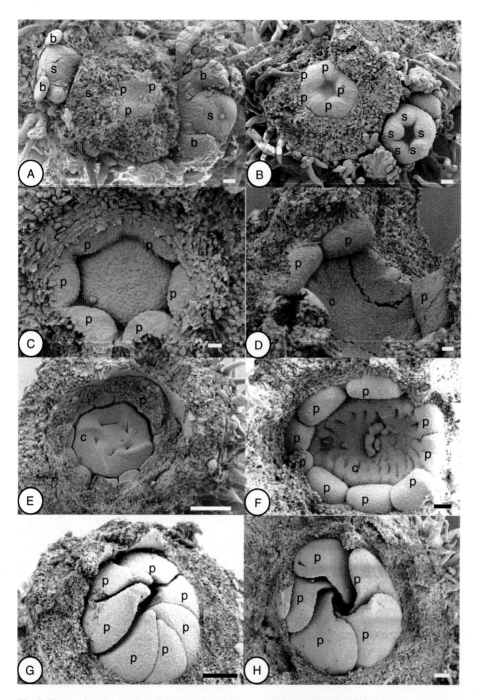

Fig 9.2 Development of sepals, petals and carpels in *Conostegia*. (A), (E) *C. icosandra*; (B) *C. xalapensis*; (C) *C. pittieri*; (D), (F) *C. macrantha*; (G) *C. rhodopetala*; (H) *C. centronioides*. (A) Young inflorescence with older apical flower and two lateral flowers with bracteoles. (B) Partial view of inflorescence with apical flower showing early petal initiation and lateral flower with sepal initiation. (C) Early initiation of gynoecial dome

vary in number in the same plant. Flowers of all species of *Conostegia* have two bracteoles, although in some cases these are very early caduceus, while in others they are persistent, even until fruit maturation. Flowers of *Conostegia* generally have five sepals (Figs 9.1A–C, 9.2B,C) united in a calyptra. In *C. icosandra* we occasionally observed a tetramerous calyx subtended by two bracteoles. The calyptra is highly variable in size, colour, texture, amount of pubescence and in the shape of its tip (Fig 9.1A–D): for example in *C. monteleagreana*, *C. pittieri* and *C. subcrustulata* the calyptra is pyriform, while it is elongated in *C. rhodopetala*, pointed in *C. xalapensis*, and more or less globose in *C. oerstediana*. In some species, such as *C.* aff. *montana* and *C. oerstediana*, sclereids form a layer within the parenchymatic tissue making the calyptras hard. Free sepals are only visible during the early stages of flower development before they fuse into a calyptra. Sepal development occurs in a rapid 1/2 or 2/5 sequence (Fig 9.2A–B) with the lobes becoming rapidly subequal. While the sepal lobes close at the top, they are lifted by extensive basal growth. At anthesis, the calyptra drops from the flower, breaking circumscissily along the line of petal insertion and it is usually not possible to distinguish the individual sepals at this stage (Fig 9.1C).

The petals of *Conostegia* are free, as in other Melastomataceae, glabrous, usually white or pink, and rounded. They are also asymmetrical as a consequence of their strongly convolute aestivation (Figs 9.1C–D, 9.2G–H). Petals are inserted more or less horizontally on top of the hypanthium, putting pressure on the rest of the flower (Figs 9.1A, 9.2G). Scars caused by the pressure of the petals are visible on the stamens and may also cause deformation of the style (Figs 9.3B,E–H, 9.4D,E). The number of petals has been reported to be 4 to 12 (Schnell, 1996). We were able to confirm a very variable number of petals for *Conostegia* (Table 9.2) and found, for example, six to eight petals in *C. pittieri* (Fig 9.2C), five in *C. monteleangreana*, *C.* aff. *montana*, *C. subcrustulata*, *C. setosa*, *C. centronioides* (Fig 9.2H) and eight to nine slightly asymmetrical petals with a retuse apex in *C. macrantha* (Fig 9.2C,F), as well as six to seven large petals in *C. rhodopetala* (Fig 9.2G). Petals are initiated in a rapid sequence on the sides of an invaginating floral apex. The further development of petals is often unidirectional from the abaxial to the adaxial side in all species examined (Fig 9.2C,F).

Caption for figure 9.2 continued
and unilateral development of five or six petals. (D) Early initiation of numerous carpel primordia on the hypanthial slope. (E) Development of five carpel primordia and early initiation of stamens. (F) Development of several carpels and petals before initiation of the stamens. (G) Convolute aestivation of seven petals. (H) Convolute aestivation of five petals. The calyx has been removed in most figures. Abbreviations: b = bracteole; c = carpel primordium; p = petal; s = sepal. Scale bars: (A)–(D) = 20 μm, (E)–(H) = 100 μm.

Fig 9.3 Development and morphology of androecium and gynoecium in *Conostegia*.
(A)–(B); *C. rhodopetala*; (C) *C. xalapensis*; (D)–(E) *C. macrantha*; (F) *C. icosandra*; (G)
C. montana; (H) *C. pittieri*. (A) Young flower with developing stamens and gynoecium;
B Mature flower with well-developed pistil. (C) Flower with six petals and eleven stamens
(one alternipetalous stamen considerably larger shown by asterisk). (D)–(E) Flower buds
with development of petals, single alternipetalous and groups of antipetalous stamens.
(F) Lateral view with development of globular stigma. (G) Apical view of flower with
14 stamens and 5 carpels; note the contortion of the carpel lobes. (H) Apical view
with unequal groups of antepetalous stamens and stigma with numerous lobes.

Table 9.2 Meristic variation in *Conostegia*. Taxon names with collector numbers and details on the numbers of floral parts according to: S = calyx; P = corolla; A = androecium; G = gynoecium.

Species	Coll. Number	S	P	A	G
C. aff. *montana*	1520	5	5	10–14	5–6
C. cf. *centronioides* var. *lancifolia*	1857	5	4–6	12–15	5
C. *icosandra*	1445	4	4–6	13–22	11
C. *macrantha*	1545	n/a	8–9	33–40 (+)	20–22
	1738	n/a	8–9	36	19–20
C. *monteleagreana*	1719a	n/a	5	11–14	7–9
C. *oerstediana*	1486	n/a	6–9	24–26	13–15
C. *pittieri*	1488	n/a	6–8	23	9–10
C. *rhodopetala*	1505	n/a	6–7	12–13	5
C. *setosa*	1760	n/a	5	10–13	5–6
C. *subcrustulata*	1727	5	5	10	5
C. *xalapensis*	1444	5	4–5	10–12	4–10
	1485	n/a	5–6	10–14	6
	1758	5	5	10	5–6
	1742	n/a	6	10–12	5–6

The androecium in *Conostegia* develops at a rather late stage after the initiation of the gynoecium (Figs 9.2D,E, 9.3D). The stamens in *Conostegia* are always regularly arranged in a single whorl (Figs 9.3A-G, 9.4A,C,D,F), but the number of stamens is highly variable across species. Stamen number ranges between 8 and 52, according to Schnell (1996). Some species, such as *C. monteleagreana* and *C. rhodopetala* have a relatively low number of stamens (10-14) while others, such as *C. macrantha* and *C. oerstediana* have more than 20-30 stamens per flower (Figs 9.3E, 9.4D). Between these extremes, we observed a range of intermediate numbers in the species studied. The basic arrangement is with ten stamens, as in other Melastomataceae (e.g. *C. setosa*, *C. subcrustulata*: Fig 9.4F), with five

Caption for figure 9.3 continued
Abbreviations: as = alternipetalous stamens; ap = antepetalous stamens; g = gynoecium; p = petal; st = stigma; sy = style. Scale bars: (A)–(C), (G)–(H) = 100 μm; (D)–(F) = 200 μm.

Fig 9.4 Development of androecium and gynoecium in *Conostegia*. (A) *C. monteleagrana*; (B)–(C) *C.* aff. *montana*; (D) *C. oerstediana*; (E), (F), (H) *C. subcrustulata*; (G) *C. pittieri*. (A) Flower bud showing the relative distribution of the corolla and of the androecium. Note the stamen pair with filaments in close proximity (asterisks). (B) Flower bud with capitate stigma, showing incurved stamens and ovules. (C) Apical view of flower bud with ten stamens; one filament removed showing two thecae. (D) Apical view of flower bud with several carpels and stamens; note the scars formed by the petals. (E) Details of contorted

opposite the sepals and five opposite the petals. However, most flowers deviate from this arrangement, with stamens often in pairs (Figs 9.3C, 9.4C). In other species, stamen numbers were much higher, sometimes in correlation with higher petal and carpel numbers, but not always. When two stamens are found opposite a petal (Figs 9.3B, 9.4A,C), they are usually smaller and appear closer to each other than the other stamens. In *C. macrantha* there are three to five stamens opposite each petal (Fig 9.3D–E) and some of them can be smaller or fused to each other (Fig 9.3D). In *C. pittieri* two to three stamens occur opposite the petals, and are of different sizes, with the central one being the largest (Fig 9.3H). The number of stamens per flower varies between species, and in some species even between different flowers of a single inflorescence. Stamens are initiated on a narrow ledge often described as a torus (e.g. Penneys and Judd, 2005) between the developing gynoecium and petals, with little space for development. Growth of the hypanthium allows the stamens to bend inwards and fill the space between the hypanthium and the style. As a result, the anthers are inverted and adherent to the flattened filament over a considerable distance. The filament has a characteristic hook-like shape (Fig 9.4B,G). The anthers dehisce by a single, apical pore. In section the anthers appear inverted.

In *Conostegia*, the pistil comprises a narrow style, in many species considerably thinner than the expanded stigma, which is either lobed, and sometimes crateriform in the centre (Figs 9.3F,H; 9.4C–D,G), or unlobed (Figs 9.3B–C, 9.4F). The stigma is densely covered with long papillae (Fig 9.4A–C). The arrangement of these papillae follows the aestivation of the petals (Figs 9.3G,H, 9.4A,E). The base of the style is surrounded by an expansion of the ovary wall forming a crown-like ring, which is prominent in some species (Figs 9.3B,E,F, 9.4D), but becomes mostly hidden at the bottom of the hypanthium (Fig 9.4B,C,F,G). The inferior ovary generally consists of four to six carpels (Fig 9.4H), which are opposite the stamens if they are equal in number. However, in some species the carpels are increased in number to nine to ten units (*C. pittieri*: Fig 9.4G) or even more (*C. macrantha*: Fig 9.2D,F). Ovules are small and numerous per locule on an intruding axile placenta (Fig 9.4H). During the early developmental stages, the top of the gynoecium has a crateriform shape with carpels arising as slits on the periphery (Figs 9.2D–F, 9.3D–E). By expansion of

Caption for figure 9.4 continued
stigmatic papillae. (F) Apical view of flower with narrow style and capitate stigma; (G) Lateral view of nearly mature flower with almost erect stamens and several carpellary lobes. (H) Cross section of the ovary showing the placenta and the ovules. Abbreviations: as = alternipetalous stamen; ap = antepetalous stamen; an = anther; f = filament; h = hypanthium; ov = ovules; p = petal; pl = placenta; st = stigma; sy = style; w = ovary wall. Scale bars: (A)-(B), (F)-(H) = 200 μm; (C)-(D) = 100 μm; (E) = 20 μm.

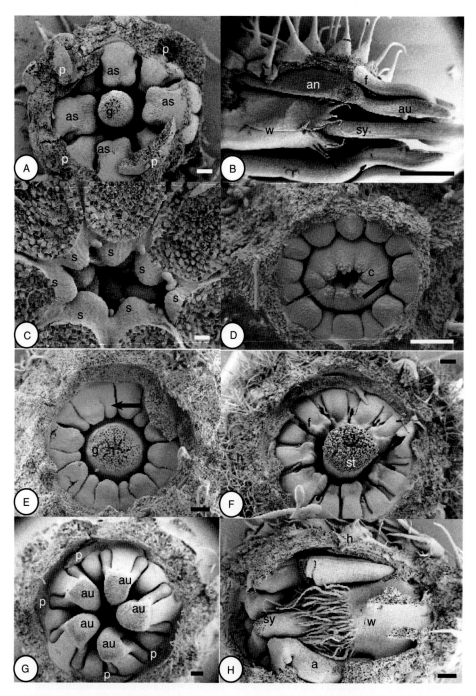

Fig 9.5 Flower morphology of *Heterocentron elegans* (A)–(B), *Clidemia octona* (C)–(F) and of *Dissotis rotundifolia* (G)–(H). (A) Mature flower bud with four petals (removed) and eight stamens. (B) Detail of flower showing a style surrounded by an extension of the ovary wall at the base. The anthers are provided with basal auriculate appendages. (C) Development of sepal lobes surrounding petal primordia. (D) Initiation of stamens and carpels. Note the

the lower part of the ovary, the slits become confined to the top of the stigma and differentiate into papillate lobes (Figs 9.3E–F,H, 9.4A–D,F).

9.3.2 *Heterocentron elegans, Clidemia octona* and *Dissotis rotundifolia*

Heterocentron elegans has tetramerous flowers. The androecium is diplostemonous and the stamens of the external, alternisepalous whorl are larger than the antepetalous ones (Fig 9.5A). The anthers are appendaged at the base and the connective is thickened and geniculate. The overall shape of the gynoecium is elongate and slender. The crown that the ovary wall forms around the style culminates with a fringe of hair-like formations (Fig 9.5B).

Clidemia octona is very similar to *Conostegia* in flower morphology. The seven sepals develop sequentially (Fig 9.5C) and differentiate in two distinct parts: a lower horizontally growing calyx lobe covering the lower formed organs and the external calyx tooth, an upper apical extension having a similar shape as a calyptra. The apical appendages are covered in trichomes like the rest of the bud. These multicellular trichomes can also be found on internal tissues of the flower bud, between sepals, petals and stamens (Fig 9.5F). Petal lobes have a strongly contorted aestivation. In *Clidemia octona* the stamens number 11–15 and differ in size, with some of them in pairs and others much larger, with a slit along the filament (Fig 9.5D–E). The anthers of *C. octona* are short and reach the basal part of the ovary during development (Fig 9.5F). Alternating with the stamens are smaller appendages formed lower on the hypanthium. They resemble staminodes, but are probably trichomes arising late during development. The ovary has seven to nine locules and each carpellary lobe is opposite a couple of stamens. The upper part of each carpel differentiates papillate cells, which develop as stylar lobes (Fig 9.5D–E). A short style extends above the rest of the ovary, which develops external lobes fitting between the anthers. The superior ovary develops several locules divided by narrow partitions and with an axile intruding placenta departing in the middle of each locule.

Caption for figure 9.5 continued
double stamen primordium (arrow). (E) Older bud with early stigma differentiation. Note the numerous stamen primordia of unequal size. The arrow points to basally connected primordia. (F) Nearly mature bud; note the grouped stamens and small appendages in between. (G) Mature flower bud with five petals and two pentamerous whorls of stamens with auriculate anthers. (H) Longitudinal section of flower bud showing the wing-like extension of the ovary wall surrounding the style base. The attachment of the ovary wall to the hypanthium is also visible. Abbreviations: an = anther; ap = antepetalous stamen; as = alternipetalous stamens; au = auriculate appendage; c = carpels; f = filament; g = gynoecium; h = hypanthium; p = petals; s = sepals; sy = stigma; w = ovary wall. Scale bars: (A) = 100 μm, (B) = 1 mm, (C) = 20 μm, (D)–(H) = 200 μm.

As in *Conostegia*, anthers are bent inwards and fit between the hypanthium and ovary.

Dissotis rotundifolia is pentamerous and diplostemonous (Fig 9.5G). Its stamens are biauriculate with the outer whorl stamens shorter than the inner whorl stamens, which extend almost to the base of the superior ovary (Fig 9.5H). The upper part of the ovary wall, which surrounds the style base is covered in long hairs (Fig 9.5H).

9.4 Discussion

In order to better understand how flowers of Melastomataceae vary in their morphology and number of organs, we have examined the development of flowers of different species of the genus *Conostegia*. For comparison and for producing a more complete picture of the morphological variation of the family we have also investigated flowers of three other species of Melastomataceae pertaining to two tribes: *Clidemia octona* (Miconieae), *Heterocentron elegans* (Melastomeae) and *Dissotis rotundifolia* (Melastomeae). Table 9.2 summarizes the numbers of different flower parts in all the taxa examined here.

To a certain extent floral development was highly comparable between different species and genera, reflecting a developmental syndrome common to Melastomataceae, though a broader survey across this highly diverse family is necessary. Flowers are characterized by the development of a deep hypanthium with petals and stamens inserted below the upper rim. Petals are strongly contorted with horizontal insertion and asymmetric shape, strongly compressing underlying organs. Stamens are incurved with the anthers inverted early in development, but flexing upwards at anthesis. The gynoecium is superior to inferior, well differentiated in a globular ovary, with axile and intruding placentation, and an elongate style with extended, papillate stigmatic lobes corresponding to the number of carpels. No nectary disc is formed.

Compared to *Dissotis* and *Heterocentron*, merism in *Conostegia* and *Clidemia octona* is highly variable. In general, all floral organ whorls in *Conostegia* can develop additional parts when space is available, e.g. on the other side of a previous member of the same whorl and alternate to that of the inner and/or outer whorl. We could distinguish several flower forms in *Conostegia*: (1) pentamerous flowers without differences in merism between whorls or with lower carpel number: (e.g. *C. subcrustulata, C. setosa*); (2) flowers in which sepals and petals are not affected, while androecium and gynoecium have more parts (e.g. *C. xalapensis, C. monteleagreana*); (3) flowers where all organs, except sepals, are affected and show a multiplication of parts (e.g. *Clidemia octona, Conostegia pittieri, C. oerstediana, C. macrantha*).

The variable number of sepals and petals in *Conostegia* appears to be linked to the variable width of the hypanthium. Most Melastomataceae have the same number of sepals and petals. In *Conostegia*, there is the occasional decoupling of sepal and petal numbers with sepals keeping the original merism and petals being variable. This occurs only in species where the number of stamens is higher than ten. The number of stamens varies considerably due to a secondary increase of the androecium. We observed that the increase of the stamens in the species of *Conostegia* examined could at least follow two different staminal developmental patterns:

(1) In one developmental pattern, a single large stamen is positioned in each alternipetalous position, and several smaller and closely linked stamens occur in antepetalous position (e.g. *C. oerstediana*, *C. rhodopetala*, *C. pittieri*, *C. monteleagreana* and *C. macrantha*). The size of the antepetalous stamens, which were sometimes partially fused, suggests that the group of stamens opposite each petal might be derived from a single primordium. A comparable stamen increase has been found in some Myrtaceae with a single row of several stamens opposite the petals (e.g. *Callistemon*, *Melaleuca*, *Hypocalyma*: Orlovich et al., 1999; Carrucan and Drinnan, 2000) and in *Plethiandra*, another member of Melastomataceae with a higher number of stamens (Kadereit, 2005).

(2) In flowers of *Conostegia rhodopetala*, *C. aff. montana*, *C. xalapensis* and *Clidemia octona*, the secondary increase of the stamens can also occur by a clear process of dédoublement, a phenomenon implying that each or most of the stamens in a flower splits into two parts. Paired stamens are closely adjacent, occasionally laterally fused and occur in antepetalous positions. While dédoublement generally results in two equal stamens, the strong contortion of the petals leads to an unequal development. The smaller stamen in *Clidemia octona* was always found on the left side of the larger one, and this regular pattern was observed in all flowers. In *Clidemia*, this phenomenon seems to take place at a later stage than observed in *Conostegia* and may occur in both antepetalous and alternipetalous stamens. A comparable process of dédoublement has also been described for some species of *Passiflora* leading to a fluctuation in the number of stamens (Krosnick et al., 2006). A paired arrangement of either antepetalous or antesepalous stamens was also found in some Lythraceae (e.g. *Ginoria*, *Crenea*: Tobe et al., 1998) and Myrtaceae (e.g. *Thryptomene*: Carrucan and Drinnan, 2000) of Myrtales.

Diplostemony has been considered to be the plesiomorphic condition in *Conostegia*, since it is widespread in other Melastomataceae and Myrtales (Dahlgren and Thorne, 1984; Goldenberg et al., 2008). In some species we observed ten stamens, five alternipetalous and the other five antepetalous, although all stamens appeared to be arranged in a single whorl. Clear diplostemony was observed in *Heterocentron elegans* and in *Dissotis rotundifolia* with stamens arranged at two different levels. Although the stamens in *Conostegia* are always inserted in a single series it is reasonable to assume that diplostemony represents an ancestral condition.

A clear correlation was found between a secondary stamen increase and a higher number of carpels in *C. macrantha*, *C. oerstediana* and *C. icosandra*. This pattern of increase is comparable to coordinated multiplications of carpels and stamens in Araliaceae (e.g. *Tupidanthus*: Philipson, 1970; Sokoloff et al., 2007), Rhizophoraceae (e.g. *Crossostylis*: Setoguchi et al., 1996), Sapotaceae (Hartog, 1878; Pennington, 2004) and Actinidiaceae (van Heel, 1987). Other examples are given in Ronse De Craene and Smets (1998) and Endress (2006). The secondary increase involves both androecium and gynoecium, and the way it occurs was described as a 'unicyclic simultaneous pleiomery' (Sokoloff et al., 2007), i.e. organs of each whorl arise simultaneously and later differentiate in size. A clear symmetrical arrangement of alternating stamens and carpels becomes obscured by insertion of supernumerary stamens and carpels in a limited area leading to a distortion of the initial structure. In the Rhizophoraceae, numbers of stamens vary between 8 and 30 and carpels between 2 and 20 (Juncosa, 1988). Species of *Crossostylis* have up to 30 stamens arising simultaneously in a single girdle and roughly the same number of alternating carpels, indicating a coordinated increase of stamens and carpels (Juncosa, 1988; Setoguchi et al., 1996). At a later stage of development, four antepetalous stamens differentiate from the others by increased growth indicating that stamen groups are antesepalous (Juncosa, 1988). In Sapotaceae changes in merism of petals, androecium and gynoecium are common and resemble *Conostegia* (e.g. *Labourdonnaisia*, *Mimusops*, *Letestua*: Baillon, 1890; Pennington, 2004), or increases in merism may only affect androecium and perianth (e.g. *Burckella*: Pennington, 2004). This was also observed in *Lafoensia* of Lythraceae and *Bruguiera* of Rhizophoraceae (Juncosa, 1988; Tobe et al., 1998). Endress (2006) discussed the spatial problems of having more carpels in one whorl, leading to displacement of carpels, irregular growth of carpel margins and irregular closure. While carpel proliferation in *Conostegia* affects the gynoecium in early stages of development, the growth of a common style and the inferior position of the ovary lead to a more regular arrangement of carpels at anthesis.

Carpels initiate simultaneously in *Conostegia*, just after petal initiation and before stamen initiation (Fig 9.2C,D). The early initiation of the gynoecium before the androecium is unusual, but has also been reported in *Phyllagathis* (Eberwein, 1993) and may be more common in the family. However, Melastomataceae remain ontogenetically underinvestigated. While this initiation sequence is uncommon in angiosperms, it has been reported for some other families (Chapters 6 and 12, this volume). In *Conostegia*, the delayed initiation of stamens might be related to the strong pressure of petal lobes and limited space before hypanthial growth. In regularly diplostemonous flowers, such as those of *C. xalapensis*, *C. setosa*, and in *Heterocentron* and *Dissotis*, carpels alternate with the inner whorl of stamens and petals.

In flowers with a secondary increase of the androecium, the respective positions of petals, stamens and carpels tend to be less clear and any relationship between parts seems to be lost. In *Conostegia macrantha* (Fig 9.3E), carpels tend to be opposite the alternipetalous stamens, but the different numbers of stamens and carpels blur any general pattern. Within *Conostegia* the number of carpels varies between 4 and 22 (25 according to Schnell, 1996). In most of the species the number of carpels (nearly) equals the number of petals, although there are some species, such as *C. macrantha*, which have many more carpels than petals.

9.5 Conclusions

The present investigation has clearly shown the occurrence of a secondary increase in some of the floral organs of species of *Conostegia*. The androecium is especially affected and we were able to observe two patterns of stamen increase in *Conostegia*. Although diplostemony has been claimed to be the original configuration in the genus (Dahlgren and Thorne, 1984), in most of the species examined some antepetalous stamen primordia divide before developing, thus producing a variable number of stamens.

Merism is extremely variable in *Conostegia*. There is little constancy in the number of parts of the flower, even within the same species. Pairwise correlations in the corolla, androecium and gynoecium show that a secondary increase may affect all organs except sepals, although to a different extent. The range of organ increases is manifold and comprises different combinations of either petals with stamens, carpels with stamens, petals with carpels, petals, stamens and carpels, or no combinations at all (Schnell, 1996). The spatial organization of the developing androecia and gynoecia of the flowers of *Conostegia* allows for some grouping of species. This grouping shows some similarities with the clades of a recent phylogenetic tree based on molecular data (Michelangeli et al., 2008), and will be further studied in the context of a Planetary Biodiversity Inventory investigation on the Miconieae (Michelangeli et al., ongoing). Furthermore, the patterns followed by the secondary increase seem to be more comparable to other Myrtalean families, e.g. Lythraceae and Myrtaceae, rather than with other groups of angiosperms, with the exception of Rhizophoraceae and Sapotaceae. This may be correlated with similarities in development of the hypanthium and perianth, influencing the mode of development of stamens and carpels. Patterns of morphological character evolution and distribution in the Melastomataceae, such as inflorescence architecture, calyx form, pleiostemony, anther appendages, ovary position, seed types, etc., are currently being examined in the context of a multigene, family-level, phylogenetic analysis (Penneys, in prep.).

Acknowledgements

We thank Frieda Christie for assistance with the SEM. Funding for this research was provided to DSP by the National Science Foundation (grants DEB-0508582 and DEB-0515665). The morphological and ontogenetic studies on *Conostegia* here included were performed by CP as part of her MSc dissertation in 2007. We acknowledge fruitful comments by Fabián Michelangeli and Favio González.

9.6 References

Almeda, F. (2008). Melastomataceae. pp. 164–338 in Davidse, G., Sousa-Sanchez, M.,Knapp, S. and Chiang, F. (eds.), *Flora Mesoamericana*, Vol. 4. Mexico City: Universidad Nacional Autónoma de México, St Louis, MO: Missouri Botanical Garden, London: The Natural History Museum.

Baillon, H. (1890). Observations sur les Sapotacées de la Nouvelle-Calédonie. *Bulletin Mensuel de la Société Linnéenne de Paris*, **2**, 881–912, 915–920, 922–926, 941–949.

Briggs, L. A. S. and Johnson, B. G. (1984). Myrtales and Myrtaceae – a phylogenetic analysis. *Annals of the Missouri Botanical Garden*, **71**, 700–756.

Carrucan, A. E. and Drinnan, A. N. (2000). The ontogenetic basis for floral diversity in the *Baeckea* sub-group. *Kew Bulletin*, **55**, 593–613

Dahlgren, R. and Thorne, R. F. (1984). The order Myrtales: circumscription, variation, and relationships. *Annals of the Missouri Botanical Garden*, **71**, 633–699.

Don, D. (1823). An illustration of the natural family of plants called Melastomataceae. *Memoirs of the Wernerian Natural History Society*, **4**, 276–329.

Eberwein, R. W. (1993). Ontogenese der Blüten von *Phyllagathis magnifica* (Melastomataceae). p. 61 in Fürnkranz, D. and Schantl, V. (eds.), *Kurzfassungen 11 Symposium Morphologie, Anatomie, Systematik Salzburg*, Salzburg: University of Salzburg.

Endress, P. K. (2006). Angiosperm floral evolution: morphological developmental framework. *Advances in Botanical Research*, **44**, 1–61.

Goldenberg, R., Penneys, D. S., Ameda, F., Judd, W. S. and Michelangeli, F. A. (2008). Phylogeny of *Miconia* (Melastomataceae): patterns of stamen diversification in a megadiverse neotropical genus. *International Journal of the Plant Sciences*, **169**, 963–979.

Hartog, M. H. (1878). On the floral structure and affinities of Sapotaceae. *Journal of Botany*, **16**, 65–72.

Judd, W. S. and Skean, J. D. (1991). Taxonomic studies in the Miconieae (Melastomataceae). IV. Generic realignments among terminal-flowered taxa. *Bulletin of the Florida Museum of Natural History*, **36**, 25–84.

Juncosa, A. M. (1988). Floral development and character evolution in Rhizophoraceae. pp. 83–101 in Leins P., Tucker S. C. and Endress P. K. (eds.), *Aspects of Floral Development*. Berlin, Stuttgart, Vaduz, Liechtenstein: Cramer.

Kadereit, G. (2005). Revision of *Plethiandra* Hook.f.: a polystaminate, East Asian genus of Melastomataceae. *Edinburgh Journal of Botany*, **62**, 127–144.

Krosnick, S. E. Harris E. M. and Freudenstein, J. V. (2006). Patterns of anomalous floral development in the Asian *Passiflora* (subgenus *Decaloba*: supersection *Disemma*). *American Journal of Botany*, **93**, 620–636.

Mabberley, D. J. (1997). *The Plant Book. A Portable Dictionary of the Vascular Plants*, 2nd edn. Cambridge: Cambridge University Press.

Michelangeli, F. A., Judd, W. S., Penneys, D. S. et al. (2008). Multiple events of dispersal and radiation of the tribe Miconieae (Melastomataceae) in the Caribbean. *Botanical Reviews*, **74**, 53–77.

Michelangeli, F. A., Penneys, D. S., Giza, J. et al. (2004). A preliminary phylogeny of the tribe Miconieae (Melastomataceae) based on nrITS sequence data and its implications on inflorescence position. *Taxon*, **53**, 279–290.

Orlovich, D. A., Drinnan, A. N. and Ladiges, P. Y. (1999). Floral development in *Melaleuca* and *Callistemon* (Myrtaceae). *Australian Systematic Botany*, **11**, 689–710.

Penneys, D. S. and Judd, W. S. 2005. A cladistic analysis and systematic revision of *Charianthus* (Miconieae: Melastomataceae) using morphological and molecular characters. *Systematic Botany*, **30**, 559–584.

Penneys, D. S., Judd, W. S., Michelangeli, F. A. and Almeda, F. (2010). Henriettieae: A new berry-fruited tribe of neotropical Melastomataceae. *Systematic Botany*, **35**, 783–800.

Pennington, T. D. (2004) Sapotaceae. pp. 390–421 in Kubitzki, K. (ed.), *The Families and Genera of Vascular Plants VI*. Berlin: Springer.

Philipson, W. R. (1970). Constant and variable features of the Araliaceae. *New Research in Plant Anatomy. Supplement to the Botanical Journal of the Linnean Society*, **63**, 87–100.

Ronse De Craene, L. P. and Smets, E. (1998). Meristic changes in gynoecium morphology. pp. 85–112 in Owens, S. J. and Rudall, P. J. (eds.), *Reproductive Biology*. Kew: Royal Botanic Gardens, Kew.

Schnell, C. E. (1996). The genus *Conostegia* (Melastomataceae). PhD thesis. Cambridge, MA: Harvard University.

Setoguchi, H., Ohba, H. and Tobe, H. (1996). Floral morphology and phylogenetic analysis in *Crossostylis* (Rhizophoraceae). *Journal of Plant Research*, **109**, 7–19.

Sokoloff, D. D., Oskolski, A. A., Remizowa, M. V. and Nuraliev, M. S. (2007). Flower structure and development in *Tupidanthus calyptratus* (Araliaceae): an extreme case of polymery among asterids. *Plant Systematics and Evolution*, **268**, 209–234.

Tobe, H. Graham, S. A. and Raven, P. H. (1998). Floral morphology and evolution in Lythraceae *sensu lato*. pp. 329–344 in Owens, S. J. and Rudall, P. J. (eds.), *Reproductive Biology*. Kew: Royal Botanic Gardens, Kew.

Van Heel W. A. (1987). Androecium development in *Actinidia chinensis* and *A. melanandra* (Actinidiaceae). *Botanische Jahrbücher für Systematik, Pflanzengeschichte und Pflanzengeographie*, **109**, 17–23.

10

Ontogenetic and phylogenetic diversification in Marantaceae

ALEXANDRA C. LEY AND REGINE CLAßEN-BOCKHOFF

10.1 Introduction

The Marantaceae Petersen (31 genera; ~530 ssp.: Andersson, 1998) are a pantropically (80% America; 11% Asia; 9% Africa: Kennedy, 2000) distributed family of perennial herbs and lianas found in the understory of tropical lowland rainforests. They are characterized by a unique pollination mechanism combining secondary pollen presentation with an explosive style movement (Kunze, 1984; Claßen-Bockhoff, 1991; Claßen-Bockhoff and Heller, 2008a). The specific pollen transfer mechanism is found in conjunction with a high synorganization of morphologically modified floral elements and has been postulated to be a key innovation responsible for the radiation of the Marantaceae (Kennedy, 2000).

Flowers in Marantaceae are trimerous, with inconspicuous sepals and petals and extremely modified elements in the two androeceal whorls (Fig 10.1). In the outer whorl one or two petaloid 'outer staminodes' act as the showy organs of the flowers. The three elements of the inner whorl are functionally differentiated into: (1) a single (monothecate) anther, (2) a 'fleshy (callose) staminode' and (3) a 'hooded (cucullate) staminode' (Kunze, 1984; Claßen-Bockhoff, 1991). These organs closely interact with the style resulting in secondary pollen presentation, set-up of tension and finally the explosive pollination mechanism (e.g. Gris, 1859; Delpino, 1869; Schumann, 1902; Yeo, 1993; Claßen-Bockhoff and Heller, 2008a, b;

Flowers on the Tree of Life, ed. Livia Wanntorp and Louis P. Ronse De Craene. Published by Cambridge University Press. © The Systematics Association 2011.

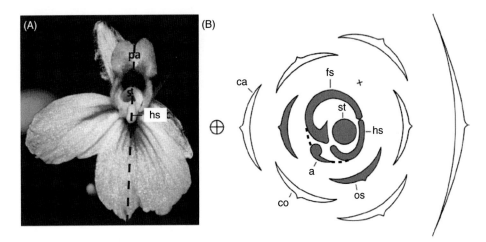

Fig 10.1 *Maranta leuconeura.* (A) Frontal view of a flower showing the petaloid outer staminodes and the monosymmetry of the flower centre (see A: axis, dashed line). (B) Floral diagram showing the asymmetry of the flower and the primary position of the floral elements. a, anther; ca, calyx; co, corolla; fs, fleshy staminode; hs, hooded staminode; os, outer staminode; pa, petaloid appendage of the fleshy staminode; st, style; x, missing organ in the inner androeceal whorl (after Pischtschan and Claßen-Bockhoff, 2008).

Ley, 2008; Pischtschan and Claßen-Bockhoff, 2008; Fig 10.2). As the style movement demands a high degree of synorganization of floral parts and synchronization with the pollinator and as the movement is irreversible, providing the flowers with a single opportunity for pollination, one should expect rather uniform structures across the whole family, as slight morphological deviations might result in a loss of operability. However, the high degree of floral diversity in the Marantaceae contradicts this expectation (Kunze, 1984; Kennedy, 2000; Claßen-Bockhoff and Heller, 2008a; Ley, 2008). It instead raises the questions: how far are elements of a functional unit allowed to vary without jeopardizing the reproductive success, and has the variation of the flowers influenced speciation in the family?

In the present chapter we summarize recent phylogenetic and floral structural findings for the Marantaceae. By using additional ecological and geographical data, we reconstruct the major character transformations per node and test the hypotheses of the pollination mechanism being: (1) a key innovation for speciation or (2) exclusively a tool for optimization of the mating system.

10.2 Phylogeny of the Marantaceae

Sequences for more than 130 of the known ~530 species of Marantaceae are so far available and used in either family-level phylogenies (Andersson and Chase, 2001:

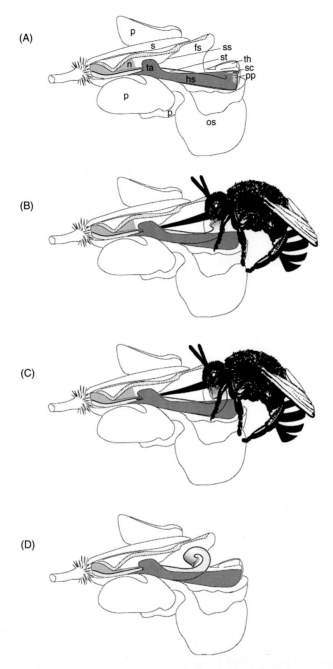

Fig 10.2 *Marantochloa purpurea* visited by *Amegilla vivida*. (A) Flower construction; note the functional unit composed of the style (st, unreleased) which is hidden behind the hooded staminode (hs; shaded) and fleshy staminode (fs); the trigger appendage (ta) and the stiff swelling (ss) together restrict access to nectar (n). (B) As soon as the pollinator deflects the trigger appendage with its mouthparts, the style tension is released. (B), (C) Pollen is exchanged by the rapid style movement (pollen uptake (B), pollen deposition (C)). (D) This mechanism is irreversible. os, outer staminode; p, petal; pp, pollen plate; s, sepal; sc, stigmatic cavity; th, theca (empty) (from Ley and Claßen-Bockhoff, 2009).

*rps*16 intron; Prince and Kress, 2006a: *matK* and 3′ intergenic spacer region and the *trnL-F* intergenic spacer region, 2006b: chloroplast: *matK*, *ndhF*, *rbcL*, *rps*16 intron and *trnL–trnF* intergenic spacer; mitochondrion: *cox*1; nuclear: ITS region and the 59-end of 26S) or to obtain more fine-scale resolution of distinct geographical clades (Asia: Suksathan et al., 2009: *rps*16 intron and nuclear internal transcribed spacer (ITS)1 and 5.8S-non-transcribed spacer (NTS), Africa: Ley and Claßen-Bockhoff, 2011: nuclear internal transcribed spacer (ITS) and *trnL–trnF* intergenic spacer; America: L. S. Suarez and F. Borchsenius, pers. com.; S. Vieira, pers. com.). The topologies of these different family-level phylogenies are so far congruent and thus the family Marantaceae has been subdivided (Fig 10.3) into the first-branching African *Sarcophrynium* clade as sister to all other Marantaceae, subsequently followed by the American *Calathea* clade (with the African *Haumania* in basal position) which is sister to an African-Asian-American clade. In the latter, the *Donax* clade splits first, while the *Maranta* and the *Stachyphrynium* clades are sisters to each other (Prince and Kress, 2006a).

However, despite this congruence between family-level phylogenies resolution along the backbone of the Marantaceae phylogenetic tree is a problem. For this reason, biogeographic interpretations must be handled with care. The present data support an African origin for the family, followed by a minimum of two dispersal events to the New World tropics and four or more dispersal events to the Asian tropics. The low number of extant species in tropical Africa is tentatively attributed to high extinction rates resulting from shrinking lowland forests during the Tertiary (Prince and Kress, 2006b).

10.3 General floral ontogeny in Marantaceae

Aspects of floral ontogeny have been investigated by Kirchhoff (1983), Kunze (1984, 2005), Claßen-Bockhoff and Heller (2008a), Ley (2008) and Pischtschan et al. (2010). Primordia initiation in Marantaceae flowers show an unusual sequence of development, with the complex of petals and inner whorl androeceal elements developing before the sepals and the outer staminodial whorl (see Kirchhoff, 1983). The lack of one or two outer staminodes and the individual differentiation of the organs of the inner androeceal whorl result in floral asymmetry, which is maintained in the adult flower by the outer staminodes (Fig 10.1A). In contrast, the functional 'pollination complex', which includes the inner androeceal whorl plus the style, is almost monosymmetric in the open flower. This is due to its secondary arrangement in a vertical position (Kunze, 2005) which provides the open flower with a roof (fleshy staminode) and a floor (hooded staminode). The hooded staminode together with one of the outer staminodes acts as a landing platform in bee pollinated flowers (Figs 10.1, 10.2).

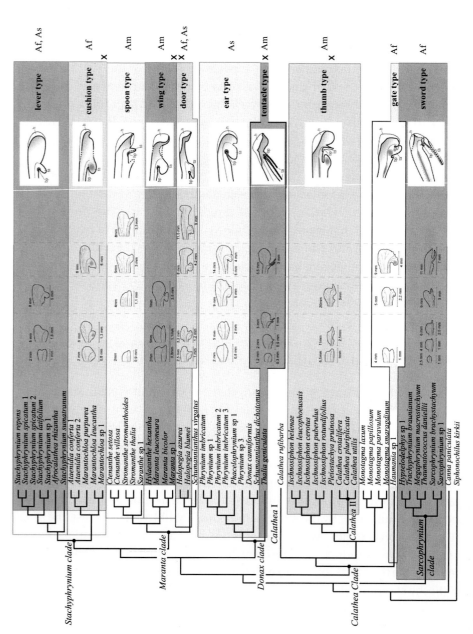

Fig 10.3 Ontogeny and distribution of the ten types of hooded staminodes within the Marantaceae. Each type is represented by one exemplary ontogeny. Ontogenetic stages corresponding in relative bud size are located in the same column. Note that the hooded staminodes are clade and continent specific and that they differ in the relative proportions of their subunits and in the shaping of the trigger appendages and basal plates. The asymmetry of the hooded staminode is initiated during ontogeny by the turning of the midrib. The trigger appendage develops close to the tip of the hood, whereas the hood proliferates on the opposite site and overtops the trigger appendage at a clade-specific ontogenetic stage. The degree of overtopping causes the final position (proximal or distal) of the trigger appendage alongside the hooded staminode. Af, Africa; Am, America; As, Asia; bp, basal plate; h, hood of the hooded staminode; ta, trigger appendage; x, autogamous species (phylogeny after Prince and Kress, 2006a; ontogenies and types of hooded staminodes after Pischtschan et al., 2010). Numbers above ontogenetic stages length of bud.

During early ontogeny, the theca of the anther is the dominant element becoming mature before the staminodes of the inner and outer whorl and the style complete their development (Fig 10.4A). At a later stage, when the fleshy and hooded staminodes have obtained the same length as the fertile anther they tightly enwrap the style and the anther (Fig 10.4B) forcing the pollen grains to be squeezed out of the theca by the elongating style (Fig 10.4C). By further style elongation pollen is fixed onto the pollen plate at the back of the style head (Fig 10.4D) from where it is secondarily presented (Claßen-Bockhoff and Heller, 2008a). After pollen transfer to the pollen plate (usually 7–12 hours before anthesis), the bud doubles its length. The style elongates more than the hooded staminode, which still enwraps the style, resulting in tension built up as the style grows against the hood (Fig 10.4E, F: backwards bending of the style; Pischtschan and Claßen-Bockhoff, 2008). Comparative morphological studies elucidate that it is not the hood per se that holds the tension but rather a distinct 'pressure point' beneath the pollen plate where the style comes into contact with the hooded staminode for the first time (Ley, 2008; Fig 10.5: black stars). This finding explains the so far inexplicable experiments in which large parts of the hood in *Maranta leuconeura*, *Calathea undulata* (Kunze, 1984), *Thalia geniculata* (Claßen-Bockhoff, 1991) and *Hylaeanthe hoffmannii* (Pischtschan and Claßen-Bockhoff, 2008) could be removed without releasing style tension. In these, as in all other tested species, the style rapidly springs forward as soon as the pressure point is tampered with (Ley, 2008).

In the adult flower, the fleshy staminode is characterized by a stiff swelling on its adaxial plane (Fig 10.4B: ss), while the hooded staminode bears a basal plate and trigger appendage at its adaxial margin (with reference to its mother axis; Fig 10.4C, D: bp, ta, th). These structures are essential for the style release mechanism, as together they narrow the access passage to the nectar, thereby forcing contact between the nectar-searching pollinator and the trigger mechanism, which mediates the release of tension. During the style movement, pollen from another flower is scraped off the arriving pollinator by the stigmatic cavity and the flower's own pollen from the pollen plate is subsequently glued onto the insect or bird (Fig 10.2B, C: sc, pp). This trigger movement is irreversible and one of the fastest in the plant kingdom (~0.03 s *Thalia geniculata*, see Claßen-Bockhoff, 1991).

10.4 Specific ontogenetic diversification and functional consequences

Though the processes of bud development, pollen transfer to the pollen plate, secondary pollen presentation, set-up of style tension and pollination by means of the explosive style movement are found in all Marantaceae, adult flowers vary in

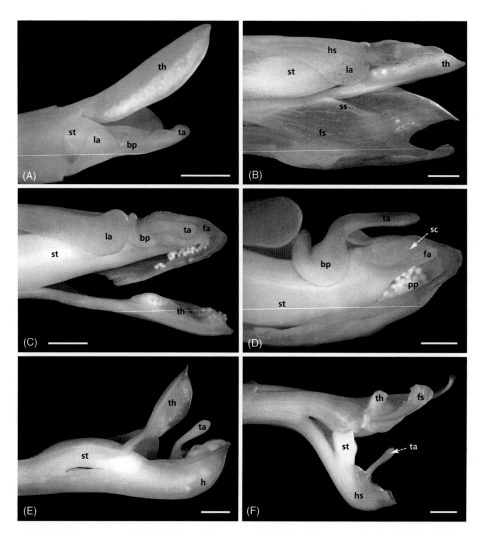

Fig 10.4 Floral ontogeny and secondary pollen presentation in *Calathea platystachya*. (A) The theca (th) is the dominant structure in a bud of 12 mm. (B) Theca and style (st) are both enwrapped by the hooded (hs) and fleshy (fs) staminodes; note the stiff swelling (ss) on the fleshy staminode (bud size: 16 mm; fleshy staminode slightly separated from the remaining parts). (C) In a bud of 24 mm size, the lateral appendage (la), basal plate (bp), trigger appendage (ta) and frontal appendage (fa) of the hooded staminode are well developed; pollen is squeezed out of the theca (th) (slightly removed) by the elongating style. (D) Pollen is finally fixed onto the pollen plate (pp) opposite to the stigmatic cavity (sc) (same size as in (C)). (E) After pollen transfer to the style the theca (th) is empty and fading; the style (st) elongates more than the hooded staminode and sets up tension. (F) In the adult stage, the style (st) is bent backwards and the trigger appendage (ta) restricts access to nectar. Scale bars: 1 mm ((A)–(C), (E)), 500 μm (D), 2 mm (F) (after Claßen-Bockhoff and Heller 2008a).

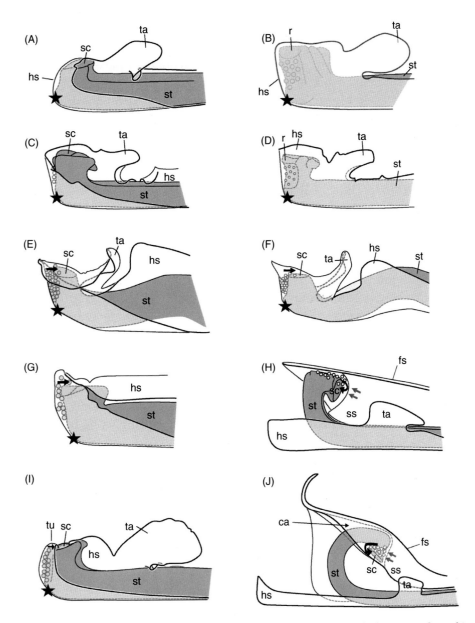

Fig 10.5 Diversity of the style head and the distal part of the hooded staminode and its influence on the mating system. (A), (B) *Maranta leuconeura*. (C), (D) *Donax canniformis*. (E), (F) *Pleiostachya pruinosa*. (G), (H) *Marantochloa leucantha*. (I) *Maranta noctiflora*. (J) *Halopegia azurea*. The style head is more or less covered by the hood of the hooded staminode; at the contact point (black star in (A)–(G), (I)) the style tension is set up. Note that in allogamous flowers (A)–(D) the pollen grains are separated from the stigmatic cavity by a high rim with abundant glandular material. In the autogamous species (E)–(J) the pollen grains (circles) reach the stigmatic cavity by being pushed through the space between style head and hood of the hooded staminode (black arrow). Glandular material is reduced. In *Halopegia azurea* (H) and *Marantochloa leucantha* (J) additional pollen

the relative position, proportions and specific shaping of their parts. As a consequence, different kinds of mating systems (allo- or autogamy), of barriers within the floral tube (trigger appendages and stiff swellings), of pollen deposition sites (lower side of the head or onto the mouthparts) and of floral constructions (for bee or bird pollination) appear.

10.4.1 Style head

The shape and size of the style head (with the pollen grains on the back) and the hood of the hooded staminode vary considerably between species (Fig 10.5). This consequently alters the relative position of both organs to each other and thereby influences the degree of autogamous versus allogamous pollination (Fig 10.5). While in most species the pollen grains are strictly separated from the stigmatic cavity by rims around the pollen plate or above the stigmatic cavity (Fig 10.5A–D) some species show a modified morphology allowing self-pollination (Fig 10.5E–J). Autogamy has arisen several times independently in the Marantaceae (Fig 10.3) probably representing 8% of the total number of species in the Marantaceae (Kennedy, 2000). In *Halopegia azurea* (Fig 10.5J) the proximal rim of the stigmatic cavity is lowered and in *Pleiostachya pruinosa* (Fig 10.5E, F) and *Marantochloa leucantha* (Fig 10.5G, H) the bulge of glandular material at the upper rim diminished. Furthermore, in each of these flowers the hooded staminode leaves a space, where, for example in *Maranta noctiflora*, a narrow tunnel (Fig 10.5I: tu is formed), which allows single pollen grains to move freely into the stigmatic cavity. Dependent on the specific flower construction, self-pollination takes place either before (e.g. *Maranta noctiflora*) or during the release of the style movement (e.g. *Pleiostachya pruinosa, Marantochloa leucantha, Halopegia azurea*) (see Ley, 2008). A common trait of all autogamous species is self-triggering of the style at the end of the day.

10.4.2 Hooded staminodes

The morphological diversity of the hooded staminode is based on its variable proportions, including the relative position of its trigger appendage (Fig 10.3). They are caused by differential growth during ontogeny. While the midrib of the young staminode is straight, its tip is shifted to the abaxial side in later ontogenetic stages due to the promoted growth of the hood side. The more it is overtopped by the hood,

Caption for figure 10.5 continued
grains are pushed into the stigmatic cavity by the collision of the style head with the stiff swelling of the fleshy staminode (grey arrow). ca, cavity formed by the fleshy staminode; fs, fleshy staminode; hs, hooded staminode; r, rim with glandular material; sc, stigmatic cavity; ss, stiff swelling of the fleshy staminode; st, style; ta, trigger appendage; tu, tunnel. (A), (C), (E), (G), (I): adaxial view; (B), (D), (F), (H), (J): abaxial view (after Ley, 2008).

the more proximal is the final position of the trigger appendage (Pischtschan et al., 2010). By comparing the ontogenies of the different types of staminodes, a gradual change from more distal (e.g. sword, gate and thumb type) to more proximal (e.g. door and cushion type) positioned trigger appendages is apparent in basal branching versus derived clades, respectively (Fig 10.3).

10.4.3 Fleshy staminodes

The fleshy staminode also shows morphological diversity. Its stiff swelling is either elongated or bipartite and it always develops independently of the midrib, from a secondary growth centre. Its adaptive significance is closely linked with the trigger appendage and the pollination success, as the swelling constricts the floral tube and forces the pollinators to deflect the trigger appendage and release the style movement. Different handling requirements (e.g. strength and behaviour of the pollinator) for different trigger appendages (e.g. more force for the thick gate type than for the soft sword type) might therefore lead to mechanical isolation by excluding certain pollinators.

Variation in the relative length of the fleshy staminode to the complex of style and hooded staminode leads to different pollen deposition sites onto the pollinator (Fig 10.6). In cases where the fleshy staminode is shorter than the complex of style and hooded staminode, the pollinator's head will be positioned directly above the style head, resulting in pollen deposition on its lower side (Fig 10.6A). If the fleshy staminode is of equal length to the 'complex', the style movement takes place inside the floral tube and as the pollinator is only able to insert its mouthparts and not its head into the floral tube, pollen is deposited onto the tip of its proboscis (Fig 10.6B) (see Ley, 2008). Kennedy (2000) postulated that differential pollen deposition might have caused reproductive isolation, but to properly estimate its evolutionary significance it has to be viewed in a phylogenetic context.

10.4.4 Flower types

Differential growth and development of the various floral parts result in adult flowers differing in their overall size, tube length, colour and position within the inflorescence. In African species, the specific combinations of these characters appear to correlate with ecological data such as nectar sugar concentration. These differences have led to the description of five flower types (small, medium-sized, large, locked (horizontal) and (large) vertical; Ley and Claßen-Bockhoff, 2010). The flower types closely correlate with different pollinator guilds (small, medium-sized, large bees and birds) indicating close floral adaptations to pollinators' body size, colour perception and foraging behaviour. A comparison of the pollinators of sister species might indicate whether this could have contributed to species divergence.

(A)

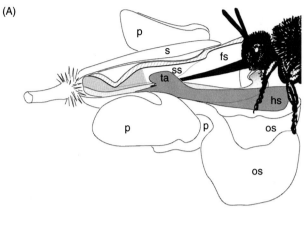

(B)

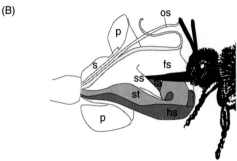

Fig 10.6 Differential pollen deposition in Marantaceae. (A) Pollen deposition underneath the head of *Amegilla vivida* in *Hypselodelphys* spp.; note that the fleshy staminode is shorter than the style and the hooded staminode. (B) Pollen deposition on the mouthparts of *Amegilla vivida* in *Megaphrynium* spp.; note that the fleshy staminode is as long as the style and the hooded staminode. fs, fleshy staminode; hs, hooded staminode; os, outer staminode; p, petal; s, sepal; ss, stiff swelling of the fleshy staminode; st, style; ta, trigger appendage.

10.5 Forces triggering speciation in the Marantaceae

In order to reconstruct character transformations in the evolution of the Marantaceae we used the most comprehensive family molecular phylogeny (Prince and Kress, 2006a), together with the existing phenotypic data (Fig 10.7A) (Dhetchuvi, 1996; Ley and Claßen-Bockhoff, 2010). Special attention is given to the African species, which are best investigated as to their phylogeny, morphology and ecology (Ley and Claßen-Bockhoff, 2011) (Fig 10.7B, C).

10.5.1 Mechanical isolation

If the trigger mechanism has indeed influenced speciation (key innovation hypothesis), then mechanical isolation should be expected by either differential

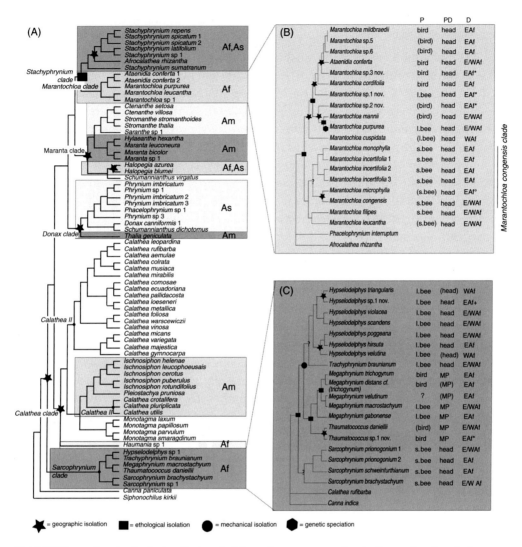

Fig 10.7 Reproductive isolation factors in Marantaceae mapped onto the molecular phylogeny of Marantaceae with special reference to the African species. (A) Phylogeny after Prince and Kress, 2006a, based on *matK* and 3′ intergenic spacer region and the *trnL-F* intergenic spacer region. (B), (C) Phylogeny after Ley and Claßen-Bockhoff, 2011, based on nuclear internal transcribed spacer (including ITS1, 5.8S and ITS2) and *trnL-F* intergenic spacer. Af, Africa; EAf, East Africa; WAf, West Africa; Am, America; As, Asia; D, distribution; P, pollinator guild; PD, site of pollen deposition; MP, mouth parts; s.bee, small bee; l.bee, large bee; values in brackets indicate assumed character states; collection sites in Gabon from Ley, 2008: °, Massif du Chaillu; *, Cristal Mountain; +, Lope).

pollen deposition onto the same pollinator, or by adaptation to different pollinators (Grant, 1994).

The first option has so far only been found once in the family (Fig 10.7B, C: PD), i.e. at the split of the two co-occurring African sister clades including the genera *Hypselodelphys/Trachyphrynium* and *Megaphrynium* (Dhetchuvi, 1996; Prince and Kress, 2006a; Ley and Claßen-Bockhoff, 2011). All three genera are pollinated by *Amegilla vivida*, but due to different flower proportions, pollen is deposited underneath the pollinator's head in *Hypselodelphys/Trachyphrynium* and onto its mouthparts in *Megaphrynium* (for details see Ley, 2008). Kennedy (2000) already mentioned different sites of pollen deposition in *Stromanthe/Ctenanthe* (*Maranta* clade) and *Calathea* (*Calathea* clade), which are, however, not closely related to each other (Fig 10.7A). She also observed different pollen deposition sites within the genus *Maranta*. This could well represent a further case of speciation by mechanical isolation, but it must be verified on a species level phylogeny.

The second option, i.e. the exclusion or allowance of specific pollinators due to the construction of the style release mechanism, has found no support as yet within the family. Instead, pollination studies in African species with different hooded staminode types elucidated that there is no correlation between certain pollinators and staminode types (Ley and Claßen-Bockhoff, 2010) (Fig 10.7B, C: P).

Based on the current knowledge there is no indication that the explosive pollination mechanism has acted as an important isolation factor during the evolution of the Marantaceae. The key innovation hypothesis is thus rejected, giving rise to the question of which other forces may have driven speciation.

10.5.2 Ethological isolation

Adaptation to different pollinators usually results from a complete change in the flower construction, including size, shape and colour, nectar peculiarities and exposition within the inflorescence (e.g. *Impatiens*/Balsaminaceae and *Pelargonium*/Geraniaceae in Vogel, 1954; *Disa*/Orchidaceae in Johnson and Steiner, 1995; *Petunia*/Solanaceae in Ando et al., 2001; *Tritoniopsis*/Iridaceae in Manning and Goldblatt, 2005; *Ruellia*/Acanthaceae in Tripp and Manos, 2008). This can be clearly confirmed by the African Marantaceae, in which different flower types are correlated with different pollinator guilds (Ley, 2008; Ley and Claßen-Bockhoff, 2010). Mapping these data onto the phylogeny of Marantaceae reveals that species of a given clade usually correspond in their flower type, but differ from the species of their sister clade (Fig 10.7B, C: P). Thus, they have obviously diverged through the adaptation to different pollinators (Ley and Claßen-Bockhoff, 2011) as was already shown in many other plant groups (e.g. Stebbins, 1970; Grant, 1994; Johnson et al., 1998; Specht, 2001; Wilson et al., 2004).

In the American genus *Calathea*, which is the largest in the family (~300 spp., Andersson, 1998) and characterized by a single type of hooded staminode

(Pischtschan et al., 2010), floral tube length shows a high degree of variation. Field observations confirm, that, for example, *C. crotalifera* subspecies, which differ in their flower tube length, are pollinated by different pollinators (Kennedy, 2000), thus providing an example of potential radiation in progress.

10.5.3 Geographical isolation

In the Marantaceae, larger sister clades are typically confined to different continents (Fig 10.7A) indicating that geographical isolation has been a major factor in the initial evolution of the family. According to the molecular clock approach conducted by Kress and Specht (2006), the radiation of the Marantaceae dates back to 63 ± 5 mya so that their pantropical distribution can only be explained by long-distance dispersal events from Africa to both the New World and Asia (Prince and Kress, 2006b). The complete interruption of gene flow through geographic isolation may have led to the substantial flower morphological differences, including the different types of hooded staminodes observed today. The general absence of transoceanic distribution patterns of genera (except *Thalia* and *Halopegia*) suggests a long period of divergence with no recent dispersal events.

But also within the continents, geographical isolation may have played an important role. In Africa, pairs of sister species are distributed in West and Central Africa (Fig 10.7B, C: D) separated by the Dahomey gap in Togo and Benin (see Booth, 1958; Hepper, 1968; Maley, 1996; Dupont et al., 2000; Salzman and Hoelzmann, 2005; Jongkind, 2008), or they occur as endemics in montane areas of central Africa and are separated from each other by mountain ridges and deep valleys (Dhetchuvi, 1996; Fjeldsa and Lovett, 1997; Barthlott et al., 1999). For the neotropics it could be shown that such topographically rich areas are centres of species richness (see Gentry and Dodson, 1987; Kessler, 1995; Ibisch, 1996; Ley et al., 2004).

Occasionally, extant distribution patterns may hide historical events of geographic isolation. A potential scenario in Africa includes geographic isolation in Pleistocene rainforest refuges (see Maley, 1996). This scenario has already been proposed for other African genera, such as *Aframomum* K. Schum. (Harris et al., 2000), *Begonia* L. (Plana et al., 2004) and *Renealmia* L. (Särkinen et al., 2007). During dry periods, African tropical lowland rainforest retracted to montane areas (Maley, 1996) so that long periods of geographic isolation led to their genetic incompatibility (compare de Nettancourt, 1977; Orr and Presgraves, 2000; Skrede et al., 2006). Only the extant distribution patterns of the three *Haumania* species (see Dhetchuvi, 1996) still exhibit a high conformity, each with a different postulated refuge area supporting this hypothesis. Their large and heavy seeds might have prevented extensive range expansions up to today (Fig 10.8 and see Dhetchuvi, 1996). All other Marantaceae species exhibit highly overlapping distribution patterns probably due to more efficient dispersal mechanisms, thereby potentially masking past geographic isolation (see Comes and Abbott, 2001).

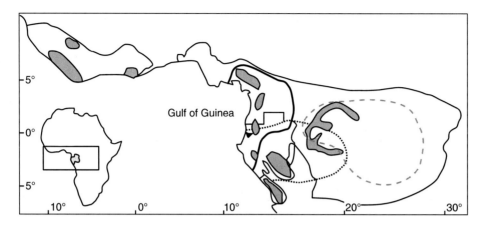

Fig 10.8 Distribution of *Haumania* species (after Dhetchuvi, 1996). Thick line, limit of current rainforest distribution; thin line, distribution area of *Haumania danckelmanniana*; dotted grey line, distribution area of *Haumania liebrechtsiana*; dashed grey line, distribution area of *Haumania leonardiana*; dark grey areas, postulated refugial areas (from Leal, 2004, simplified after Maley, 1996); light grey area, Gabon.

Phylogeographic investigations are now underway to elucidate this hypothesis (Ley and Hardy, ULB, Belgium: http://ebe.ulb.ac.be/ebe/ebe-Welcome.html).

Similar speciation events might be hypothesized for America. Data from related Zingiberales families suggest an increase in the diversification rate with the orogeny of the Andes (e.g. Kennedy, 2000; Kay et al., 2005; Särkinen et al., 2007). In Asia, the distribution of sister species often correspond with the scattered island landscape (for Marantaceae, see Suksathan et al., 2009; see also Cannon and Manos, 2003; Mausfeld and Schmitz, 2003; Outlaw and Voelker, 2007), again pointing to geographical isolation.

10.5.4 Speciation via hybridization

Occasionally, genetic factors might have played a role in the speciation of Marantaceae. Phylogeographic studies in *Haumania* (Ley and Hardy, 2010) and topological incongruencies in phylogenetic trees based on cp- and nr-DNA in *Marantochloa* and *Megaphrynium* suggest several hybridization events (see Ley and Claßen-Bockhoff, 2011). However, more detailed genetic data is still needed to define whether these processes actually contributed to speciation (see also Rieseberg, 1997).

10.6 Conclusion

The study exemplarily demonstrates how the combination of different data sets (ecological, molecular, morphological) can reveal a comprehensive view on the evolution of a given taxon. The radiation of the Marantaceae includes a variety of possible isolation factors. Beginning with the backbone of the phylogenetic tree, geographic isolation on distant continents might have forced primary separation into major clades. Following which, the adaptation to different pollinators may have given rise to large clades within continents. The divergence of extant closely related species is probably mainly shaped by past and present geographic isolations due to climate changes, island and mountainous landscapes and, only occasionally, due to mechanical isolation and hybridization events.

The diversity of different types of hooded staminodes might have also arisen solely by chance in geographic isolation (see Fig 10.3). Only in a few cases, sister clades with different types of hooded staminodes can be found co-occurring on the same continent. Here, the adaptation to different pollinators might indeed have played a role. For the African 'gate' and 'sword' types differential pollination is already proven (Fig 10.7, Ley and Claßen-Bockhoff, 2011) and for the American 'thumb' and 'tentacle' types, which, however, do not appear in sister groups, it is highly probable (see Davis, 1987; Kennedy, 2000). Only the sympatric occurrence of the American 'spoon' and 'wing' types still remains to be investigated.

At least once different trigger appendages (cushion and sword type, see Ley and Claßen-Bockhoff, 2010) have developed in parallel adaptations to the same pollinators in Africa. This is probably due to the same selection pressures imposed on both clades (*Marantochloa* and *Sarcophrynium* clade) by identical local environments (Perret et al., 2007). Thereby the genetic distance between the clades should have successfully impeded hybridization. Instead, the benefits from using the same pollinators probably outweighed the advantage of diverging to reduce competition, especially as all these species are sparsely flowering (see also Schemske, 1981). However, whether the species in the two clades evolved consecutively or simultaneously is still a matter of debate.

When one counts the number of nodes at which certain isolation factors may have acted, both geographic isolation and adaptation to different pollinators occur most frequently. Only very few isolation events could be attributed to the diversification of the trigger mechanism and for this reason it is rejected as a key innovation forcing speciation within the Marantaceae for the Marantaceae. The surprising result is that the diversity of the floral elements intimately involved in the reproductive success does not appear to influence radiation. Its existence should instead be interpreted as an optimization of the reproductive system guaranteeing successful pollen transfer – a new hypothesis to be tested.

Acknowledgements

We thank Louis Ronse de Craene and Livia Wanntorp for organizing the symposium on the 'Flowers on the Tree of Life' and two anonymous reviewers for their comments on the manuscript. The first author thanks the Systematics Association for financing the trip to the Systematics 2009 conference in Leiden, August 2009.

10.7 References

Andersson, L. (1998). Marantaceae. pp. 278–293 in Kubitzki, K. (ed.), *The Families and Genera of Vascular Plants.* Vol. IV. *Flowering Plants, Monocotyledons, Alismatanae and Commelinanae (except Gramineae).* Berlin: Springer.

Andersson, L. and Chase, M. W. (2001). Phylogeny and classification of *Marantaceae. Botanical Journal of the Linnean Society,* **135**, 275–287.

Ando, T., Nomura, M. and Tsukahara, J. et al. (2001). Reproductive isolation in a native population of Petunia sensu Jussieu (Solanaceae). *Annals of Botany,* **88**, 403–413.

Barthlott, W., Kier, G. and Mutke, J. (1999). Biodiversity – The uneven distribution of a treasure. *NNA-Report,* Special Issue **2**, 18–28.

Booth, A. H. (1958). The Niger, the Volta and the Dahomey Gap as geographic barriers. *Evolution,* **12**(1), 48–62.

Cannon, C. H. and Manos, P. S. (2003). Phylogeography of the Southeast Asian stone oaks (*Lithocarpus*). *Journal of Biogeography,* **30**, 211–226.

Claßen-Bockhoff, R. (1991). Untersuchungen zur Konstruktion des Bestäubungsapparates von *Thalia geniculata* (Marantaceae). *Botanica Acta,* **74**, 183–193.

Claßen-Bockhoff, R. and Heller, A. (2008a). Floral synorganisation and secondary pollen presentation in four Marantaceae from Costa Rica. *International Journal of Plant Sciences,* **169**, 745–760.

Claßen-Bockhoff, R. and Heller, A. (2008b). Style release experiments in four species of Marantaceae from the Golfo Dulce area, Costa Rica. *Stapfia,* **88**, 557–571.

Comes, H. P. and Abbott, R. J. (2001). Phylogeography, reticulation and lineage sorting in Mediterranean *Senecio* sect. Senecio (Asteraceae). *Evolution,* **55**(10), 1943–1962.

Davis, M. A. (1987). The role of flower visitors in the explosive pollination of *Thalia geniculata* (Marantaceae), a Costa Rican marsh plant. *Bulletin of the Torrey Botanical Club,* **114**, 134–138.

Delpino, F. (1869). Breve cenno sulle relazioni biologiche e genealogiche delle Marantacee. *Nuovo Giornale Botanico Italiano,* **1**, 293–306.

Dhetchuvi, J. B. (1996). Taxonomie et phytogéographie des Marantaceae et des Zingiberaceae de l'Afrique Centrale (Gabon, Congo, Zaire, Rwanda et Brundi). PhD Thesis, Université Libre de Bruxelles, Belgique.

Dupont, L. M., Jahns, S., Marret, F. and Ning, S. (2000). Vegetational change in equatorial West Africa: time-slices for the last 150ka. *Palaeogeography, Palaeoclimatology and Palaeoecology,* **155**, 95–122.

Fjeldsa, J. and Lovett, J. C. (1997). Geographical patterns of old and young species in African forest biota: the significance of specific montane areas as evolutionary centres. *Biodiversity and Conservation,* **6**, 325–346.

Gentry, A. H. and Dodson, C. H. (1987). Diversity and biogeography of neotropical vascular epiphytes. *Annals of the Missouri Botanical Garden*, **24** (2), 205–233.

Grant, V. (1994). Modes and origins of mechanical and ethological isolation in angiosperms. *Proceedings of the National Academy of Sciences USA*, **91**, 3–10.

Gris, A. (1859). Observations sur la fleur des Marantées. *Annales des Sciences naturelles*, **IV, 12**, 193–219.

Harris, D. J., Poulsen, A. D., Frimodt-Moller, C., Preston, J. and Cronk, Q. C. B. (2000). Rapid radiation in *Aframomum* (Zingiberaceae): evidence from nuclear ribosomal DNA internal transcribed spacer (ITS) sequences. *Edinburgh Journal of Botany*, **57**, 377–395.

Hepper, F. N. (1968). Marantaceae. pp. 79–89 in Hutchinson, J. and Dalziel, J. M. (eds.), *Flora of Tropical West Africa*. London: Crown Agents for Oversea Governments and Administrations.

Ibisch, P. L. (1996). *Neotropische Epiphytendiversität – das Beispiel Bolivien. Archiv naturwissenschaftlicher Dissertationen*. Wiehl: Matina Galunder-Verlag.

Johnson, S. D. and Steiner, K. E. (1995). Long-proboscid fly pollination of two orchids in the Cape Drakensberg mountains, South Africa. *Plant Systematics and Evolution*, **195**, 169–175.

Johnson, S. D., Linder, H. P. and Steiner, K. E. (1998). Phylogeny and radiation of pollination systems in *Disa* (Orchidaceae). *American Journal of Botany*, **85**(3), 402–411.

Jongkind, C. C. H. (2008). Two new species of *Hypselodelphys* (Marantaceae) from West Africa. *Adansonia*, **30** (1), 57–62.

Kay, K. M., Reeves, P. A., Olmstead, R. G. and Schemske, D. W. (2005). Rapid speciation and the evolution of hummingbird pollination in neotropical *Costus* subgenus *Costus* (Costaceae): evidence from nrDNA ITS and ETS sequences. *American Journal of Botany*, **92**(1), 1899–1910.

Kennedy, H. (2000). Diversification in pollination mechanisms in the Marantaceae. pp. 335–343 in Wilson K. L. and Morrison, D. A. (eds.), *Monocots: Systematics and Evolution*. Melbourne, CSIRO.

Kessler, M. (1995). *Polylepis*-Wälder Boliviens: Taxa, Ökologie, Verbreitung und Geschichte. *Dissertationes Botanicae*, **246**. J. Cramer, Berlin-Stuttgart.

Kirchoff, B. K. (1983). Allometric growth of the flowers in five genera of the Marantaceae and in Canna (Cannaceae). *Botanical Gazette*, **144** (1), 110–118.

Kress, J. and Specht, C. D. (2006). The evolutionary and biogeographic origin and diversification of the tropical monocot order Zingiberales. *Aliso*, **22** (1), 619–630.

Kunze, H. (1984). Comparative studies in the flower of Cannaceae and Marantaceae. *Flora*, **175**, 301–318.

Kunze, H. (2005). Musterbildungsprozesse in der Blütenontogenese der Zingiberales. Werte und Grenzen des Typus in der botanischen Morphologie. pp. 189–200 in Harlan, V. (ed.), *Wert und Grenzen des Typus in der botanischen Morphologie*. Nümbrecht: Martin Galunder-Verlag.

Leal, M. E. (2004). *The African Rainforest During the Last Glacial Maximum: An Archipelago of Forests in a Sea of Grass*. Wageningen: Department of Plant Science, Biosystematics Group.

Ley, A. C. (2008). Evolutionary tendencies in African Marantaceae – evidence from floral morphology, ecology and phylogeny. PhD thesis. Mainz, Germany: University of Mainz.

Ley, A. C. and Claßen-Bockhoff, R. (2009). Pollination syndromes in African Marantaceae. *Annals of Botany*, **104**(1), 41–56.

Ley, A. C. and Claßen-Bockhoff, R. (2010). Parallel evolution in plant-pollinator interaction in African Marantaceae. pp. 847–854 in van der Burgt, X., van der Maesen, J. and Onana, J.-M. (eds.), *Systematics and Conservation of African Plants*. Kew: Royal Botanic Gardens, Kew.

Ley, A. C. and Claßen-Bockhoff, R. (2011). Evolution in African Marantaceae – evidence from phylogenetic, ecological and morphological studies. *Systematic Botany*. In press.

Ley, A. C., Nowicki, C., Barthlott, W. and Ibisch, P. L. (2004). Biogeography and spatial diversity. pp. 492–571 in Vasquez, Ch. R. and Ibisch, P. L. (eds.), *Orchids of Bolivia – Diversity and Conservation Status, Laeliinae, Polystachinae, Sobraliinae with updates of the Pleurothallidinae*, Vol. 2. Santa Cruz de la Sierra, Bolivia: Editorial FAN.

Ley, A. C., Hardy, O. J., (2010). Species delimitation in the Central African herbs Haumania (Marantaceae) using georeferenced nuclear and chloroplastic DNA sequences. *Molecular Phylogenetics and Evolution* **57**, 859.

Maley, J. (1996). The African rain forest – main characteristics of changes in vegetation and climate from the Upper Cretaceous to the Quaternary. *Proceedings of the Royal Society of Edinburgh*, **104B**, 31–73.

Manning, J. C. and Goldblatt, P. (2005). Radiation of pollination systems in the cape genus *Tritoniopsis* (Iridaceae:

Crocoideae) and the development of bimodal pollination strategies. *International Journal of Plant Sciences*, **166** (3), 459–474.

Mausfeld, P. and Schmitz, A. (2003). Molecular phylogeography, intraspecific variation and speciation of the Asian scincid lizard genus *Eutropis* Fitzinger, 1843. (Squamata: Reptilia: Scincidae): taxonomic and biogeographic implications. *Organisms, Diversity and Evolution*, **3** (3), 161–171.

Nettancourt, D. de (1977). *Incompatibility in Angiosperms*. Berlin, Heidelberg and New York: Springer Verlag.

Orr, H. A. and Presgraves, D. C. (2000). Speciation by postzygotic isolation: forces, genes and molecules. *BioEssays*, **22**, 1085–1094.

Outlaw, D. C. and Voelker, G. (2007). Pliocene climatic change in insular Southeast Asia as an engine of diversification in *Ficedula* flycatchers. *Journal of Biogeography*, **35** (4), 739–752.

Perret, M., Chautems, R., Spichiger, R., Barraclough, T. G. and Savolainen, V. (2007). The geographical pattern of speciation and floral diversification in the neotropics: the Tribe Sinningiaeae (Gesnericaeae) as a case study. *Evolution*, **61** (7), 1641–1660.

Pischtschan, E. and Claßen-Bockhoff, R. (2008). The setting-up of tension in the style of Marantaceae. *Plant Biology*, **10**, 441–450.

Pischtschan, E., Ley, A. C. and Claßen-Bockhoff, R. (2010). Ontogenetic and phylogenetic diversification of the hooded staminode in Marantaceae. *Taxon*, **59**, 1111–1125.

Plana, V., Gascoigne, A., Forrest, L. L., Harris, D. and Pennington, R. T. (2004). Pleistocene and pre-Pleistocene *Begonia* speciation in Africa. *Molecular Phylogenetics and Evolution*, **31**, 449–461.

Prince, L. M. and Kress, W. J. (2006a). Phylogenetic relationships and classification in Marantaceae: insights from plastid DNA sequence data. *Taxon*, **55** (2), 281–296.

Prince, L. M. and Kress, W. J. (2006b). Phylogeny and biogeography of the prayer plant family: getting to the root problem in Marantaceae. *Aliso*, **22** (1), 643–657.

Rieseberg, L. H. (1997). Hybrid origins of plant species. *Annual Review of Ecology and Systematics*, **28**, 359–389.

Salzmann, U. and Hoelzmann, P. (2005). The Dahomey Gap: an abrupt climatically induced rain forest fragmentation in West Africa during the late Holocene. *The Holocene*, **15** (2), 190–199.

Särkinen, R. E., Newman, M. F. and Maas, P. J. M. et al. (2007). Recent oceanic long-distance dispersal and divergence in the amphi-Atlantic rain forest genus *Renealmia* L.f. (Zingiberaceae). *Molecular Phylogenetics and Evolution*, **44**, 968–980.

Schemske, D. W. (1981). Floral convergence and pollinator sharing in the bee-pollinated tropical herbs. *Ecology*, **62** (4), 946–954.

Schumann, K. (1902). Marantaceae. pp. 1–184 in Engler, A. (ed.), *Das Pflanzenreich IV*. Leipzig: W. Engelmann.

Skrede, I., Eidesen, P. B., Portela, R. P. and Brochmann, C. (2006). Refugia, differentiation and postglacial migration in arctic-alpine Eurasia, exemplified by the mountain avens (*Dryas octopetala* L.). *Molecular Ecology*, **15** (7), 1827–1840.

Specht, C. D. (2001). Phylogenetics, floral evolution and rapid radiation in the tropical monocotyledon family Costaceae (Zingiberales). pp. 29–60 in Sharma, A. K. and Sharma, A. (eds.), *Plant Genome – Biodiversity and Evolution*. Ensfield, NH: Science Publishers, Inc.

Stebbins, G. L. (1970). Adaptive radiation of reproductive characteristics in angiosperms, I: pollination mechanisms. *Annual Review of Ecology and Systematics*, **1**, 307–326.

Suksathan, P., Gustafsson, M. H. and Borchsenius, F. (2009). Phylogeny and generic delimitation of Asian Marantaceae. *Botanical Journal of Linnean Society*, **159**, 381–395.

Tripp, E. A. and Manos, P. S. (2008). Is floral specialization an evolutionary dead-end? Pollination system transitions in *Ruellia* (Acanthaceae). *Evolution*, **62**, 1712–1737.

Vogel, S. (1954). *Blütenbiologische Typen als Elemente der Sippengliederung*. Jena: Gustav Fischer Verlag.

Wilson, P., Castellanos, C., Hogue, J. N., Thomson, J. D. and Armbruster, W. S. (2004). A multivariate search for pollination syndromes among penstemons. *Oikos*, **104**, 345–361.

Yeo, P. (1993). Secondary pollen presentation. Form, function and evolution. *Plant Systematics and Evolution (Supplement)*, **6**, 1–268.

11

Floral ontogeny of *Acacia celastrifolia*: an enigmatic mimosoid legume with pronounced polyandry and multiple carpels

GERHARD PRENNER

11.1 Introduction

The genus *Acacia* is among the largest plant genera. It was recently treated either in a broad sense with *c.* 1450 species (Lewis, 2005) or in a strict sense (*Acacia* s.s. with *c.* 987 species). The latter follows the re-typification of *Acacia* with an Australian type (Orchard and Maslin, 2003; see also Murphy, 2008). According to Maslin (1995), the Australian species *A. celastrifolia* belongs to the '*Acacia myrtifolia* group' and is most closely related to *A. myrtifolia*. In molecular studies, only *A. myrtifolia* was sampled, which is sister to *A. pulchella* in the Pulchelloidea clade (e.g. Miller and Bayer, 2001; Miller et al., 2003; Murphy et al., 2010). Molecular sampling of the hitherto unsampled *A. celastrifolia* is highly desirable in order to verify the hypothesized close relationship with *A. myrtifolia* (e.g. Maslin, 1995).

Flowers of the genus *Acacia* s.l. are always found in globular heads or spikes. The flowers are (3–)4–5(–6)-merous, with free or united sepals and small reduced petals, which are postgenitally fused and which split open at anthesis. The androecium is

Flowers on the Tree of Life, ed. Livia Wanntorp and Louis P. Ronse De Craene. Published by Cambridge University Press. © The Systematics Association 2011.

composed of many free stamens (i.e. a polyandrous androecium) and the flower is normally terminated by a single superior carpel.

Acacia celastrifolia deviates from this bauplan in that it has 3–7 carpels. Furthermore its exceptionally high polyandry, with more than 500 stamens, is noteworthy. Flowers of the majority of Leguminosae are terminated by a single carpel and multicarpelly in Fabales is restricted to Polygalaceae, Quillajaceae and Surianaceae. Thus, a comparative study of legume flowers with more than one carpel per flower is of special interest for a better understanding of legume flowers and their evolution.

11.2 Materials and methods

Floral buds and inflorescences of *Acacia celastrifolia* Benth. were collected from the Botanic Garden Graz (University of Graz, Austria) and immediately fixed in FAA (five parts formalin:five parts 100% acetic acid:90 parts 70% ethanol). Storage and further dissection was in 70% ethanol. For scanning electron microscopy (SEM), material was dehydrated through an alcohol series to absolute alcohol and critical-point dried using an Autosamdri-815B critical-point dryer. Dried material was mounted onto specimen stubs using nail polish, coated with platinum using an Emitech K550 sputter coater and examined using a Hitachi cold field emission SEM S-4700-II. More than 300 SEM micrographs were analysed and figures were processed using Adobe Photoshop CS.

11.3 Results

11.3.1 Habitus (Fig 11.1A–D)

Acacia celastrifolia is a glabrous shrub of 1–3 m. Flowers are clustered in 10–20-headed racemes (Fig 11.1A). Phyllodes show a distinct midrib and a prominent gland 0.5–2 cm above the pulvinus (Fig 11.1B,C). Individual flowers form (1–) 2–3-flowered heads on 3–10 mm long peduncles. In the mature flowers the recurved petals and the highly polyandrous androecium are the most distinct characters (Fig 11.1D). Styles of the three to seven carpels are longer than the stamens and arch clearly above the level of the anthers (Fig 11.1D).

11.3.2 Flower formation (Fig 11.2A–D)

On the inflorescence axis (1–)2–3-flowered heads are formed in a spiral (Fig 11.1A–B, Fig 11.2A). The flowers are formed in a quick succession on a short roundish peduncle (Fig 11.2A–B) and flower subtending bracts are either formed or missing (Fig 11.2A–F). No flower preceding bracteoles are formed. Multicellular and club-shaped hairs are rarely found at the base of the floral heads (Fig 11.2E) and in groups between the individual flowers (Fig 11.2F,G).

Fig 11.1 *Acacia celastrifolia*. (A) Flowering branch. (B) Inflorescence with phyllode at its base. (C) Detail of phyllode with nectary and drop of nectar. (D) Flower with recurved petals, highly polyandrous androecium and four styles which are protruding above the level of the anthers. Scale bar = 1 cm in all. For colour illustration see plate section.

11.3.3 Calyx (Fig 11.3A–F)

Soon after flower primordia become visible, a calyx is produced in the form of a closed ring primordium (Fig 11.3A–B). The shape of the primordium is uniform at the beginning and only later in ontogeny do four distinct lobes become discernible. The two lobes in the median position (one pointing to the main floral axis and one opposite) become larger than the two lateral ones, which become the inner sepals in later developmental stages (Fig 11.3C–D). Sepals are congenitally fused with only the upper lobes free (Fig 11.3D–E). Later in ontogeny, sepal aestivation becomes open, due to the enlarging bud and petals take over bud protection (Fig 11.3F).

11.3.4 Corolla (Fig 11.4A–G)

Distinctly after the sepals are formed (i.e. after a longer plastochron during which sepals are formed and enlarge), formation of four petals commences in a simultaneous pattern (Fig 11.4A). The four petals are formed apart from each other and only in some flowers are two petals formed in a very close proximity. The petals

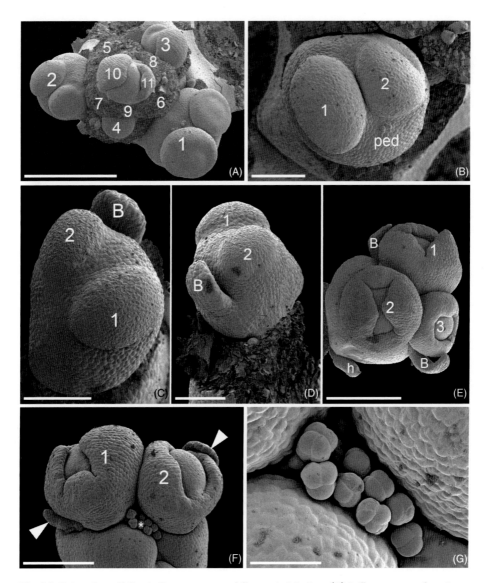

Fig 11.2 *A. celastrifolia*, inflorescence and flower initiation. (A) Inflorescence showing spiral initiation of partial inflorescences (1–4) and their subtending bracts respectively (5–11). (B) Partial inflorescence with two flowers initiated in a short sequence on short roundish peduncle (ped). (C), (D) Two-flowered partial inflorescence showing initiation of two flowers and one flower subtending bract which can be assigned to the second flower. (E) Three-flowered partial inflorescence. Note that there is a developmental difference between the three flowers. Two flowers are subtended by bracts and at the base of the entire structure one club-shaped multicellular hair (h) is found. (F) Two-flowered partial inflorescence, each of the flowers associated to a flower subtending bract (arrowheads). Note the hairs at the base of the two flowers (asterisk). (G) Detail of (F) showing multicellular hairs. B = flower subtending bract, h = multicellular hair, ped = peduncle. Scale bars: (A) = 400 µm; (B)–(D) = 100 µm; (E) = 300 µm; (F) = 200 µm; (G) = 50 µm.

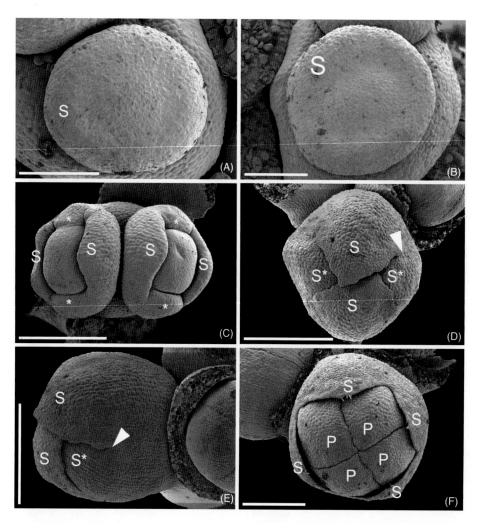

Fig 11.3 *A. celastrifolia*, sepal initiation and development. (A) Young flower in which the sepaloid ring primordium becomes just visible. (B) Somewhat later stage showing distinct ring primordium. (C) Two-flowered partial inflorescence. The sepals in median position are larger than the lateral sepals (asterisks). (D) Frontal view of older flower with closed calyx. The sepals in median position are covering the lateral sepals (S*). Sepals are fused (arrowhead) and only the upper lobes remain free. (E) Lateral view of older flower showing fused calyx and free sepal lobes. (F) Frontal view of older flower in which the calyx is open and protection of the sexual organs is by the valvately closed petals. P = petal, S = sepal. Scale bars: (A), (B) = 100 μm; (C) = 200 μm; (D)–(F) = 500 μm.

enlarge quickly and fuse postgenitally due to interlocking epidermal papillae, which become most prominent at the tips of the petals (Fig 11.4B–G). Petal aestivation is valvate and due to the enlarging corolla the calyx is squeezed open (Fig 11.3F). At this developmental stage petals take over the protection of the developing

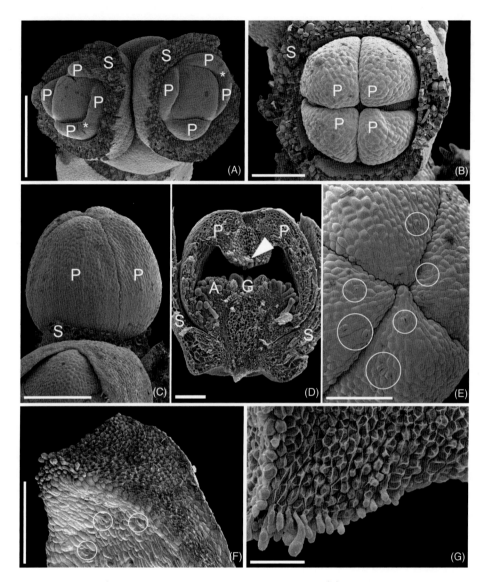

Fig 11.4 *A. celastrifolia*, petal initiation and development. (A) Two flowers in which the sepals are removed and four petals are formed simultaneously. (B) Frontal view of young flower (sepals removed). Four petals are still free from each other, petal aestivation becomes valvate. (C) Side view of older flower bud (sepals removed). Note the distinct suture between adjacent petals. (D) Median section through flower bud with short sepals in open aestivation, valvate petals with distinct papillae on the inner tips (arrowhead), androecium and gynoecium. (E) Outer petal surface of young floral bud showing many stomata (circles). (F) Stomata on the inner petal surface (circles). (G) Petal tip with distinct papillate epidermal cells and epidermal ridges in the regions of postgenital fusion. A = androecium, G = gynoecium, P = petal, S = sepal/calyx. Scale bars: (A), (F) = 300 μm; (B), (D), (E) = 200 μm; (C) = 500 μm; (G) = 100 μm.

androecium and gynoecium inside the floral bud. Both on the outer and on the inner surface of the mature petals abundant stomata are found (Fig 11.4E–F).

11.3.5 Androecium (Figs 11.5A–F, 11.6A–E)

Androecium formation starts after a longer plastochron, during which the petals enlarge distinctly and the young floral meristem is protected by the corolla. At this developmental stage the floral meristem becomes square in shape and a distinct ring primordium becomes visible (Fig 11.5A). Formation of individual stamens starts with four larger primordia which arise at the four corners of the square alternating with the four petals (Fig 11.5A–D). Stamen formation continues in lateral and centripetal direction. Individual stamen primordia are *c.* 20 μm in diameter (Fig 11.5D–F). In the mature flower more than 500 individual stamens are found, which arise on a distinct androecial bulge (i.e. reminder of the ring primordium) at the base of the flower (Fig 11.6A–B). The anthers of *A. celastrifolia* show two thecae separated by a broad connective (Fig 11.6C). Each theca is parted into two locules (i.e. microsporangia) and each locule is again divided by a transverse septum. The inner surface of the locules is covered with orbicules which makes their appearance granulate (Fig 11.6D). In each locule two eight-grained polyads are formed (Fig 11.6D–E). This makes a total of eight polyads and 64 pollen grains per anther. The pollen grains are heteromorphic within each polyad. Two central pollen grains can be clearly distinguished from six lateral grains (Fig 11.6E).

11.3.6 Gynoecium (Figs 11.5E–F, 11.7A–H, 11.8A–E)

Carpel formation starts before the last stamens are formed (Fig 11.5F). At this developmental stage there is a distinct separation of the androecial ring-wall and the central floral region where the carpels are formed. Following the somewhat square-shaped androecial ring-wall, the central region of the flower also becomes square shaped and in some instances at each corner of the square one carpel is formed (Fig 11.5E–F). However, in other flowers only three or up to seven carpels are formed. Carpel initiation appears, in some flowers, simultaneous (Figs 11.5F, 11.7A–B), while in others it is erratic with no distinct discernible sequence (Fig 11.7C–H). Common to all studied flowers is: (1) that the floral centre remains organ free, (2) that the carpels are formed only around this centre and (3) that the cleft of the carpels always points towards the floral centre. The mature carpel shows a long style and a small stigmatic region on top (Fig 11.8A–B). This can harbour only one single polyad (Fig 11.8B). Abundant stomata are located on the surface of the carpel and at the base a short gynophore is formed (Fig 11.8A, C–D). A maximum of eight ovules are formed within each carpel and in each ovule initially only the outer integument is visible (Fig 11.8D–E). The inner integument is formed later in ontogeny (not shown).

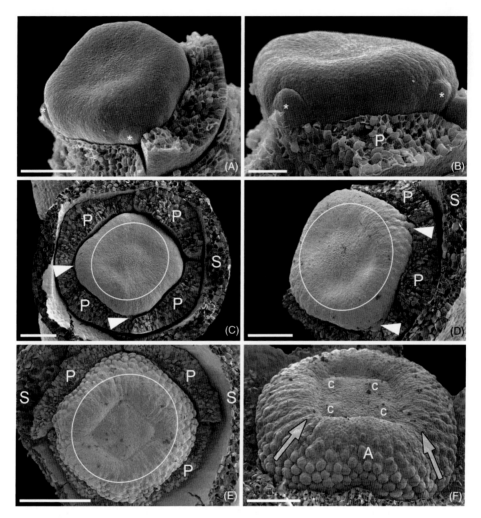

Fig 11.5 *A. celastrifolia*, stamen initiation. Sepals and petals removed in all. (A) Frontal view of androecial ring-wall with first stamens formed (asterisks) alternating with the petals. (B) Lateral view of a bud with first formed stamens (asterisks). (C) Androecial ring-wall (circle) and first formed stamens (arrowheads) alternating with the petals. (D) Side view of somewhat older bud. Stamen primordia are formed in a centripetal pattern. (E) Older floral bud with massive androecial ring-wall (circle) distinctly separated from inner part of the flower. (F) Side view of androecial ring-wall. Arrows show the direction of stamen initiation. In the centre of the flower four carpel primordia are just formed (c). A = androecium, c = carpel primordium, P = petal, S = sepal/calyx. Scale bars: (A), (C), (D), (F) = 100 μm; (B) = 50 μm; (E) = 200 μm.

Fig 11.6 *A. celastrifolia*, mature androecium and pollen. (A) Part of the androecium of a mature flower. Note the small anthers sitting on long filaments. (B) Base of mature flower, stamens removed leaving only a basal ring-wall of androecial tissue. (C) Versatile anther with two thecae attached to a broad connective. Note a polyad looking out of one theca (arrowhead). (D) Opened theca showing a locule which is divided by a central transverse tissue (arrowheads). Note the inner surface of the locule which is covered with orbicules (granular surface). (E) Eight-grained polyad with two central grains surrounded by six lateral grains. Note the difference of the central and the lateral grains (i.e. heteromorphic pollen). A = androecium, Co = connective, C = carpel, L = locule, P = petal, Th = theca. Scale bars: (A), (B) = 500 μm; (C) = 50 μm; (D) = 20 μm; (E) = 10 μm.

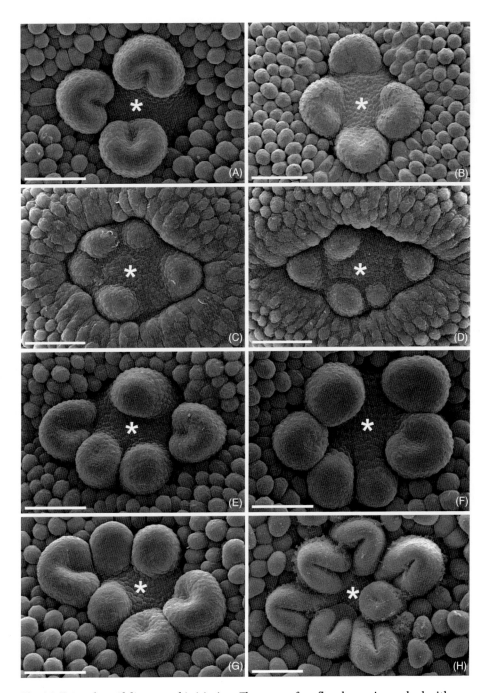

Fig 11.7 *A. celastrifolia*, carpel initiation. The organ-free floral apex is marked with an asterisk in all. Note that in all samples the carpellary cleft (if already present) is pointing towards the floral centre. (A) Three very similar carpels suggesting simultaneous initiation. (B) Four carpels initiated and very similar in size. (C)–(E). Five carpels formed in no defined sequence. (F) Six carpels formed apparently in anticlockwise sequence. (G) Six carpels formed in no discernible sequence (i.e. erratic). (H) Seven carpels formed, six are of similar developmental stage, one lags behind. Scale bars = 100 μm in all.

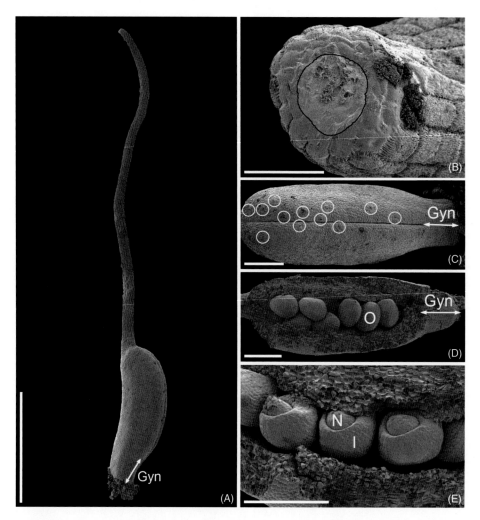

Fig 11.8 *A. celastrifolia*, mature carpel. (A) Mature carpel with a gynophore, the ovary and a long style. (B) Stigmatic area of mature carpel with the outline of a polyad projected onto its surface. Note that the polyad covers almost the entire receptive surface of the style. (C) Mature ovary with many stomata on its surface (circles) and with a distinct basal gynophore. (D) Dissected ovary with seven ovules. (E) Detail of ovules. Note that only the outer integument and the nucellus are visible. Gyn= gynophore, I = outer integument, N = nucellus, O = ovule. Scale bars: (A) = 1 mm; (B) = 30 μm; (C)-(E) = 200 μm.

11.4 Discussion

11.4.1 Inflorescence and floral orientation

In *A. celastrifolia* the flower number per head is reduced to (1–)2–3 flowers, which contrasts with most other acacias (*Acacia* s.l.), in which the heads are many-flowered. Guinet et al. (1980) highlight that *Acacia gilbertii* is highly unusual because of the 2–8 relatively large tetramerous flowers per head; this feature is otherwise unknown in Pulchellae but 'identical' with that of the phyllodinous species *A. myrtifolia* and *A. celastrifolia*. Floral orientation in Mimosoideae differs from Caesalpinioideae and Papilionoideae in that one sepal is oriented towards the main axis in mimosoids, while one petal is oriented this way in the other two subfamilies. In this respect, flowers of *A. celastrifolia* are orientated in the same way as in most Mimosoideae, with one sepal towards the main axis (Fig 11.9; see also Eichler, 1878; Tucker, 1987, 2003; Prenner, 2004a; Ronse De Craene, 2010). In other mimosoid taxa with only four sepals, a shift in orientation has occurred and a petal is pointing towards the main axis (e.g. *Mimosa pudica*, Eichler, 1878). Flower subtending bracts are found infrequently and flower preceding bracteoles are entirely missing (see Prenner, 2004c for a discussion of the occurrence of bracteoles in Leguminosae).

It is interesting to note that within a single inflorescence, organ formation starts sequentially and sepals are therefore at different developmental stages within one

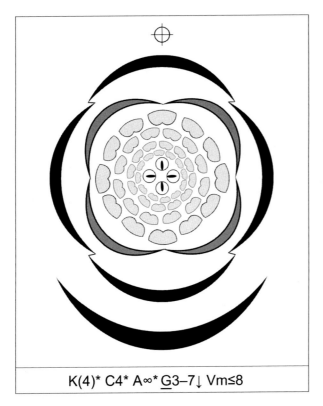

$$K(4)^* \ C4^* \ A\infty^* \ \underline{G}3\text{–}7\downarrow \ Vm\leq8$$

Fig 11.9 Floral diagram and floral formula. Bract and sepals in black, petals in dark grey, stamens in bright grey. Only a reduced number of stamens is shown and four carpels are shown representing the multicarpellate gynoecium which consists of 3–7 carpels. The format of the floral formula follows Prenner et al. (2010).

inflorescence (see Fig 11.2E–F). This contrasts with other Mimosoideae, where within one inflorescence organ formation commences simultaneously (i.e. the floral primordia stay organ free until the last primordium is formed and only then does organ formation start in all meristems simultaneously) (Tucker, 1992; Prenner, 2004a). In Papilionoideae and Caesalpinioideae, organ formation always starts immediately after the floral primordia are formed and therefore different ontogenetic stages can be found within a single inflorescence (c.f. Prenner, 2004d; Prenner and Klitgaard, 2008).

11.4.2 Calyx

While sepal orientation follows the common mimosoid pattern (see discussion above), sepal initiation in the form of a ring meristem is a rarely documented phenomenon for the subfamily. It was only reported from the genus *Mimosa* (Gemmeke, 1982; Ramírez-Domenech and Tucker, 1989, 1990) and this is the first report for *Acacia* s.l. All other acacias studied so far show sepal initiation in a rapid sequence, in a spiral, bidirectional and reversed unidirectional sequence from the adaxial towards the abaxial side of the flower (Gemmeke, 1982; Ramírez-Domenech and Tucker, 1990; Derstine and Tucker, 1991; Gómez-Acevedo et al., 2007). Putting these data together, it is clear that sepal initiation is variable within *Acacia* s.l. However, only a small fraction of species have been studied and broader conclusions can only be made on the basis of more studied taxa. Analogous to Mimosoideae, calyx initiation in Papilionoideae is also more variable than previously thought (Prenner, 2004c).

11.4.3 Corolla

The simultaneous petal formation found in *A. celastrifolia* is among the most stable characters in mimosoid floral development (Gemmeke, 1982; Ramírez-Domenech and Tucker, 1990; Prenner, 2004a; Gómez-Acevedo et al., 2007). Besides simultaneous initiation, Prenner and Klitgaard (2008) highlighted that rapid petal enlargement and early closure of the corolla is a common character in Mimosoideae. As shown here for *A. celastrifolia*, the valvate corolla takes over the protection of the sexual organs in early floral development. While in Papilionoideae petal growth is always distinctly retarded, both early and late petal enlargement can be found in Caesalpinioideae (see Prenner and Klitgaard, 2008). The petals of *A. celastrifolia* and in many other mimosoid legumes are reduced in size and therefore play a minor role in pollination biology. Nevertheless, they perform an important protective function during early floral ontogeny. In Mimosoideae, the showy function of the petals is frequently transferred to the androecium, which consists of many coloured stamens (see next section). The presence of stomata on both the inside and outside of the petals is a rare character and was found only once in *A. berlandieri*. Gómez -Acevedo et al. (2007) speculated that in *A. berlandieri* this is nectariferous tissue which is involved in pollination biology. No such observation was made for *A. celastrifolia*.

11.4.4 Androecium

Acacia celastrifolia is remarkable because of its extreme polyandry, with more than 500 stamens in a single flower. In an ontogenetic context the androecium of *A. celastrifolia* (and that of many other polyandrous Mimosoideae) can be seen as a stamen fascicle in which the ring meristem acts as the primary meristem, which gives rise to secondary stamen primordia (for a review of stamen fascicles see Prenner et al., 2008). This type of secondary polyandry differs from the primary polyandry of early-divergent angiosperms where stamens arise directly from primary stamen primorda (which are frequently formed in a spiral or whorled pattern). Similar to the conditions in *A. baileyana* (studied by Derstine and Tucker, 1991), stamen formation in *A. celastrifolia* starts in an antesepalous position and proliferates in a lateral direction. However *A. baileyana* shows only moderate polyandry (30–40 stamens per flower) and does not form a pronounced ring-wall, as found in *A. celastrifolia*.

Anther morphology of *A. celastrifolia* is similar to that of *Calliandra* and shows four divided locules of which each harbours a single polyad (Prenner and Teppner, 2005; Teppner and Stabentheiner, 2007). The finding of orbicules on the inner surface of the locules is another proof of this character in Mimosoideae (c.f. Huismans et al., 1998).

11.4.5 Pollen and pollination

Another interesting and rather uncommon feature of *A. celastrifolia* is its eight-grained polyads. This is only half the grain number commonly found in *Acacia* s.l. (see Kenrick and Knox, 1982). Other acacias with eight grains per polyad are *A. hispidula*, *A. lineata*, *A. paradoxa* and *A. rupicola* (see Kenrick and Knox, 1982) and there is only one report of four grains per polyad in *A. baueri*. Kenrick and Knox (1982) showed that grain number per polyad is correlated with seed number per pod. Species with eight grains per polyad show two to ten ovules per pod and never more than eight seeds per mature fruit. This fits with the present observations of *A. celastrifolia*, which never showed more than eight ovules per carpel or seeds per fruit (present study; Prenner, G., pers. obs.). This balance of grain number and ovule number also makes sense considering the fact that the stigmatic surface of *A. celastrifolia* can only provide space for a single polyad.

A similar pattern of eight-grained polyads and a maximum of eight ovules per carpel were found in *Calliandra* (Prenner, 2004a; Prenner and Teppner, 2005; Teppner and Stabentheiner, 2007; Santos, 2008). However, in this genus, the correlation of the size of the stigma and the polyad is lost and the massive stigma can hold several polyads, though the number of ovules (i.e. eight) is still correlated with the eight pollen grains per polyad. It seems possible that stigma size is here enlarged because of the special mode of pollination found in *Calliandra* (bird, bat and moth pollination are reported for the genus).

Guinet et al. (1980) discussed the rather unexpected similarities of the eight-celled polyads of *A. celastrifolia* and *A. newbeyi*, of which the latter is otherwise not closely related to the series Pulchellae. Regarding the morphology of the polyad, *A. mitchellii* also closely resembles *A. celastrifolia* (see Guinet et al., 1980 and the present study). However, this strong similarity is not reflected in recent molecular phylogenies; *A. mitchellii* is found in the Botrycephalae subclade, which is not closely related to the Pulchelloidea subclade to which *A. celastrifolia* belongs (Miller et al., 2003, Murphy et al., 2010).

Stone et al. (2003) thoroughly reviewed the pollination ecology of *Acacia* s.l. They highlighted that all members of subgenus Phyllodineae offer only pollen as a floral reward and that the most important pollinators are bees. In specific cases, other insects and nectar-feeding birds are mentioned as important pollinators. The authors also highlight the need of more taxa to be studied prior to more general assumptions and conclusions on the pollination biology in *Acacia*.

Sargent (1909, 1918) mentioned that *Acacia celastrifolia* is pollinated by 'Silvereyes' (*Zosterops gouldi*; Passeriformes, Zosteropidae) and other honeyeaters. The birds visit the extrafloral nectaries at the base of the phyllodes and thereby come into contact with the flowers. In this case, these extrafloral nectaries can be seen as nuptial (i.e. they are involved in pollination). Knox et al. (1985) and Vanstone and Paton (1988) showed a similar condition for bird pollination of *A. terminalis* and *A. pycnantha*. Bird pollination is not exclusive and the inflorescences are also visited by insects; Bernhardt (1987) showed a broad range of bees and wasps as effective pollinators of some Australian acacias. Concerning the observation of bird-pollinated acacias, the genus appears to be a good example for a taxon where the floral morphology (nectarless, small and yellow flowers aggregated to inflorescences) does not 'fit' the actual observed syndrome of bird pollination (cf. Fenster et al., 2004).

Sargent (1909) even suggested the possibility of wind pollination in *A. celastrifolia*. However, there are no studies providing evidence for this hypothesis and the presence of pollenkitt on the surface of polyads in *Acacia* (Teppner, 2009) makes wind rather improbable as a major pollinating agent for this species.

11.4.6 Gynoecium

Perhaps the most remarkable floral feature of *A. celastrifolia* is its multicarpellate gynoecium that consists of three to seven free carpels (Figs 11.5F, 11.7.). Tucker (1987, p. 207) mentioned legume flowers with multiple carpels in some species of the papilionoid genus *Swartzia* (see also Cowan, 1968, 1981) and in *Archidendron* and *Affonsea* (= *Inga*) of the mimosoid tribe Ingeae (see also van Heel, 1993;

Pennington, 1997). It is intriguing that there is no report of multicarpellate species in Caesalpinioideae; this paraphyletic subfamily has the highest amount of floral diversity among Leguminosae, and Papilionoideae and Mimosoideae are nested within it (e.g. Wojciechowski et al., 2004). Within Fabales, flowers with more than one carpel are common among Polygalaceae (e.g. Prenner, 2004e; Bello et al., 2010), Quillajaceae and Surianaceae (e.g. Bello et al., 2007). Among rosids, taxa with more than five carpels in a single whorl are restricted to Myrtales, Malpighiales, Brassicales, Malvales and Sapindales (c.f. Endress and Matthews, 2006).

Van Heel (1983) studied the ontogeny of free carpels from a wide range of distantly related plants. For *Amherstia nobilis* (Caesalpinioideae) he showed abnormal gynoecia with a second smaller carpel formed in an adaxial position. One sample (his Fig 117) shows the early initiation of such an aberrant carpel situated at the proximal end of the adaxial cleft (i.e. it resembles part of the well-formed carpel rather than an entirely independent organ; see also van Heel, 1993). Besides these abnormalities, there are no well-documented instances of regularly multicarpellate flowers in Caesalpinioideae. Sattler (1973) documented one anomalous flower of the mimosoid *Albizia lophanta* with two young carpels with their sutures facing each other. While in *Albizia* this pattern must be seen as aberrant, there are other mimosoid taxa that show more than one carpel as a stable character in all flowers.

Archidendron lucyi is one such example of a mimosoid with five carpels that was studied in detail by van Heel (1993). In this species, carpel formation is more or less simultaneous and the five carpels are distributed more or less regularly and in most cases alternating with the petals. In contrast to *A. celastrifolia*, carpel formation in *Archidendron* starts much earlier and almost simultaneously with the formation of the first stamen primordia. The border between the androecial ring-wall and the central part of the flower on which the carpels are formed is not as sharp as in *A. celastrifolia*. It is formed only later in ontogeny when the young carpels are distinctly visible.

The discovery that in *A. celastrifolia* carpel initiation starts before the last stamens are formed, confirms Endress' (1994, p. 103) observation that early carpel initiation is a characteristic of species in which stamens are formed on a 'ring-like androecial mound'. However, early carpel formation is also frequently found in Papilionoideae with only ten stamens in two whorls (e.g. *Daviesia cordata*, Prenner, 2004d) and in Caesalpinioideae (c.f. Prenner and Klitgaard, 2008).

It is intriguing that early developmental stages in *A. celastrifolia* closely resemble the distantly related *Nelumbo* (Nelumbonaceae) (Hayes et al., 2000). In particular, the clear-cut border between androecial ring-wall and central floral meristem on which the carpels will be formed is striking. In contrast, in polyandrous and multicarpellate Ranunculaceae carpel initiation commences more or less gradually after the stamens are formed and there is no clear-cut boundary between the zone of stamen formation and that of carpel initiation (e.g. Ren et al., 2009, 2010).

Orientation of the carpels with the cleft pointing towards the centre of the flower is similar in *Archidendron* and *A. celastrifolia*. Furthermore, in both species the central part of the flower remains organ free. Most other legumes show in this position (i.e. terminating the flower) a single carpel which is in most cases oriented with the carpellary cleft pointing towards the flower's main axis (cf. Tucker, 1987, 2003; Prenner, 2004a). However, there are exceptions, as in *A. baileyana*, where in some flowers the carpel cleft is tilted out of the median plane (Derstine and Tucker, 1991). In *A. berlandieri*, Gómez-Acevedo et al. (2007) found the position of the cleft highly variable, and in the caesalpinioid *Gleditsia* the carpel is even inverted in the flowers of some inflorescences (i.e. the cleft lies at the abaxial side of the flower) (Tucker, 1991). In papilionoid flowers there are two patterns of cleft orientation (normal and oblique) which are linked to androecial symmetry (Prenner, 2004b). In species with an asymmetric androecium (i.e. the adaxial stamen of the inner whorl is formed off the median plane) the cleft is frequently turned either to the left or to the right of the median axis. In species with a symmetric androecium (i.e. the adaxial inner stamen lies exactly in the median plane) the carpel cleft always points in exactly the adaxial direction towards the main axis.

Another interesting feature of the mature carpels in *A. celastrifolia* is the abundance of stomata on the surface. Since no secretory products were found within the flowers it seems more plausible that the stomata are important for photosynthesis. Galen et al. (1993) investigated the function of carpels in *Ranunculus adoneus* and found that due to carpellate assimilation the net carbon rates of flowers rise from negative during bud expansion to positive during early fruit growth. This variation of carbon rates in the course of anthesis is probably an understudied aspect of flowering. Recently, Earley et al. (2009) found that the inflorescences of *Arabidopsis thaliana* contribute more to lifetime carbon gain than its rosettes. This seems to suggest that flowering is not necessarily a costly investment for the plant (i.e. negative energy bill), but that reproductive structures and its surroundings can also significantly contribute to the plants carbon rate balance.

Sinha (1971) highlighted that in *A. nilotica* the integuments start to develop only when the seed formation starts. A similar instance is found in *A. celastrifolia* in which in young ovules only the outer integument is visible and the formation of the inner integument is delayed.

It would be interesting to compare the flower development of *A. celastrifolia* with that of its putative close relative *A. myrtifolia*. Similar to *A. celastrifolia*, this species also has a highly polyandrous androecium, but only one carpel per flower. To verify the close relationship of these two species, molecular sampling should be a goal for the near future.

11.4.7 Is multicarpelly a 'primitive' (plesiomorphic) condition in Leguminosae?

Arber and Parkin (1907) argued that apocarpous and superior carpels represent the primitive gynoecial conditions in angiosperms. This hypothesis was supported by Endress and Doyle (2009) who reconstructed more than one carpel as ancestral and showed that a reduction to one carpel per flower has occurred several times independently.

Within Fabales, the families Polygalaceae, Quillajaceae and Surianaceae are multicarpellate. Thus, the question arises whether among legumes multicarpelly represents the 'primitive' condition from which unicarpellate flowers evolved. So far, there is no evidence for this assumption. Multicarpellate taxa are not closely related and multicarpelly apparently arose several times independently in the legume radiation. Considering the occurrence of multicarpellate taxa in related families, this character could also be interpreted as atavism (i.e. reversion to a previous evolutionary state). More obvious than a phylogenetic signal is the fact that multicarpelly seems correlated with a higher degree of polyandry. In legumes, multicarpellate flowers could therefore be a secondary phenomenon resulting from an enlarged floral meristem. The blueprint for multicarpelly can be found in the related families of Fabales. Possibly due to the extended size of the meristem there is more space available for more than one carpel (see also Endress, 1994). The study of more multicarpellate legume taxa will help to better understand this phenomenon in Leguminosae.

Another independent theory for multicarpelly in Leguminosae is that multicarpellate flowers are the result of fasciation and are in fact 'hidden inflorescences'. Extreme polyandry and moderate multicarpelly of *A. celastrifolia* can be linked with reduced flower number per head. The floral heads of most other acacias are many-flowered and the individual flowers frequently show moderate polyandry and only a single terminal carpel. In *A. retinodes*, flower heads each consist of 18–50 flowers and each flower possesses up to 50 stamens, giving a total of 1500 (30 × 50) stamens per head (in a 30-flowered head). Because of its extreme polyandry with more than 500 stamens, the (1–)2–3-flowered heads of *A. celastrifolia* show potentially the same number of 1500 (3 × 500) stamens, indicating that reduction in flower number is matched by increase in stamen number per flower. Albertsen et al. (1983) reported and analysed fasciation in soybeans (*Glycine max*). They described the fasciation of inflorescences and mentioned that flowers with multiple carpels were common. Recently Sinjushin and Gostimskii (2008) investigated the genetic control of fasciation in pea (*Pisum sativum*).

Instances of pseudanthial 'flowers' are scattered throughout angiosperms. Rozefelds and Drinnan (1998) interpreted the polyandrous flowers of *Lophozonia* (Nothofagaceae) as pseudanthia. Sokoloff et al. (2007) showed that 'flowers' of *Tupidanthus calyptratus* (Araliaceae) are probably the result of fasciation. Strong tendencies to fasciation are also found in the reproductive units of the early branching angiosperm *Trithuria* (Hydatellaceae) (Rudall et al., 2009). These examples demonstrate that fasciation could be widespread and that there may be more 'flowers' that in fact are of a pseudanthial nature.

11.5 Conclusions

Together with earlier ontogenetic studies, the present study shows that even though flowers of *Acacia* s.l. are superficially uniform they show a wide range of different ontogenetic patterns. Sepal initiation is the most flexible pattern, not only in *Acacia*, but in Mimosoideae in general (e.g. Gemmeke, 1982; Ramírez-Domenech and Tucker, 1990; Prenner, 2004a; Gómez-Acevedo et al., 2007). This is similar to Papilionoideae, in which Prenner (2004c) showed a wide range of ontogenetic patterns of sepal initiation. The reason for this variability might be that sepals are not crucial for the function of these flowers during anthesis and therefore sepal initiation and sepal aestivation are of minor importance and more flexible. In contrast to the sepals, petal formation in *Acacia* and other Mimosoideae appears to be uniformly simultaneous. The reason for this could be that simultaneous organ formation is a prerequisite for the valvate petal aestivation frequently found in Mimosoideae. Regarding the androecium, secondary polyandry with primary primordia giving rise to secondary stamen primordia is a common pattern in *Acacia*. So far, a range of developmental types from 'sectorial initiation' (in *A. baileyana*, Derstine and Tucker, 1991) to ring primordia (e.g. present study; Gemmeke, 1982; Gómez-Acevedo et al., 2007) have been reported in *Acacia* s.l. The gynoecium in Leguminosae is typically very conservative and formed of a single terminal carpel, but some species of Mimosoideae–Ingeae and Papilionoideae–Swartzieae show deviations towards multicarpelly. It is possible that this character evolved secondarily and apparently linked with polyandry. However, the possibility that the correlated features of high polyandry and moderate multicarpelly are the result of fasciation should be considered in future studies.

Acknowledgements

Sincere thanks go to Herwig Teppner who drew my attention to the enigmatic flowers of *Acacia celastrifolia* and to legumes in general, Paula Rudall for reading the manuscript and for very helpful suggestions, Bruce Maslin for valuable comments and help with some literature, and two anonymous referees for their helpful criticism and remarks on the manuscript.

11.6 References

Albertsen, M. C., Curry, T. M., Palmer, R. G. and Lamotte, C. E. (1983). Genetics and comparative growth morphology of fasciation in soybeans (*Glycine max* [L.] Merr.). *Botanical Gazette*, **144**, 263–275.

Arber, E. A. N. and Parkin, J. (1907). On the origin of angiosperms. *Journal of the Linnean Society*, **38**, 29–80.

Bello, M. A., Hawkins, J. A. and Rudall, P. J. (2007). Floral morphology and development in Quillajaceae and Surianaceae (Fabales), the species-poor relatives of Leguminosae and Polygalaceae. *Annals of Botany*, **100**, 1491–1505.

Bello, M. A., Hawkins, J. A. and Rudall, P. J. (2010). Floral ontogeny in Polygalaceae and its bearing on the origin of keeled flowers in Fabales. *International Journal of Plant Sciences*, **171**, 482–498.

Bernhardt, P. (1987). A comparison of the diversity, density and foraging behavior of bees and wasps on Australian *Acacia*. *Annals of the Missouri Botanical Garden*, **74**, 42–50.

Cowan, R. S. (1968). *Swartzia* (Leguminosae, Caesalpinioideae, Swartzieae). pp. 1–228 in *Flora Neotropica*, Monograph, number 1. New York: Hafner.

Cowan, R. S. (1981). Swartzieae. pp. 209–212 in Polhill, R. M. and Raven, P. R. (eds.), *Advances in Legume Systematics*. Kew: Royal Botanic Gardens, Kew.

Derstine, K. S. and Tucker, S. C. (1991). Organ initiation and development of inflorescences and flowers of *Acacia baileyana*. *American Journal of Botany*, **78**, 816–832.

Earley, E. J., Ingland, B., Winkler, J. and Tonsor, S. J. (2009). Inflorescences contribute more than rosettes to lifetime carbon gain in *Arabidopsis thaliana* (Brassicaceae). *American Journal of Botany*, **96**, 786–792.

Eichler, A. W. (1878). *Blüthendiagramme, 2. Teil*. Leipzig: Wilhelm Engelmann.

Endress, P. K. (1994). *Diversity and Evolutionary Biology of Tropical Flowers*. Cambridge: Cambridge University Press.

Endress, P. K. and Doyle, J. A. (2009). Reconstructing the ancestral angiosperm flower and its initial specializations. *American Journal of Botany*, **96**, 22–66.

Endress, P. K. and Matthews, M. L. (2006). First steps towards a floral structural characterization of the major rosid subclades. *Plant Systematics and Evolution*, **260**, 223–251.

Fenster, C. B., Armbruster, W. S., Wilson, P., Dudash, M. R. and Thomson, J. D. (2004). Pollination syndromes and floral specialization. *Annual Review of Ecology, Evolution, and Systematics*, **35**, 375–403.

Galen, G., Dawson, T. E. and Stanton, M. L. (1993). Carpels as leaves: meeting the carbon cost of reproduction in an alpine buttercup. *Oecologia*, **95**, 187–193.

Gemmeke, V. (1982). Entwicklungsgeschichtliche Untersuchungen an Mimosaceen-Blüten. *Botanische Jahrbücher für Systematik*, **103**, 185–210.

Gómez-Acevedo, S. L., Magallón, S. and Rico-Arce, L. (2007). Floral development in three species of *Acacia* (Leguminosae, Mimosoideae). *Australian Journal of Botany*, **55**, 30–41.

Guinet, P., Vassal, J. Evans, C. S. and Maslin, B. R. (1980). *Acacia* (Mimosoideae): composition and affinities of the series Pulchellae Bentham. *Botanical Journal of the Linnean Society*, **80**, 53–68.

Hayes, V., Schneider, E. L. and Carlquist, S. (2000). Floral development in *Nelumbo nucifera* (Nelumbonaceae). *International Journal of Plant Sciences*, **161**, S183–S191.

Huismans, S., El-Ghazaly, G. and Smets, E. (1998). Orbicules in angiosperms: morphology, function, distribution and relation with tapetum types. *The Botanical Review*, **64**, 240–272.

Kenrick, J. and Knox, R. B. (1982). Function of the polyad in reproduction of *Acacia. Annals of Botany*, **50**, 721–727.

Knox, R. B., Kenrick, J., Bernhardt, P. et al. (1985). Extrafloral nectaries as adaptations for bird pollination in *Acacia terminalis. American Journal of Botany*, **72**, 1185–1196.

Lewis, G. (2005). Tribe Acacieae. pp. 187–191 in Lewis, G., Schrire, B., Mackinder, B. and Lock, M. (eds.), *Legumes of the World*. Kew: Royal Botanic Gardens, Kew.

Maslin, B. R. (1995). *Acacia* Miscellany 12. *Acacia myrtifolia* (Leguminosae: Mimosoideae: section *Phyllodineae*) and its allies in Western Australia. *Nuytsia*, **10**, 85–101.

Miller, J. T. and Bayer, R. J. (2001). Molecular phylogenetics of *Acacia* (Fabaceae: Mimosoideae) based on the chloroplast *matK* coding sequence and flanking *trnK* intron spacer regions. *American Journal of Botany*, **88**, 697–705.

Miller, J. T., Andrew, R. and Bayer, R. J. (2003). Molecular phylogenetics of the Australian acacias of subg. Phyllodineae (Fabaceae: Mimosoideae) based on trnK intron. *Australian Journal of Botany*, **51**, 167–177.

Murphy, D. J. (2008). A review of the classification of *Acacia* (Leguminosae, Mimosoideae). *Muelleria*, **26**, 10–26.

Murphy, D. J., Brown, G. K., Miller, J. T. and Ladiges, P. Y. (2010) Molecular phylogeny of *Acacia* Mill. (Mimosoideae: Leguminosae): Evidence for major clades and informal classification. *Taxon*, **59**, 7–19.

Orchard, A. E. and Maslin, B. R. (2003). Proposal to conserve the name *Acacia* (Leguminosae: Mimosoideae) with a conserved type. *Taxon*, **52**, 362–363.

Pennington, T. D. (1997). *The Genus Inga. Botany*. Kew: Royal Botanic Gardens, Kew.

Prenner, G. (2004a). Floral development in *Calliandra angustifolia* (Leguminosae-Mimosoideae) and its systematic implications. *International Journal of Plant Sciences*, **165**, 417–426.

Prenner, G. (2004b). The asymmetric androecium in Papilionoideae (Leguminosae): definition, occurrence, and possible systematic value. *International Journal of Plant Sciences*, **165**, 499–510.

Prenner, G. (2004c). New aspects in floral development of Papilionoideae: initiated but suppressed bracteoles and variable initiation of sepals. *Annals of Botany*, **93**, 537–545.

Prenner, G. (2004d). Floral development in *Daviesia cordata* (Leguminosae: Papilionoideae: Mirbelieae) and its systematic implications. *Australian Journal of Botany*, **52**, 285–291.

Prenner, G. (2004e). Floral development in *Polygala myrtifolia* (Polygalaceae) and its similarities with Leguminosae. *Plant Systematics and Evolution*, **249**, 67–76.

Prenner, G., Bateman, R. M. and Rudall, P. J. (2010). Floral formulae updated for routine inclusion in formal taxonomic descriptions. *Taxon*, **59**, 241–250.

Prenner, G., Box, M. S., Cunniff, J. and Rudall, P. J. (2008). The branching stamens of *Ricinus* and the homologies of the angiosperms stamen fascicle. *International Journal of Plant Sciences*, **169**, 735–744.

Prenner, G. and Klitgaard, B. B. (2008). Towards unlocking the deep nodes of Leguminosae: floral development and morphology of the enigmatic *Duparquetia orchidacea* (Leguminosae, Caesalpinioideae). *American Journal of Botany*, **95**, 1349–1365.

Prenner, G. and Teppner, H. (2005). Anther development, pollen presentation and pollen adhesive of parenchymatous origin in *Callinadra angustifolia* (Leguminosae-Mimosoideae-Ingeae). *Phyton (Horn, Austria)*, **45**, 267–286.

Ramírez-Domenech, J. I. and Tucker, S. C. (1989). Phylogenetic implications of inflorescence and floral ontogeny of *Mimosa strigillosa*. *American Journal of Botany*, **76**, 1583–1593.

Ramírez-Domenech, J. I. and Tucker, S. C. (1990). Comparative ontogeny of the perianth in mimosoid legumes. *American Journal of Botany*, **77**, 624–635.

Ren, Y., Chang, H-L. and Endress, P. K. (2010). Floral development in Anemoneae (Ranunculaceae). *Botanical Journal of the Linnean Society*, **162**, 77–100.

Ren, Y., Chang, H.-L., Tian, X.-H., Song, P. and Endress, P. K. (2009). Floral development in Adonideae (Ranunculaceae). *Flora*, **204**, 506–517.

Ronse De Craene, L. P. (2010). *Floral Diagrams. An Aid to Understanding Flower Morphology and Evolution.* Cambridge: Cambridge University Press.

Rozefelds, A. C. and Drinnan, A. N. (1998). Ontogeny and diversity in staminate flowers of *Nothofagus* (Nothofagaceae). *International Journal of Plant Sciences*, **159**, 906–922.

Rudall, P. J., Remizowa, M. V., Prenner, G. et al. (2009). Nonflowers near the base of extant angiosperms? Spatiotemporal arrangement of organs in reproductive units of Hydatellaceae and its bearing on the origin of the flower. *American Journal of Botany*, **96**, 67–82.

Santos, F. A. R. and Romano, C. O. (2008). Pollen morphology of some species of *Calliandra* Benth. (Leguminosae – Mimosoideae) from Bahia, Brazil. *Grana*, **47**, 101–116.

Sargent, O. H. (1909). Biological notes on *Acacia celastrifolia*. *Journal of Western Australian Natural History*, **6**, 38–44.

Sargent, O. H. (1918). Fragments of the flower biology of Westralian plants. *Annals of Botany*, **32**, 215–231.

Sattler, R. (1973). *Organogenesis of Flowers.* Toronto: University of Toronto Press.

Sinha, S. C. (1971). Floral morphology of Acacias. *Caribbean Journal of Science*, **11**, 137–153.

Sinjushin, A. A. and Gostimskii, S. A. (2008). Genetic control of fasciation in pea (*Pisum sativum* L.). *Russian Journal of Genetics*, **44**, 702–708.

Sokoloff, D. D., Oskolski, A. A., Remizowa, M. V. and Nuraliev, M. S. (2007). Flower structure and development in *Tupidanthus calyptratus* (Araliaceae): an extreme case of polymery among asterids. *Plant Systematics and Evolution*, **268**, 209–234.

Stone, G. N., Raine, N. E., Prescott, M. and Willmer, P. G. (2003). Pollination ecology of acacias (Fabaceae, Mimosoideae). *Australian Systematic Botany*, **16**, 103–118.

Teppner, H. (2009). The easiest proof for the presence of pollenkitt. *Phyton (Horn, Austria)*, **48**, 169–328.

Teppner, H. and Stabentheiner, E. (2007). Anther opening polyad presentation, pollenkitt and pollen adhesive in four *Calliandra* species (Mimosaceae-Ingeae). *Phyton (Horn, Austria)*, **47**, 291–320.

Tucker, S. C. (1987). Floral initiation and development in legumes. pp. 183–293 in Stirton, C. H. (ed.), *Advances in Legume Systematics*. Kew: Royal Botanic Gardens, Kew.

Tucker, S. C. (1991). Helical floral organogenesis in *Gleditsia*, a primitive caesalpinioid legume. *American Journal of Botany*, **78**, 1130–1149.

Tucker, S. C. (1992). The role of floral development in studies of legume evolution. *Canadian Journal of Botany*, **70**, 692–700.

Tucker, S. C. (2003). Floral development in legumes. *Plant Physiology*, **131**, 911–926.

van Heel, W. A. (1983). The ascidiform early development of free carpels. A S.E.M.-investigation. *Blumea*, **28**, 231–270.

van Heel, W. A. (1993). Floral ontogeny of *Archidendron lucyi* (Mimosaceae) with remarks on *Amherstia nobilis* (Caesalpiniaceae). *Botanische Jahrbücher für Systematik*, **114**, 551–560.

Vanstone, V. A. and Paton, D. C. (1988). Extrafloral nectaries and pollination of *Acacia pycnantha* Benth. by birds. *Australian Journal of Botany*, **36**, 519–531.

Wojciechowski, M. F., Lavin, M. and Sanderson, M. J. (2004). A phylogeny of legumes (Leguminosae) based on analysis of the plastid *matK* gene resolves many well-supported subclades within the family. *American Journal of Botany*, **91**, 1846–1862.

12

Floral development of *Napoleonaea* (Lecythidaceae), a deceptively complex flower

LOUIS P. RONSE DE CRAENE

12.1 Introduction

Napoleonaea is a small genus with about eight to ten species mainly restricted to west and central Africa and extending into southern Africa (Liben, 1971; Frame and Durou, 2001). The genus was initially described by Palisot de Beauvois in 1804 and dedicated to Napoleon Buonaparte (Thompson, 1922; Liben, 1971), but became often misspelled as *Napoleona* in later publications.

Thompson (1922) reviewed the early classification of the genus. A close relationship with Myrtaceae was put forward by Bentham and Hooker (1867) on the belief that the corona of *Napoleonaea* represents sterile outer stamens, as found in some Myrtaceae and in the genera now placed in Lecythidaceae (e.g. *Grias, Couroupita, Lecythis*). The interpretation of the corolla is central in the discussion of affinities, as most authors accepted a Myrtalean affinity of *Napoleonaea* and Lecythidaceae (e.g. Masters, 1869; Baillon, 1875; Thompson, 1922, 1927). Later authors removed Lecythidaceae from Myrtales because of important morphological distinctions (see Dahlgren and Thorne, 1984). Recent molecular phylogenies have placed Lecythidaceae (including *Napoleonaea*) in Ericales (e.g. Morton et al., 1997; Schönenberger et al., 2005; APG, 2009).

Flowers on the Tree of Life, ed. Livia Wanntorp and Louis P. Ronse De Craene. Published by Cambridge University Press. © The Systematics Association 2011.

Mature flowers of *Napoleonaea* appear extremely complex at a glance; they resemble the French 'cocarde', a tricoloured cockade or rosette, with a reflexed outer tube lacking distinctive appendages, a corona of two series of threads (horizontal staminodes) and laminate appendages (vertical staminodes) and an inner whorl of 20 half stamens and staminodes inserted below an umbrella-shaped style (Fig 12.2A–C; see also Fig 3 in Mori et al., 2007). The flower of *Napoleonaea* has often been described as apetalous (e.g. Masters, 1869; Niedenzu, 1893; Knuth, 1939; Prance and Mori, 1979; Endress, 1994; Morton et al., 1998; Prance, 2004) and the outer series was interpreted as part of the staminodial corona. At maturity there is little distinction between the different whorls of staminodial appendages and the outer fused corolla with many-ribbed margins. Despite the persistent belief that *Napoleonaea* is apetalous, a number of earlier and more recent morphological studies have claimed the contrary, i.e. that a true corolla is present in the flower (e.g. Baillon, 1875; Thompson, 1922, 1927; Liben, 1971; Takhtajan, 1997; Frame and Durou, 2001).

Napoleonaea has been associated with two other genera *Asteranthos* and *Crateranthus* in a subfamily Napoleonoideae (Niedenzu, 1893) or in a family Napoleonaeaceae (Prance, 2004). *Asteranthos* is a Neotropical genus, while both *Crateranthus* and *Napoleonaea* are from W Africa. *Asteranthos brasiliensis* Desf. bears a circular membranous cup closely linked to the androecium, which is pleated in bud and unfolds like an umbrella (Thompson, 1927; Appel, 1996). This cup has been variously interpreted as a staminodial corona or a true corolla. *Crateranthus talbotii* Bak. Fil. bears an urn-shaped fleshy rim associated with a multistaminate androecium of about ten cycles (Thompson, 1927). Again, the interpretation of the rim is controversial. Appel (1996) recognized that *Asteranthos* is highly similar to Scytopetalaceae. Genera of Scytopetalaceae (e.g. *Oubanguia*, *Scytopetalum*) have showy petaloid appendages linked to the stamen bases in common and these were interpreted as staminodial (Appel, 1996, 2004). The number of lobes varies from 6 to 16, or they appear completely fused. Prance (2004) implied that *Napoleonaea* (placed by him with *Crateranthus* in Napoleonaeaceae) and the related Scytopetalaceae have lost their corolla, which has been replaced by a staminodial substitute.

Recent molecular phylogenies (Morton et al., 1997, 1998; Mori et al., 2007) demonstrated the paraphyletic nature of Napoleonaeoideae and confirmed the close relationship between *Asteranthos* and Scytopetalaceae, as had been suggested on morphological evidence. Morton et al. (1998) considered Napoleonaeoideae to be a clade sister to the remaining Lecythidaceae, with two genera *Napoleonaea* and *Crateranthus*, although they did not investigate the latter. The next clade (Scytopetaloideae) consists of former Scytopetalaceae and *Asteranthos*, sister to a clade formed by Planchonioideae, Foetidioideae and Lecythidoideae. They interpreted the corolla as fundamentally absent or replaced by a staminodial corona in Napoleonaeoideae, Scytopetaloideae as well as *Foetidia* (Foetidioideae), while it is

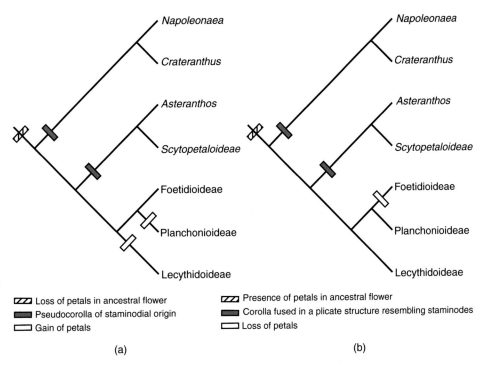

Fig 12.1 Phylogenetic tree of Lecythidaceae based on Morton et al. (1998). (a) Hypothesis of loss of corolla and elaboration of staminodial corona in Napoleonoideae. (b) Hypothesis of real corolla in Napoleonoideae.

present in other subfamilies (Fig 12.1). There remains some controversy about the circumscription of Lecythidaceae, as some authors prefer to keep Napoleonaeaceae and Scytopetalaceae as distinct from Lecythidaceae (Mori et al., 2007), while APG (2009) recognizes a single family Lecythidaceae with five subfamilies.

If the interpretation of apetaly is the correct one, based on the phylogeny of Morton et al. (1998), petals must have been lost early in the Lecythidaceae, to be regained by a transformation of outer staminodes in Napoleonaeoideae and Scytopetaloideae before being acquired again separately in Planchonioideae and Lecythidoideae, or at the base of a clade leading to Planchonoideae, Foetidioideae and Lecythidoideae (Fig 12.1A). This would indicate a loss of petals in Foetidoideae, where it represents an autapomorphy. However, if the petals of Napoleonoideae and possibly Scytopetaloideae represent a true petal whorl, they would be present in all subfamilies, except for Foetidioideae and some members of Scytopetaloideae, where they became secondarily lost (Fig 12.1B). Against this duality of interpretations it is important to verify the homology of the corona, but also to understand how the androecium develops into an extraordinarily complex system.

In this chapter the floral development of two species of *Napoleonaea* is investigated to understand the structure of the androecium and to clarify the controversy about the presence of apetaly in the family.

12.2 Materials and Methods

Floral material of two species of *Napoleonaea* (*N. vogelii* Hook.f. and Planch.; *N. imperialis* P. Beauv.) was collected in BR and RBGE. Floral buds were photographed, collected, fixed in FAA (5% acetic acid, 5% formaldehyde, 90% ethanol [70%]), and subsequently stored in 70% ethanol. Buds and mature flowers were dissected and prepared using a Wild MZ8 stereo-microscope (Leica, Wetzlar, Germany), dehydrated in an ethanol-acetone series and critical-point dried with a K850 Critical Point Dryer (Emitech Ltd, Ashford, Kent, UK). The dried material was later coated with platinum using an Emitech K575X sputter coater (Emitech Ltd, Ashford, Kent, UK) and examined with a Supra 55VP scanning electron microscope (LEO Electron Microscopy Ltd, Cambridge, UK). Reference material (in ethanol) is kept at the botanical institute in Leuven, Belgium (LM: *N. vogelii* 375 Lm) and RBG Edinburgh (*N. imperialis* 994 Led; *N. vogelii* 921 Led).

For light microscopy, flower buds were embedded in paraffin and sectioned with a Leitz Minot 1212 rotary microtome fitted with metal blade. The sections (about 8 µm thick) were stained in fastgreen and safranin, and enclosed in DMX.

12.3 Results

Flowers at maturity are deceptively complex. The polysymmetric flowers are pentamerous and have five valvate calyx lobes inserted on top of the inferior ovary. The outer corona (petals?) is reflexed and consists of a crenelated rim with pleated ribs partly enwrapping the stem (Fig 12.2A, B, D). The next (staminodial?) level forms a complex with two series of distinct appendages. The outer series consists of reflexed (horizontal), laciniate white threads (Fig 12.2A, D), while the inner series is erect (vertical), variously fused and pleated, resembling the outer reflexed whorl (Fig 12.2A, D). Both series are adherent by ventral grooves on the threads fitting in knobs formed by the erect inner series. The mature androecium has been described in detail by several authors (e.g. Thomson, 1922; Frame and Durou, 2001; Prance, 2004). I refer to figures in Thomson (1922) and Mori et al. (2007) for more clarity. The erect inner series encloses 20 incurved stamens, consisting of pairs of fertile stamens alternating with pairs of staminodes, reduced to the flattened white filaments of the fertile stamens (Fig 12.2C,E). At maturity the stamens are bent over a broad nectary and deposit their pollen under a stylar outgrowth resembling a five-angled umbrella (Fig 12.2E). The gynoecium consists of five carpels fused into an

Fig 12.2 Mature flowers of *Napoleonaea*. (A)–(B) *N. imperialis*; (C)–(F) *N. vogelii*. (A)–(D) Distal and lateral view of flower at anthesis. (E) detail of (C): central part with staminodes, stamens and ovary. (F) Floral bud prior to anthesis. Note the petals opening along the midribs. Af, fertile stamen; As, sterile stamen; Ci, inner corona; Co, outer corona; Ct, corona threads; St, stylar lobe. Scale bars = 10 mm. For colour illustration see plate section.

Fig 12.3 Initiation of floral prophylls and calyx of *Napoleonaea*. (A)–(E) *N. imperialis*;
(F)–(G) *N. vogelii*. (A) Initiation of upper prophylls. (B) Sequential initiation of calyx lobes;

inferior ovary with axile placentation bearing four ovules in two rows within each locule (cf. Liben, 1971). Differences between flowers of *Napoleonaea vogelii* and *N. imperialis* are shown in Fig 12.2A–D. Liben (1971) mentioned solitary flowers associated with a glabrous calyx for *N. vogelii* against fascicled or solitary flowers with glabrous to verrucose calyx in *N. imperialis*.

Flowers often appear to be solitary, but are grouped on short, generally cauliflorous, cymose shoots with usually a single flower reaching maturity. A series of prophylls surrounds each flower in a distichous arrangement of three to four pairs. Prophylls arise sequentially and are unequal in size (Fig 12.3A, C, G); they bear sessile terminal glands (Fig 12.3G). Lower prophylls occasionally enclose a younger flower.

Sepals initiate in a rapid, almost simultaneous, sequence or initiation is unidirectional (Fig 12.3B). However, sepals become rapidly equal in size with a valvate aestivation (Fig 12.3C–E). The apical part of the sepal is initially developed in a massive protuberance, which overtops the rest of the flower (Fig 12.3F). The broad upper section becomes covered with small trichomes on the margins (Fig 12.3E,F), spreading to the whole sepal surface (Fig 12.3G). The sepals bear one large or a pair of sessile glands on each side arising as lobes on the margins (Fig 12.3E,F). At maturity sepals are more equal in size and spread open or become reflexed.

The petals arise as distinct primordia on a peripheral ring meristem surrounding a central depression at the stage sepal lobes enclose the bud (Fig 12.4A). Triangular petal lobes are rapidly differentiated, but become basally connected as a ring (Fig 12.4B–C). The young petals develop ribs on the abaxial side in a centrifugal sequence starting with a broad midrib (Fig 12.4C–E). Lateral ribs are positioned parallel to the main rib, and as petals extend in size they meet laterally in a valvate aestivation with lateral ribs interlocking in a zig-zag pattern (Fig 12.4E–F). As the flower reaches pre-anthetic stages, different ribs become placed as separate erect appendages and the limits between different petals become unclear (Fig 12.4G–H). While the midrib is initially broader and longer, it becomes superseded before maturity by the lateral ribs lying in parallel (Figs 12.2F, 12.4G). All ribs become covered with trichomes (Fig 12.4G–H). As the flower opens, the petals tear open along the margins of the ribs, mostly at the level of the original midribs,

Caption for figure 12.3 continued
bracts removed. (C) Apical view of calyx lobes lifted by common basal growth. (D) Lateral view of floral bud. Note the development of lateral glands on calyx lobes. (E) Pre-anthetic bud. Note the valvate arrangement of calyx lobes and lateral glands. (F) Adaxial view of three calyx lobes. Note the numerous trichomes on the distal portion. (G) Lateral view of pre-anthetic bud enclosed by distichous prophylls. B, bract. Asterisks point to lateral glands on the calyx lobes. Scale bars = 100 μm; (G) = 1 mm.

Fig 12.4 Initiation of the corolla of *Napoleonaea*. (A)–(E), (H) *N. imperialis*; (F)–(G)
N. vogelii. (A) Initiation of five corolla lobes on a ring primordium; calyx removed. (B) Early
initiation of petals. Note the compression by the calyx lobes. (C) Development of lobes on
the petals. Note the basally connected area. (D) Lateral extension of lobes on each petal.

which appear to be split in half (Fig 12.2F). As the ribs become reflexed, they remain adherent by their trichomes and appear as a feathery reflexed corona (Fig Fig 12.2A–E).

Early stages of androecium and gynoecium initiation were very difficult to see. Before the differentiation of lateral lobes on petals and before any sign of the androecium, five carpel primordia initiate in a rapid sequence (Fig 12.5A–C). Initiation and development of carpels is regular, unidirectional or highly distorted by reduction of one carpel. Carpels initiate as separate lobes opposite the petals around a central depression, but rapidly develop septal branches extending towards the centre of the depression (Fig 12.5D). Carpels rapidly interconnect by development of a marginal rim around the central depression (Fig 12.5E). By more extensive intercarpellary growth, a broad stylar platform develops, with five slits running to the central depression (Fig 12.5F).

Immediately following carpel initiation a rim becomes differentiated externally of the carpel primordia. Paired primordia become visible opposite the carpel primordia, while the areas alternating with the petals show a weak swelling extending between the petal lobes (Fig 12.5C–D). As the division between the antepetalous primordia is more pronounced, each one appears as a lateral lobe of larger antesepalous primordia (Fig 12.5D). The antepetalous primordia grow horizontally and shift below the margin of the expanding carpellary lobes (Fig 12.5E). At this stage the antesepalous swellings differentiate as pairs of smaller primordia leaving a distinct gap between androecium and carpellary tissue (Fig 12.5E). As the carpellary lobes expand outwards, the antepetalous pairs may become partly hidden, while the antesepalous pairs are more distinctly erect and may become unequal in size and shape by spatial constraints (Figs 12.5F, 12.6B). Occasionally, a single primordium is formed instead of a pair. The antepetalous primordia develop as fertile stamens. Each stamen develops a ribbon-shaped filament with a single curved anther with a single theca (Fig 12.6C–E). As the stylar shield expands upwards, anthers are curved below the stylar lobes (Fig 12.6C–D). The antesepalous pair develops as two staminodes resembling the filament of the fertile stamens, with undeveloped anther tissue. They have no clear function in the flower except obstructing the passage between the carpels.

Before the inner stamens start differentiating, the androecial rim extends centrifugally, first opposite the petal lobes and later between the petals, and a whorl of

Caption for figure 12.4 continued
(E) Valvate arrangement of petals. Note the more extensively developed midrib. (F) Adaxial view of two petals. Note the basal common zone and the zig-zag connection of petal lobes. (G) Lateral view of pre-anthetic bud. The white line refers to the extent of a single petal. (H) Detail of the margins of two petals. Note the presence of interconnecting trichomes. B, bract; K, sepal; Co, petal. Scale bars = 100 µm; (G), (H) = 200 µm.

Fig 12.5 Initiation of the gynoecium and androecium of *Napoleonaea*. (A)–(E) *N. imperialis*; (F) *N. vogelii*. (A) Early initiation of carpel primordia (asterisks); petals removed. (B) Sequential initiation of carpel lobes (numbered); petals removed. (C) Initiation of paired primordia (asterisks) on androecial ring meristem. (D) Development of fertile stamen pairs opposite the petals and early initiation of common primordia alternating with the petal lobes; arrow points to dividing common primordium. (E) Development of five lobed stigma extending over the fertile stamens; differentiation of staminodial pairs. (F) Extension of the androecial ring primordium and centrifugal initiation of a second staminodial tier. Co, petal lobe; Af, fertile stamen primordium; As, sterile stamen primordium; G, carpel primordium or carpellary lobe. Scale bars = 100 μm.

Fig 12.6 Later stages of development of androecium and gynoecium of *Napoleonaea*. (A), (C), (D), (F), (H) *N. imperialis*; (B), (E), (G) *N. vogelii*. (A) Apical view at development of second staminodial tier; two petals removed. (B) Lateral view of slightly older bud showing initiation of third staminodial tier (arrowhead); two petals removed. (C) Apical view of

smaller primordia is initiated (Figs 12.5F, 12.6A–B). The about 40 organs are unequal and appear to radiate from primordia opposite the carpels towards the intercarpellary area and some appear laterally fused at initiation (Fig 12.6C). They grow into plate-like staminodes and will eventually form the inner erect corona (Fig 12.2A–E). By further expansion of the androecial ring, a third smaller whorl of primordia is initiated and develops as much shorter plate-like staminodes (Fig 12.6A, C, D). These will eventually develop as the threads in mature flowers (Fig 12.2A, B, D). During development of the anthers the area between androecium and gynoecium develops as an elevated nectary disc. Anthers are curved over the disc to fit below the stigmatic crest (Fig 12.6E–F). The ovary develops in an inferior position and is connected to the exterior by a narrow central aperture on the stigma (Fig 12.6C, E). Placentation is axile with ovules developing in two rows on each placenta in the synascidiate zone (Fig 12.6H). The roof of the ovary is not connected with the placenta and placentation becomes parietal in the symplicate zone of the ovary.

12.4 Discussion

Napolaeonaea is distinct from other Lecythidaceae in its elaborate corona that makes the flower look deceptively complex. However, several characters are common to the family, such as a multistaminate androecium with centrifugal development on a circular mound, development of outer staminodes (often forming a hood in zygomorphic genera of Lecythidoideae, broadly developed sepals and petals occasionally arising unidirectionally, inferior ovary with axile placentation and comparable early development, and occasionally tetramerous or hexamerous flowers (e.g. Endress, 1994; Tsou and Mori, 2007). However, petals are connected by a ring primordium and carpel initiation is sequential and arises before the androecium, in contrast to Lecythidoideae (Tsou and Mori, 2007). Contrary to other Lecythidaceae, petals are basally fused and are connected with

Caption for figure 12.6 continued
similar stage, showing the extent of development of stigmatic lobes together with stamen and staminodial pairs. (D) Development of anther tissue curving below the stigmatic lobes (removed). Note the development of two staminodial tiers. (E) Older bud showing arrangement of fertile and sterile stamens. (F) Transverse section of bud at a similar stage. Note the extrorse anther tissue and nectary disc. Broken line shows the connection of the filament with its respective anther. (G) Longisection of nearly mature bud. Note the anther tissue bending over the nectary disc and the developing ovules. (H) Detail of the upper part of the ovary showing five septa and locules. Note the free apical part of the placentation. Af, fertile stamen; As, sterile stamen; Ci, inner staminodial tier; Co, petal; Ct, outer threadlike staminodial tier; N, nectary disc. Scale bars = 100 µm; (D), (G) = 200 µm.

the androecium in *Napoleonaea*, *Crateranthus* and Scytopetaloideae, including *Asteranthos* (Frame and Durou, 2001; Appel, 2004; Prance, 2004).

Polyandry is reported in most genera of Lecythidaceae with multiple stamens arising centrifugally on a ring primordium (e.g. Endress, 1994; Appel, 1996; Tsou and Mori, 2007). Except for *Napoleonaea* with strongly differentiated outer stami-nodes, the androecium consists generally of fertile stamens, well delimited from the showy outer corolla.

The gynoecium is generally inferior in *Napoleonaea*, or variously inferior, half-inferior or superior in other Lecythidaceae, with axile placentation. The ovary appears partly unilocular as the connection between upper part and ovule-bearing part happens late and can sometimes be imperfect (Baillon, 1875). The early initiation of the gynoecium is unusual in angiosperms and has also been described by Rudall (2010) and in Chapters 6 and 9 of this volume.

Both Thompson (1922, Fig 2) and Frame and Durou (2001, Fig 4) thought it relevant to represent a floral diagram of *Napoleonaea vogelii*. Although the position of the carpels is wrongly depicted as antesepalous by Frame and Durou (2001), it is closer to reality than Thompson's figure, as shown in Fig 12.7. Ronse De Craene

Fig 12.7 Floral diagram of *Napoleonaea vogelii* (slightly modified from Ronse De Craene, 2010). Black arcs, bracts and sepals; white arcs, petals; ellipses, fertile stamen bases; ellipses with black dots, staminodes; grey dots represent glands; nectary shown by grey ring. Outer broken ring shows fusion of petals; thick broken lines show attachment of anther to filament; inner broken pentagone shows extent of stigmatic lobes covering anthers.

(2010) showed the fertile stamens mistakingly in alternation with the carpels. This is rectified here. Baillon (1875) rightly described the carpels as antepetalous.

It is demonstrated that the corona represents a complex structure, which is part corolla and part androecium, represented by two series of distinct staminodes. The argument that the flower is apetalous because of the strong resemblance and association of outer petaloid appendages to the stamen tube gains little support because of the initiation of petal lobes before the androecium and from the fact that petals and stamens are connected by a common stamen–petal tube, which is widespread in asterids. A floral developmental study of genera of Scytopetaloideae, *Crateranthus* and *Asteranthos* is expected to confirm the earlier initiation of the outer corolla before the centrifugal development of the androecium.

The androecium is highly complex, consisting of an inner tier of fertile half-stamens alternating with staminodial appendages, an intermediate tier of erect staminodes and an outer tier of reflexed filiform staminodes. Two possibilities can explain the existence of paired half-anthers in the inner tier:

- The inner tier consists of complex antesepalous fascicles and the half-anthers represent highly divided antesepalous stamens with elaborate connective extensions in between.
- The inner tier represents two alternating whorls, one of paired stamens and one of paired staminodes.

Baillon (1875) interpreted the androecium as consisting of antesepalous fascicles, formed by two peripheral fertile anthers and two central staminodes. Baillon viewed the outer tiers of appendages as analogous to the corona of *Passiflora* without expressing an opinion about their nature. Thompson (1922) similarly viewed the inner stamens as four fascicles, but described them as being opposite the corolla (although he placed them opposite the sepals on his floral diagram!). A similar interpretation was made by Prance (2004). The interpretation of antesepalous fascicles would be in line with the 'Alternanz' rule of Hofmeister (Kirchoff, 2000). Figure 12.5D could provide some evidence for antesepalous fascicles, with the staminode pair forming the central section of the fascicle. However, the fact that the lateral parts of the fascicle develop before the central part could only be explained by the delay in development of the staminodes and is never encountered in other plants with a lateral division of fascicles (e.g. Ronse De Craene and Smets, 1992). The arrangement of two half anthers separated by two sterile appendages is unusual and demands an interpretation of a partial sterilization of an anther of a pair or the insertion of two sterile outgrowths of connective tissue. An interpretation of highly divided antesepalous stamens with staminodes representing connective appendages seems highly unlikely, as this differentiation would occur much later in the floral development. The staminodes also have remnants of anther tissue.

My observations clearly demonstrate the sequential initiation of stamen and staminode primordia as separate pairs, not as fascicles. There is no evidence of a different growth rate between adjacent half-anthers and pairs of staminodes, suggesting that they belong together.

Unilocular anthers are relatively rare in angiosperms (see Endress and Stumpf, 1990). As the stamen primordia in *Napoleonaea* arise in pairs, it can be suggested that each primordium represents a half-stamen. Although two primordia are initiated, the process could represent a halving of an original primordium and not a secondary increase. A similar halving of primordia occurs in *Adoxa* (Erbar, 1994) or in Malvaceae, on a larger scale (van Heel, 1966). The half-anthers are basically the result of a division of single antepetalous stamens and the same applies for the paired antesepalous staminodes. The fact that the inner whorl of stamens represents ten halved units indicates that the androecium may have been ancestrally two-whorled, as is occasionally found in some Lecythidoideae (e.g. *Cariniana*: Tsou and Mori, 2007).

A comparable umbrella-like style is found in *Hura* (Euphorbiaceae) or *Sarracenia* (Sarraceniaceae) (Weberling, 1989). In Sarraceniaceae the flowers hang upside down and pollen is shed on the underside of the umbrella. As *Napoleonaea* has a comparable hanging flower it is possible that the same mechanism is responsible for pollination. However, careful field observations by Frame and Durou (2001) demonstrated that pollinators of *Napoleonaea* are probably thrips finding shelter in an inner chamber formed between the curved filaments and style. Between individual filaments a gap is present, allowing access to the inner chamber. However, the presence of a conspicuous outer staminodial chamber, a large nectary disc and other visitors might indicate that the flower has evolved strategies to attract other insect pollinators and to segregate them (Frame and Durou, 2001).

12.5 Conclusions

Floral developmental evidence supports the second hypothesis that petals are ancestral in Lecythidaceae with occasional loss in a few genera (Fig 12.1B). In Napoleonoideae and Scytopetaloideae, including *Asteranthos*, petals are transformed into a plicate corolla closely linked to the androecium. Prance (2004) grouped *Crateranthus* with *Napoleonaea* in Napoleonaeaceae. Both genera share a plicate corolla and similar wood anatomy, beside other characters (see Prance, 2004). In *Crateranthus* stamens are arranged in eight to ten girdles of fertile stamens, similarly connected to the corolla at the base.

Napoleonaea is distinctive by the combination of a highly elaborate androecium combined with the corolla. Although the corolla arises as 'normal' petals, they rapidly become so integrated in the conspicuous androecial corona as to become almost identical to staminodes in later stages of development. The fact that the corolla splits open along the midrib emphasizes this similarity. The genetic implications for these changes are exciting, as development shows a gradual reduction of genetic boundaries between androecium and perianth. This has not been equalled in the angiosperms and places *Napoleonaea* as an interesting model genus for floral evolution and for genetic studies.

Acknowledgements

I thank Elspeth Haston for providing photographs of mature flowers of *Napoleonaea vogelii* and Frieda Christie for assistance with the SEM. Financial support for travel from the Tropical Section of RBGE is acknowledged. Sincere thanks to Gerhard Prenner and Jürg Schönenberger for helpful comments.

12.6 References

Angiosperm Phylogeny Group (2009). An update of the Angiosperm Phylogeny Group classification for the orders and families of flowering plants: APG III. *Botanical Journal of the Linnean Society*, **161**, 105–121.

Appel, O. (1996). Morphology and systematics of the Scytopetalaceae. *Botanical Journal of the Linnean Society*, **121**, 207–227.

Appel, O. (2004). Scytopetalaceae. pp. 426–430 in Kubitzki, K. (ed.), *The Families and Genera of Vascular Plants*, Vol. 6. Berlin: Springer.

Baillon, H. (1875). Sur les fleurs et les fruits des *Napoleona*. *Bulletin de la Société Linnéenne de Paris*, **1** (8), 59–62.

Bentham, G. and Hooker, J. D. (1867). *Genera Plantarum*. London: Reeve and Co.

Dahlgren, R. and Thorne, R. F. (1984). The order Myrtales; circumscription, variation and relationships. *Annals of the Missouri Botanical Garden*, **71**, 633–699.

Endress, P. K. (1994). *Diversity and Evolutionary Biology of Tropical Flowers*. Cambridge: Cambridge University Press.

Endress, P. K. and Stumpf, S. (1990). Non-tetrasporangiate stamens in the angiosperms: Structure, systematic distribution and evolutionary aspects. *Botanische Jahrbücher für Systematik*, **112**, 193–240.

Erbar, C. (1994). Contributions to the affinities of *Adoxa* from the viewpoint of floral development. *Botanische Jahrbücher für Systematik*, **116**, 259–282.

Frame, D. and Durou, S. (2001). Morphology and biology of *Napoleonaea vogelii* (Lecythidaceae) flowers in relation to the natural history of insect visitors. *Biotropica*, **33**, 458–471.

Kirchoff, B. K. (2000). Hofmeister's rule and primordium shape, influences on organ position in *Hedychium coronarium* (Zingiberaceae). pp. 75–83 in Wilson, K. L. and Morrison, D. A. (eds.), *Monocots, Systematics and Evolution*. Melbourne: CSIRO.

Knuth, R. (1939). Barringtoniaceae. p. 82 in Engler, A. (ed.), *Das Pflanzenreich 105*. Leipzig: W. Engelmann.

Liben, L. (1971). Révision du genre African *Napoleonaea* P. Beauv. (Lecythidaceae). *Bulletin du Jardin Botanique National de Belgique*, **41**, 363–382.

Masters, M. T. (1869). On the structure of the flower in the genus *Napoleona*, etc. *Journal of the Linnean Society, Botany*, **10**, 492–504.

Mori, S. A., Tsou, C.-H., Wu, C.-C., Cronholm, B. and Anderberg, A. A. (2007). Evolution of Lecythidaceae with an emphasis on the circumscription of Neotropical genera: information from combined *ndhF* and *trnL-F* sequence data. *American Journal of Botany*, **94**, 289–301.

Morton, C. M., Mori, S. A. Prance, G. T., Karol, K. G. and Chase, M. W. (1997). Phylogenetic relationships of Lecythidaceae: a cladistic analysis using *rbcL* sequence and morphological data. *American Journal of Botany*, **84**, 530–540.

Morton, C. M., Prance, G. T., Mori, S. A. and Thorburn, L. G. (1998). Recircumscription of the Lecythidaceae. *Taxon*, **47**, 817–827.

Niedenzu, F. (1893). Lecythidaceae. pp. 26–41 in Engler, A. and Prantl, K. (eds.), *Die Natürlichen Pflanzenfamilien 1, 3*. Leipzig: W. Engelmann.

Prance, G. T. (2004). Napoleonaeaceae. pp. 282–284 in Kubitzki, K. (ed.), *The Families and Genera of Vascular Plants*, Vol. 6. Berlin: Springer.

Prance, G. T. and Mori, S. A. (1979). Lecythidaceae – part I. The actinomorphic-flowered New World Lecythidaceae. *Flora Neotropica, Monograph* **21**, 1–270.

Ronse De Craene, L. P. (2010). *Floral Diagrams. An Aid to Understanding Flower Morphology and Evolution*. Cambridge: Cambridge University Press.

Ronse De Craene, L. P. and Smets, E. (1992). Complex polyandry in the Magnoliatae: definition, distribution and systematic value. *Nordic Journal of Botany*, **12**, 621–649.

Rudall, P. J. (2010). All in a spin. Centrifugal organ formation and floral patterning. *Current Opinion in Plant Biology*, **13**, 108–114.

Schönenberger, J., Anderberg, A. A. and Systsma, K. J. (2005). Molecular phylogenetics and patterns of floral evolution in the Ericales. *International Journal of Plant Science*, **166**, 265–288.

Takhtajan, A. (1997). *Diversity and Classification of Flowering Plants*. New York: Columbia University Press.

Thompson, J. M. (1922). Studies in floral morphology. III. The flowering of *Napoleona imperialis*, Beauv. *Transactions of the Royal Society of Edinburgh*, **53**, 265–275.

Thompson, J. M. (1927). A study in advancing gigantism with staminal sterility with special reference to the Lecythidaceae. *Publications of the Hartley Botanical Laboratory*, **4**, 5–44.

Tsou, C.-H. (1994). The embryology, reproductive morphology and systematics of Lecythidaceae. *Memoirs of the New York Botanical Garden*, **71**, 1–110.

Tsou, C.-H. and Mori, S. A. (2007). Floral organogenesis and floral evolution of the Lecythidoideae (Lecythidaceae). *American Journal of Botany*, **94**, 716–736.

van Heel, W. A. (1966). Morphology of the androecium in Malvales. *Blumea*, **13**, 177–394.

Weberling, F. (1989). *Morphology of Flowers and Inflorescences*. Cambridge: Cambridge University Press.

Taxon index

Subject index

Systematics Association Publications

1. Bibliography of Key Works for the Identification of the British Fauna and Flora, 3rd edition (1967)†
   Edited by G.J. Kerrich, R.D. Meikie and N. Tebble

2. Function and Taxonomic Importance (1959)†
   Edited by A.J. Cain

3. The Species Concept in Palaeontology (1956)†
   Edited by P.C. Sylvester-Bradley

4. Taxonomy and Geography (1962)†
   Edited by D. Nichols

5. Speciation in the Sea (1963)†
   Edited by J.P. Harding and N. Tebble

6. Phenetic and Phylogenetic Classification (1964)†
   Edited by V.H. Heywood and J. McNeill

7. Aspects of Tethyan Biogeography (1967)†
   Edited by C.G. Adams and D.V. Ager

8. The Soil Ecosystem (1969)†
   Edited by H. Sheals

9. Organisms and Continents through Time (1973)*
   Edited by N.F. Hughes

10. Cladistics: A Practical Course in Systematics (1992)‡
    P.L. Forey, C.J. Humphries, I.J. Kitching, R.W. Scotland, D.J. Siebert and D.M. Williams

11. Cladistics: The Theory and Practice of Parsimony Analysis (2nd edition) (1998)‡
    I.J. Kitching, P.L. Forey, C.J. Humphries and D.M. Williams

† Published by the Systematics Association (out of print)
* Published by the Palaeontological Association in conjunction with the Systematics Association
‡ Published by Oxford University Press for the Systematics Association

Systematics Association Special Volumes

74. Automated Taxon Identification in Systematics: Theory, Approaches and Applications (2007)‡‡
   Edited by N. MacLeod

75. Unravelling the algae: the past, present, and future of algal systematics (2008)‡‡
   Edited by J. Brodie and J. Lewis

76. The New Taxonomy (2008)‡‡
   Edited by Q.D. Wheeler

77. Palaeogeography and Palaeobiogeography: Biodiversity in Space and Time (in press)‡‡
   Edited by P. Upchurch, A. McGowan and C. Slater

---

[a] Published by Clarendon Press for the Systematics Association
[*] Published by Academic Press for the Systematics Association
[†] Published by Oxford University Press for the Systematics Association
[**] Published by Chapman and Hall for the Systematics Association
[††] Published by CRC Press for the Systematics Association

Printed in the United States
by Baker & Taylor Publisher Services